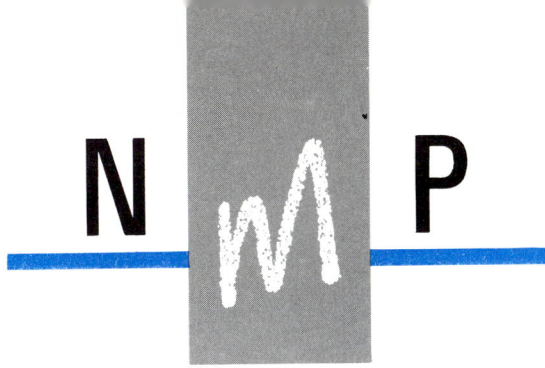

N M P

MATHEMATICS FOR **4** SECONDARY SCHOOLS

BLUE TRACK

Longman Group UK Limited
Longman House, Burnt Mill, Harlow, Essex CM20 2JE, England
and Associated Companies throughout the world.

First published 1989
ISBN 0 582 22515 9

Set in 11/13 Times

Printed in Great Britain by Scotprint Limited, Musselburgh

N M P

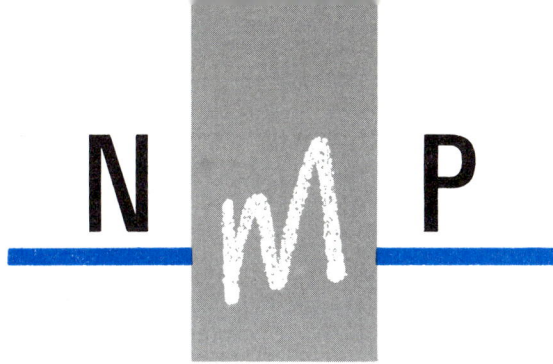

This book was written by

Eon Harper
Dietmar Küchemann
Michael Mahoney
Sally Marshall
Edward Martin
Heather McLeay
Peter Reed
Sheila Russell

NMP Director
Eon Harper

NMP Research
Edward Martin

MATHEMATICS FOR

4

SECONDARY SCHOOLS

BLUE TRACK

Longman

PREFACE

This book and the three companion books, Year 4 Red Track, Year 5 Blue Track and Year 5 Red Track cover all Levels in Key Stage 4 of the National Curriculum and provide a complete two-year Examination course for all grades of the GCSE Examination.

The course builds upon the Years 1–3 Red Track and Blue Track course of *NMP Mathematics for Secondary Schools*. Each book contains Review sections which revise the mathematics introduced in Years 1–3. The course can thus be used by schools which have hitherto not used NMP materials in the foundation years.

The Blue Track books for Years 4 and 5 provide for GCSE Grades C/D to G; the Red Track books provide for Grades A to C/D. Further details can be found in the Teachers' Handbook for each track.

NMP was founded in 1981 to develop teaching and learning materials suited to the emerging new syllabuses and National Criteria. The material was researched and evaluated at the University of Bath and written by practising teachers and professional educators.

Each text provides for pupil–pupil and pupil–teacher discussion, oral and mental work, skill and practice work, written and calculator work, problem solving, investigation and extended assignments. Each of these aspects is integrated into the text to provide the variety of learning opportunities required by the various GCSE Boards and the National Curriculum.

The text falls into two Sections, A and B, whose Chapters contain Review, Consolidation, Core and Enrichment material. The relationship of these to GCSE needs is explained in detail in the Teachers' Handbook.

CONTENTS

ACKNOWLEDGEMENTS

We are grateful to the following for permission to reproduce photographs and other copyright material:
Heather Angel, page 76 *below*; Automobile Association/*Great Britain Road Atlas*, page 97; ABI Caravans, page 106 *right*; Aerofilms, pages 34 *above right*, 56 and 264; American Museum in Britain, Bath, page 46; Andes Press Agency, page 294 *left*; Eve Arnold Magnum, page 4 *below centre*; Barnaby's Picture Library, pages 1, 34 *above left*, *below centre* and *below right*, 126, 217, 276 *right*, 279 *above* and 281 (photo Brian L. Arundale); Barratt Developments, Guildford, page 11; British Museum Connection, page 125 *above right*; British Museum, Department of Egyptian Antiquities, page 125 *above left*; British Railways Board, pages 117 (Intercity) and 136 *above*; British Waterways Board, page 122 *below*; Camera Press, page 124 *below centre* (photo Ray Hamilton); Capital and Growth Estate Agency, page 272 *above*; J. Allan Cash, pages 92, 121 *below*, 122 *above*, 132, 135, 142 *below*, 144, 147, 168 *above*, 214 and 294 *right*; Bruce Coleman, page 167 *above* (photo A.J. Mobbs); Colorific!, page 136 *below* (photo Mark Greenberg); Consumer's Association/*Which?* January 1987, page 54; Corocraft, page 226 *below* (photo Richton International); Electrolux, page 142 *above left* and *right*; Empire Mail Order, page 271 *centre*; 1989 M.C. Escher Heirs/Cordon Art, Baarn, Holland page 267; Chris Fairclough Colour Library, page 138; Format Photographers, page 21 *right* (photo Roshini Kempadoo); Geographers' A-Z Map Co. Ltd, pages 95 and 96 *left*; Geoscience Features, page 3 *centre* and *centre right* (photos Dr Basil Booth); Sally and Richard Greenhill, pages 33 *above left*, 133 *above centre*, *below left* and *below right*, 276 *left* and *centre*; William Gyoory & Co., page 272 *centre* Harlow Council Development Services, page 52; *Harlow and Epping Classified*, page 231; Holt Studios, pages 106 *left*, 130 and 271 *right*; Hoover, page 103 figure 4 *centre left*; Hulton Deutsch Collection, pages 34 *centre left*, 51, 72, 75, 93, 107, 218 and 234; Imagine Photography, pages 275 (photo Irene Lynch), 280, 300 and 311 (photos Terence White); Frank Lane Agency, pages 3 *above* and 54 *below*; London Features International, page 149 *below* (photo Kevin Mazur); Network Photographers, pages 225 (photo Mike Abrahams), 277 *below left* (photo Geoff Franklin) and 277 *below right* (photo Laurie Sparham); *The Observer Magazine*, page 24; Panasonic UK, page 101 *above*; Philips, page 103 figure 4 *left*, *centre* and *right*, figure 6 *left*; Picturepoint, pages 34 *above centre* and 133 *above right*; Pirelli, page 94 *below*; Popperfoto, pages 3 *centre left*, 32 *above left*, 74, 125 *below right* and 277 *above*; Post Office, page 213 *left*; Quadrant Picture Library, pages 21 *left* and 77; Rest Assured, page 103 figure 6 *right*; Ann Ronan Picture Library, page 245; Roy Export Company Establishment, page 4 *above left* and *above right*; Sainsbury's, page 39; Science Photo Library, page 3 *below left*, *below centre* and *below right* (photos David Scharf), 57 *above left*, *above centre* and *above right* (photos Dr. Tony Brain), 57 *centre* (photo Ferranti Electronics/A. Sternberg) and 168 *below* (photo National Cancer Institute); Seekers Estate Agency, page 272 *below*; Spectrum Yarns, page 34 *below left*; Sporting Pictures UK, page 124 *below left* and *below right*; Telefocus/British Telecom, page 125 *below left*; Thames Water, page 279 *below*; Topham, pages 4 *below left* and 282; Unigate Dairies, page 34 *centre right*; Janine Wiedel, pages 2, 60, 69 and 76 *above*; World Gold Council, page 226 *above*; Yardley of London, page 271 *left*.

The Ordnance Survey maps symbols, signs and photograph on pages 57, 120, 121 and 122 *centre* are © Crown Copyright and are reproduced by permission of Her Majesty's Stationary Office.

Photographs on pages 133 *above left* and *below centre* were taken by John Birdsall, on page 96 *above right* and *below right* by Colin Johnson, on pages 120 *below* and 309 by Martin Mulcahy and on pages 4 *below right*, 7, 9, 20, 22, 23, 32 *above right*, *centre* and *below right*, 33 *above right*, 34 *centre*, 42, 78, 79, 94 *above*, 101 *below*, 102, 103 figure 5, 109, 124 *above* and *centre*, 125 *centre left* and *below centre*, 149 *above left* and *above right*, 167 *below*, 211, 213 *right*, 222, 273 and 303 by Longman Photographic Unit.

Tables on page 63 are based on extracts taken from page 619 PEARS Cyclopaedia, 96th Edition 1987, Published by Pelham Books, Edited by C.Cook.

Illustrations by Jerry Collins, Valerie Hill (Linda Rogers Associates), Oxford Illustrators Ltd, Julie Sailing, Martin Shovel (cartoons) and John Woodcock.

A1 ENLARGING AND REDUCING

REVIEW

- In an enlargement *all* distances are increased by the same *scale factor*.

■ What can you say about the angles of a shape and the angles of a shape enlarged from it?

Enlargement scale factor × 3

These lengths are three times these lengths

A microscope enlarges by a scale factor.

This amoeba is enlarged with scale factor × 40.

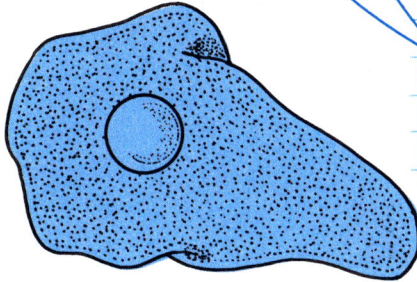

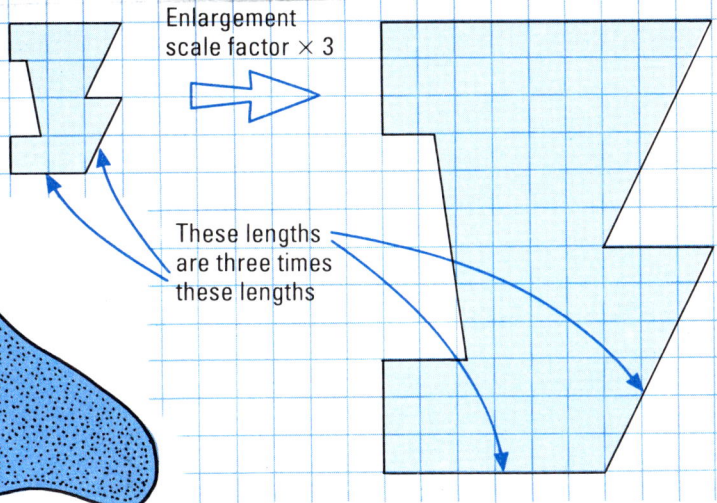

Photographic enlargements have a scale factor.

This photograph is enlarged using a scale factor × 2.

■ A photograph measures 6 cm × 8 cm. It is enlarged to 30 cm × 40 cm. What scale factor is used?

- In a reduction all distances are reduced by the same scale factor.

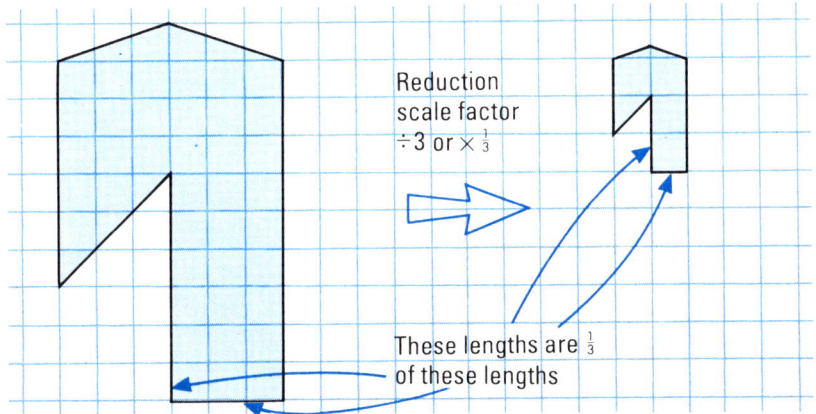

Reduction scale factor ÷3 or ×$\frac{1}{3}$

These lengths are $\frac{1}{3}$ of these lengths

A1

CONSOLIDATION

1   This is a photograph of a snipe.

a) Some of these are proper enlargements or reductions of the photograph. Which ones?

A

C

B

D

E

F

b) What is the scale factor for enlargement or reduction in (i) C, (ii) E?

2   Microscopes help us to look in detail at very small things. For example, this is a flea magnified under a microscope.

a)   Guess the magnification (enlargement) factor. Is it:

$\times\, 2$, $\times\, 5$, $\times\, 10$, or $\times\, 50$?

b)   Most fleas like the one under the microscope are about 2 mm tall. Use this to check your guess in part a).

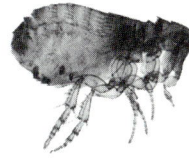

A1

3   Here are some more small things magnified under a microscope.

a)   *Guess* the magnification for each one.

Gnat

Grain of rice

Grain of sand

b)   These are the actual sizes of the things in part a).

Gnat: length 2.5 mm.   Grain of rice: width 2.7 mm, length 6 mm.   Grain of sand: width 0.5 mm.

Use them to check each of your guesses in part a). Mark each guess 'Very good', 'OK', or 'Abysmal'.

CHALLENGE

4
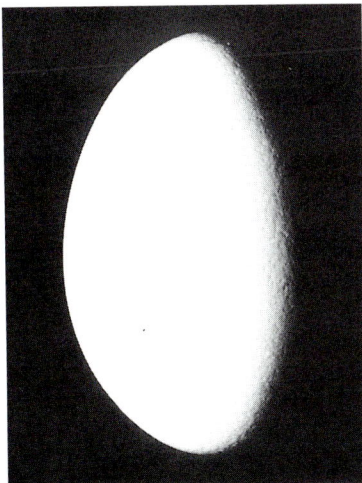

This is the crack in an eggshell magnified $\times 100$ by an electron microscope.

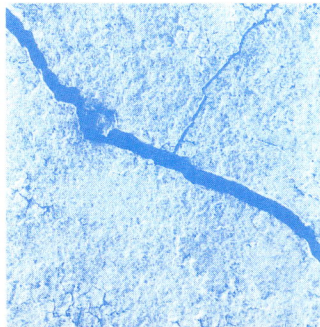

This is the same crack. Estimate the magnification for this enlargement.

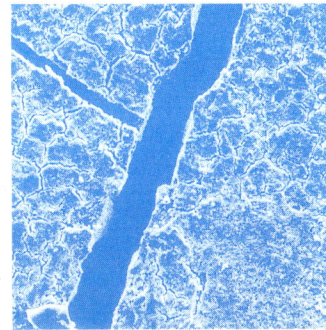

5 A photographer enlarges a photograph using a ×6 scale factor. In the original, a tiger is 5cm long and 2cm tall. How long and how tall is the tiger in the enlarged photograph?

6 This is a microfilm photograph of a sheet of A4 paper: ▭ (actual size 205mm × 295mm). Approximately, what is the scale factor for reduction?
   Write the scale factor: a) like this '÷ ▢'
                           b) like this '× $\frac{1}{▢}$'

7 This is a frame from an old movie film.

When it is projected onto a screen the picture measures 4m × 3.5m. What is the scale factor for enlargement?

========== WITH A FRIEND ==========

8 Imagine you want *life-size* enlargements made from each of these photographs. Roughly, what scale factor for enlargement would you have to use for each one? Decide between you.

ENRICHMENT

EXPLORATION: ENLARGEMENT WITH TILES AND CUBES

A1

1   You need 1 cm squared paper, and 1 cm
    isometric paper.

a) This shape is made from 1 cm square tiles.

   (i) Copy and complete the ×2 enlarged
        drawing of it.
   (ii) How many 1 cm tiles are needed for
         the enlarged version?
   (iii) Is this twice the number needed for
          the original ...
          three times the number needed
          for the original ...
          four times ...   five times ...?

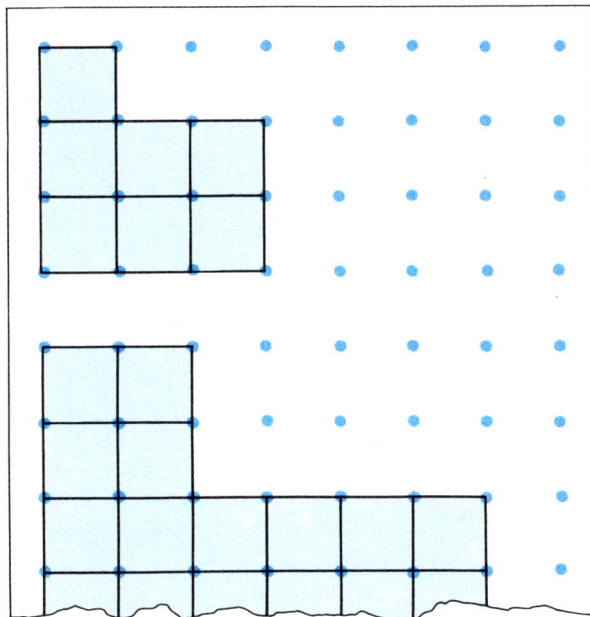

b) This shape is made from 1 cm triangle tiles.

   (i) Copy and complete the ×3
        enlargement of it.
   (ii) How many triangle tiles are needed
         for the enlarged version?
   (iii) Is this twice the number needed for
          the original ...
          three times the number ...
          four times ...   five times ...?

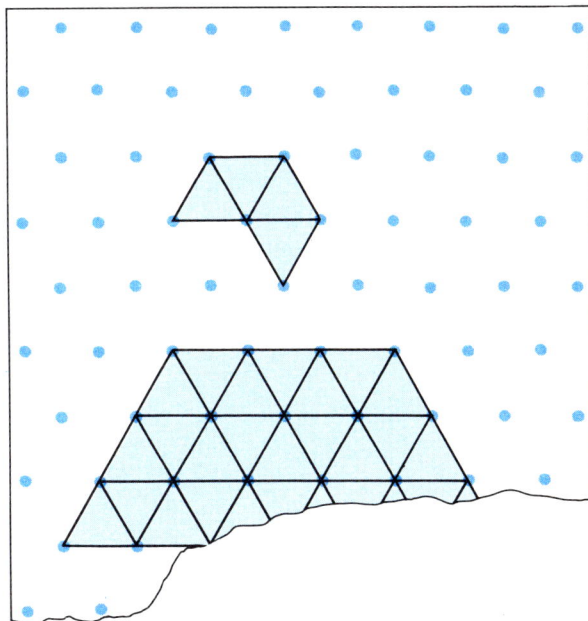

c) Investigate some more shapes and some more enlargement factors, for example ×4, ×5. Try
    to find a rule which tells you how many tiles you will need for an enlarged version of a shape.
    Write down what you discover.

A1

EXPLORATION: ENLARGING AREAS AND VOLUMES

2   You might need: some 1 cm cubes, and some isometric paper.

a)   The tower on the right is made from 1 cm cubes.
Think about a × 3 enlargement of it. (You might like to
build it.) The enlarged version will be three times as high,
three times as wide, and three times as long.

  (i)   How many cubes are needed for the enlarged version?
  (ii) Is this twice as many as the original . . .
          three times as many as the original . . .
          four times as many . . .   five times . . . ?

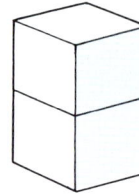

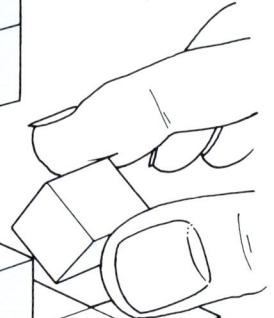

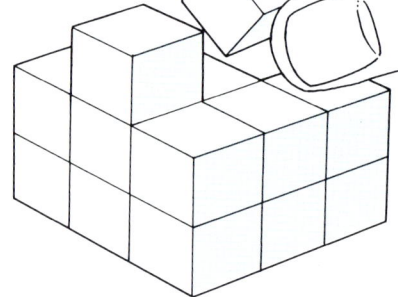

b)   This tower is made
from 1 cm cubes.

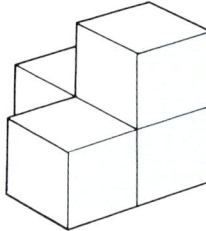

  (i)   Sketch a × 2 enlargement. (You might like to build it first.)
          Each distance on your sketch will be twice that on the original. (For example, the top
          cube will measure 2 cm × 2 cm × 2 cm.)
  (ii) How many cubes are needed to make the enlarged version?
  (iii) Is this twice as many as the original . . .
          three times as many as the original . . .
          four times as many . . .   five times . . . ?

c)   Investigate some more towers and some more enlargement factors. Try to find a rule which
tells you how many cubes you need for an enlarged version of a tower.

REVIEW

- The *area* of a shape is the amount of surface it covers.

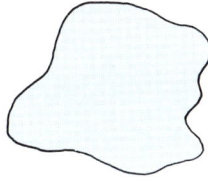

Area decreasing ⟶

- The *perimeter* of a shape is the distance around its edge.

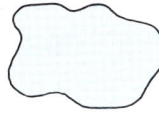

Perimeter increasing ⟶

Area is measured in:
- square millimetres (mm²)
- square metres (m²)
- square centimetres (cm²)
- hectares (ha).

Area of each shape 1 mm²

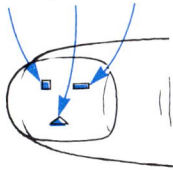

Area of each shape 1 cm²

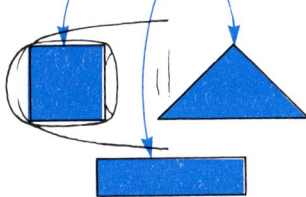

Area 1 hectare

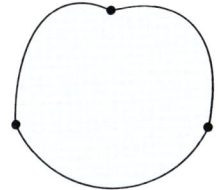

100 m

100 m

One hectare is about the same area as 1½ football pitches

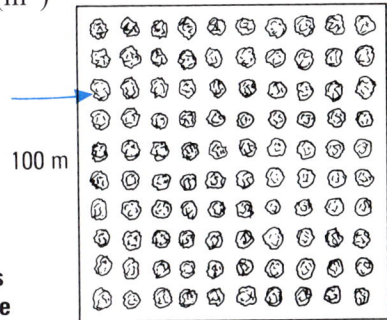

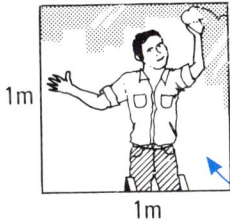

1 m

1 m

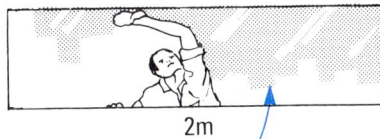

2 m

½ m

Area of glass 1 m²

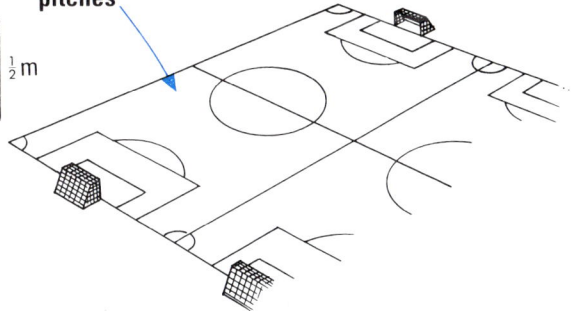

- Area of a rectangle = length × width.

Area = $2.4 \times 2\,\text{cm}^2 = 4.8\,\text{cm}^2$.

length: 5 cm

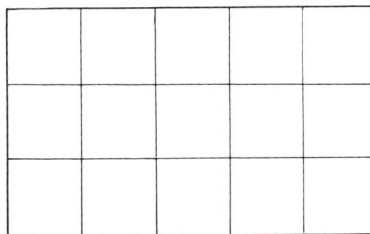

width: 3 cm

Area = $15\,\text{cm}^2 = \text{length} \times \text{width} = 5 \times 3\,\text{cm}^2$.

CONSOLIDATION

EXPLORATION

1   a) Check that the shapes have the same area, but that their perimeters gradually increase.

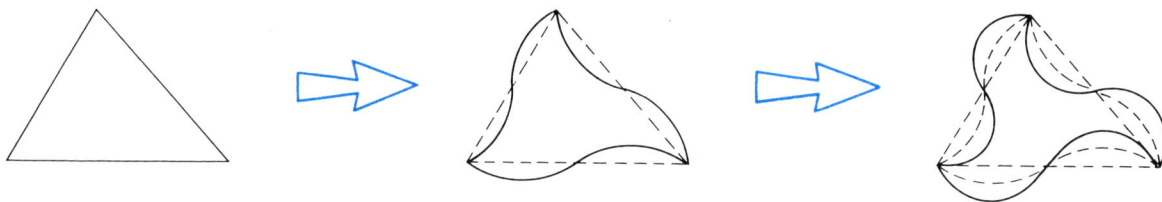

b) Start with this shape.

Draw two more shapes, to make a set of three, which have:
  ● the same area, but
  ● increasing perimeters.

c) Start with this shape.

Draw two more shapes, to make a set of three, which have:
  ● the same perimeter, but
  ● decreasing areas.

d) Start with this shape.

Draw two more shapes, to make a set of three, which have:
  ● decreasing areas, but
  ● increasing perimeters.

e) Start with this shape.

Draw two more shapes, to make a set of three, which have:
  ● increasing areas, but
  ● decreasing perimeters.

f) Investigate other possibilities. (For example, increasing areas and increasing perimeters.) Draw examples of sets of shapes to show what is possible.

2   The stamp has an area of about $5\,cm^2$. Roughly, what is the area of:

          a) this postcard   b) this envelope?

A2

3   Here are some areas: $2\,m^2$ $10\,m^2$ $50\,m^2$ $200\,m^2$ $1000\,m^2$

Which gives the best estimate for the area of:

    a) this gymnasium wall   b) this bathroom wall?

4   Some outdoor basketball courts are to be marked out like this.

    a) About how many courts will fit into a
        1 hectare square? (A 1 hectare square
        measures $100\,m \times 100\,m$.)

    b) How many m^2 is 1 hectare?

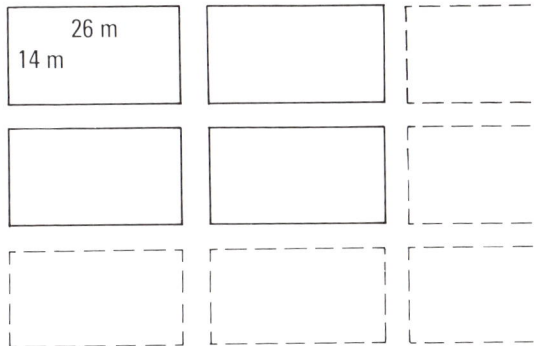

5   Use 1 mm squared paper.

    a) Roughly, what area in mm^2 does each
        coin cover?

    b) Roughly, what is the perimeter of each
        in millimetres?

━━━━━━ ACTIVITY ━━━━━━

6 You need 1 cm squared paper and scissors. Cut out the
rectangle and squares on the right.
They can be arranged in different ways so that they always
touch along a complete edge of a shape. Here are two examples.

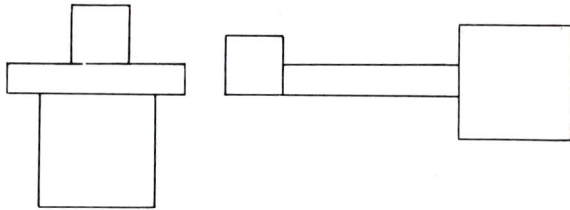

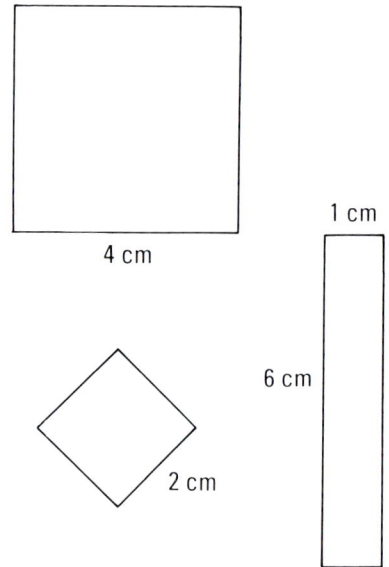

4 cm

1 cm

6 cm

2 cm

Find an arrangement for which:

a) the perimeter is a maximum

b) the perimeter is a minimum

Sketch each arrangement and write down each perimeter.

7 Helen Makepeace works in a department
store. Her job is to work out the cost of floor
covering (carpets, tiles, lino, etc.) for
customers.

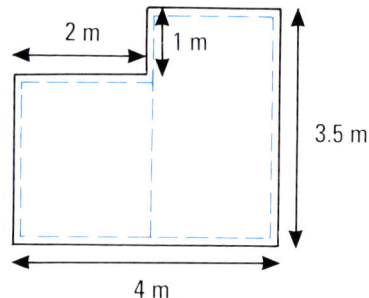

2 m   1 m   3.5 m   4 m

This is a plan of a kitchen. The customer
wants cushioned floor covering at £6 per m².

a) How many m² of floor covering is needed?

b) How much will it cost?

c) Double-sided sticky tape will be needed along the dotted lines, to stick down the floor
covering.

(i) What is the perimeter of the kitchen?
(ii) How many metres of tape are needed altogether?

d) The tape costs 75p per metre. How much will the tape and the floor covering cost altogether?

8 a) Check that the perimeter of this rectangle is 20 cm.

b) Draw two more, different, rectangles,
whose perimeter is 20 cm. Write down the
area of each of your rectangles.

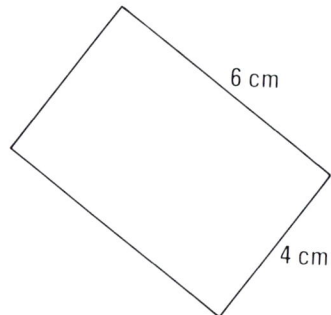

6 cm

4 cm

9 a) The perimeter of a square is 20 cm. What is its area?

b) The area of a square is 36 cm². What is its perimeter?

ENRICHMENT

1   Visit a local showroom. Choose the carpets and other floor coverings you would like for this new 'Salisbury' home. Make a list for each room. (Don't forget the stairs!) Write out an estimate of cost for the whole house, including labour.

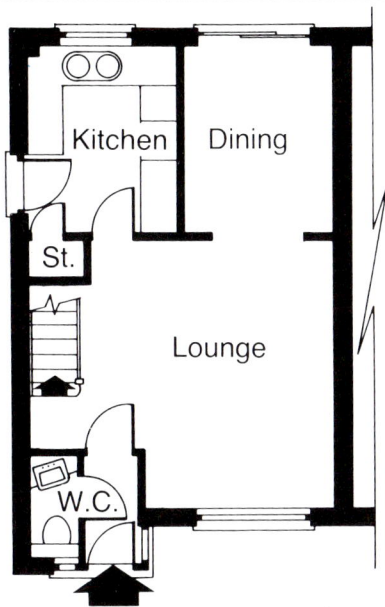

Ground Floor

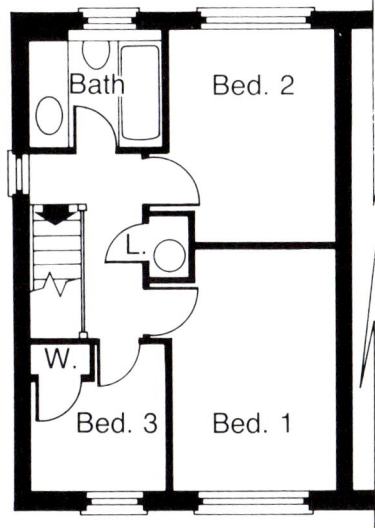

First Floor

THE SALISBURY

GROUND FLOOR	Imperial	Metric	FIRST FLOOR	Imperial	Metric
Lounge	16' 1" x 14' 1"	4900 x 4296	Bedroom 1	12'10" x 8' 0"	3900 x 2435
Dining Room	9' 8" x 7'10"	2948 x 2381	Bedroom 2	11' 3" x 8' 5"	3435 x 2559
Kitchen	9' 8" x 8' 1"	2948 x 2454	Bedroom 3	7'11" x 7'10"	2400 x 2378
Garage	8' 2" x 17' 0"	2501 x 5183	Bathroom	7' 6" x 6' 3"	2276 x 1900

THE ILLUSTRATION OVERLEAF SHOWS JUST ONE OF THE INDIVIDUAL ELEVATIONAL TREATMENTS OF THIS PREMIER COLLECTION HOUSE DESIGN.

We take every care to ensure that the information on this leaflet is correct. We hope you find it helpful, however, complete accuracy is not guaranteed and the information is expressly excluded from any contract.

B A R R A T T

Britain's Premier House Builder

REVIEW

Amounts are sometimes written

- using vulgar fractions: $3\frac{7}{10}$ cm
- using decimal fractions: 3.7 cm

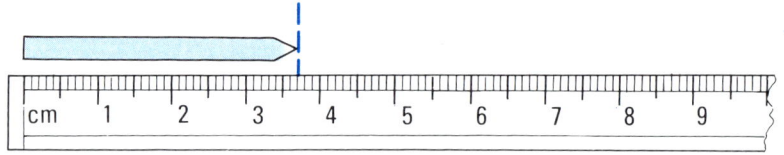

■ Which method does a calculator use, vulgar fractions or decimal fractions?

- We can get from one type to the other by thinking about column headings.

wholes . $\frac{1}{10}$ $\frac{1}{100}$

3 . 4 0 $\longrightarrow$ $3\frac{4}{10}$

wholes . $\frac{1}{10}$ $\frac{1}{100}$

$4\frac{3}{100}$ $\longrightarrow$ 4 . 0 3

$\frac{1}{5}$ can be written as $1 \div 5$.

$\boxed{C}$ $\boxed{1}$ $\boxed{\div}$ $\boxed{5}$ $\boxed{=}$ $\boxed{0.2}$

$\frac{1}{5}$ is 0.2

$$5 \overline{) \cancel{1}.^{1}0} \quad 0.2$$

■ Why has the 'l' in the division been crossed out? What does the lower '0' represent?

$\frac{1}{2}$ is 0.5

$\frac{1}{4}$ is 0.25

$\frac{3}{4}$ is 0.75

0.1 is $\frac{1}{10}$

0.7 is $\frac{7}{10}$

0.01 is $\frac{1}{100}$

0.47 is $\frac{47}{100}$

$$4 \overline{) 3.^{3}0^{2}0} \quad 0.75$$

$\boxed{C}$ $\boxed{3}$ $\boxed{\div}$ $\boxed{4}$ $\boxed{=}$ $\boxed{0.75}$

$\frac{10}{100}$

wholes. $\frac{1}{10}$ $\frac{1}{100}$

0. 4 7

CONSOLIDATION

1 a) The scales show how much water is in the kettle. Sketch the scales, side by side. Write in the missing amounts.

   b) Arrange these amounts in order, smallest first.

   1.25 tonne, $1\frac{1}{2}$ tonne, 0.5 tonne, $\frac{3}{4}$ tonne, $1\frac{3}{4}$ tonne, 2.75 tonne, $2\frac{1}{4}$ tonne.

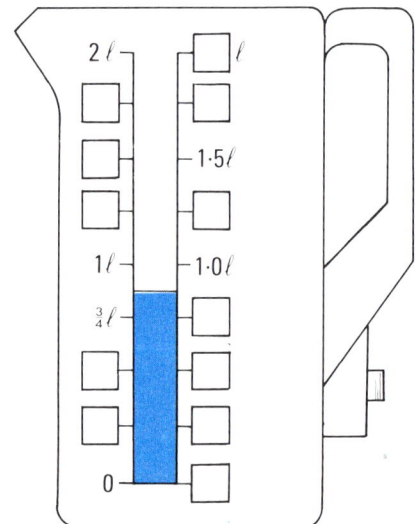

c) This 1 m ruler can be used to give measurements in decimal form or in vulgar fraction form.

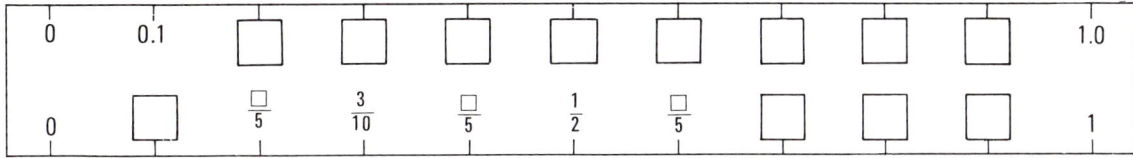

Copy the ruler. Write in the missing amounts.

d) Use your sketches in a) and c) to help you with these.

(i) Which is heavier, a 0.8 tonne horse or a $\frac{3}{4}$ tonne van?

(ii) Which is wider, a 0.7 km river crossing or a field that is $\frac{4}{5}$ km wide?

(iii) Which is worth more, a £$1\frac{1}{4}$ million order for machinery, or a £1.3 million order?

CHALLENGE

2 This weighing scale uses vulgar fractions. It rounds weights *down* to the next $\frac{1}{4}$ ounce.

This one uses decimal fractions. It rounds weights *down* to the next 0.1 ounce.

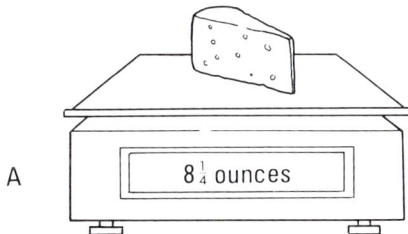

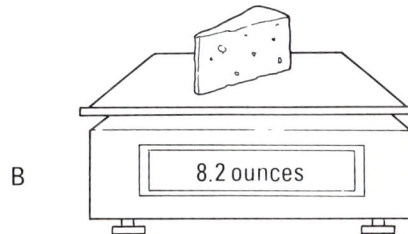

A   8$\frac{1}{4}$ ounces

B   8.2 ounces

a) Which weighing scale gives the most accurate readings, A or B? Explain your answer.

b) What would scale A show for this piece of cheese?

4.6 ounces

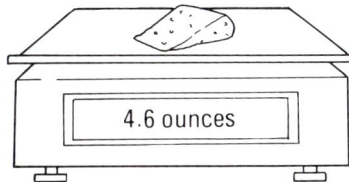

c) What would scale B show for this piece of cheese? (There are three possibilities. Give all of them.)

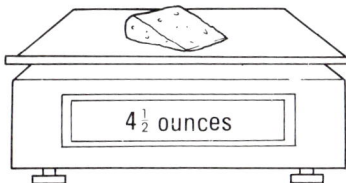

4$\frac{1}{2}$ ounces

Cutting decimals to size

1 a) The advertisement shows that three pairs of
    socks cost £2. Use your calculator to find the cost
    of one pair.

SUMMER SOCKS
SPECIAL OFFER
3 Pairs for
£2

b) In a) your calculator showed ⬛ 0.6666666
    The cost can be rounded *down* to 66p or *up* to 67p.
    Find the cost of one item in each of A, B, and C below.
    Round the cost *down* to the next penny in each case.

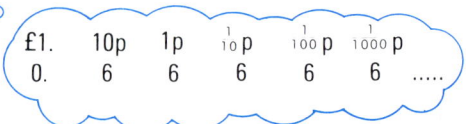

£1.   10p   1p $\frac{1}{10}$p $\frac{1}{100}$p $\frac{1}{1000}$p
0.   6   6   6   6   6   .....

A
GOLDEN OLDIES
8 Tapes for £5

B
NINE
Hot XBuns
for 98p

C
PRINTS
choose any 6 for
£8

2 a) Decide which of these two distances is the greater:

    3.416759   km

    3.4157598 km

b) Explain why you made this choice.

3 a) This shows that 5.1473 km is larger than 5.1465 km.

wholes	.	$\frac{1}{10}$	$\frac{1}{100}$	$\frac{1}{1000}$	$\frac{1}{10000}$	
5	.	1	4	7	3	km
5	.	1	4	6	5	km

same   different

Is it larger by more than $\frac{1}{1000}$ of a kilometre
or less than $\frac{1}{1000}$ of a kilometre?

b) The distance 2.1498 km is greater than another distance, by exactly $\frac{1}{1000}$ of a kilometre. What
    is the other distance?

4 a) Follow the flow chart
       for these two decimals:

       8.714906 7
       8.714905 731

   b) Choose three pairs of
       decimals of your own.
       Follow the flow chart
       for each pair. What
       does the flow chart help
       you to do?

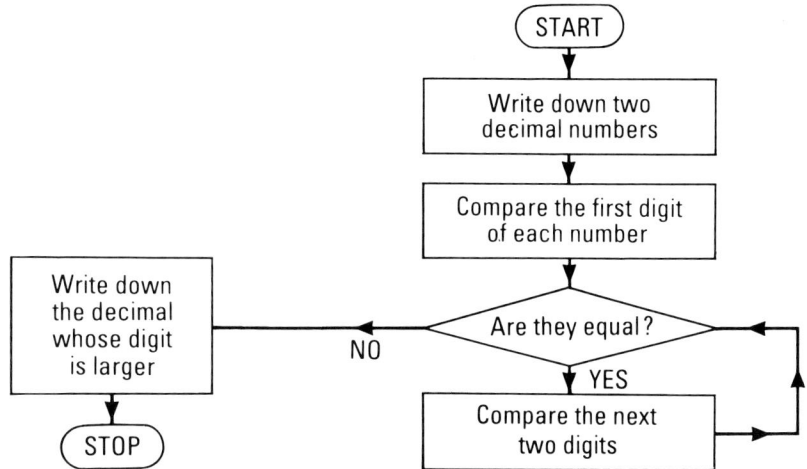

START

Write down two
decimal numbers

Compare the first digit
of each number

Are they equal?

Write down
the decimal
whose digit
is larger

NO

STOP

YES

Compare the next
two digits

5   These are the times for three athletes in a 100 m sprint:

Green 10.752 s   Cannon 10.8 s   Perry 10.76 s

Write the names of the athletes in order, the winner first.

6   Use your calculator to help you with these:

   a) Which is heavier, $\frac{14}{25}$ tonne or $\frac{3}{5}$ tonne?

   b) Which is longer, $\frac{1}{8}$ km or $\frac{5}{36}$ km?

   c) Which is the shorter time, $\frac{3}{20}$ s or $\frac{4}{25}$ s?

   d) Which is greater, $\frac{5}{9}$ or $\frac{3}{5}$?

═══════════ CHALLENGE ═══════════

7 a) Which of these
       is the best value?

BUTTER FULL CREAM   0·3kg   £1·18

BUTTER SALTED   0·33kg   £1·30

BUTTER ⅓kg   £1·13

   b) Which of these
       is the best value?

NmP SHERRY 0.6cl   £2·80

NmP SHERRY 0·66l   £3·10

NmP SHERRY ⅔l   £3·09

Thirds

1 a) Write down what your
       calculator display shows for:

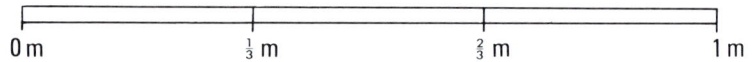

   (i) $1 \div 3$ (ii) $2 \div 3$

   b) Which is the best estimate for $\frac{1}{3}$m: 0.2m, 0.3m, or 0.4m?

   c) Which is the best estimate for $\frac{2}{3}$m: 0.4m, 0.6m, or 0.7m?

━━━━━━━━━━━━ TAKE NOTE ━━━━━━━━━━━━

These approximations are often used for $\frac{1}{3}$:   0.3 or 0.33.

These approximations are often used for $\frac{2}{3}$:   0.6 or 0.66 or 0.67 or 0.7.

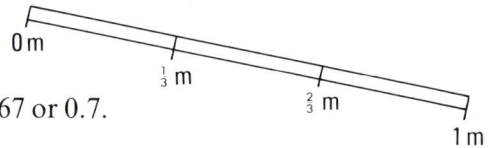

━━━━━━━━━━━━━━━━━━━━━━━━━━━━━━━━━━

   d) Which of the estimates for $\frac{2}{3}$ in the *Take note* are:

       (i) rounded down estimates
       (ii) rounded up estimates?

2   Are you closer to $1\frac{1}{3}$m tall, or to $1\frac{2}{3}$m tall?

3   Arrange these amounts in order, from smallest to largest.

    0.2 $\frac{1}{3}$ 0.6 $\frac{3}{10}$ $\frac{2}{5}$ 0.5 $\frac{2}{3}$

━━━━━━━━━━━━ WITH A FRIEND ━━━━━━━━━━━━

4 a) You should both know these facts (below). Take turns to test each other until you *do* both
       know them.
       (For example, ask 'What is 0.5 as a vulgar fraction?', 'Roughly, what is $\frac{1}{3}$ as a decimal fraction?')

    $0.5 = \frac{1}{2}$ $0.01 = \frac{1}{100}$ $0.1 = \frac{1}{10}$ $0.25 = \frac{1}{4}$

    $0.2 = \frac{1}{5}$ $\frac{2}{3} \approx 0.7$ $0.75 = \frac{3}{4}$ $\frac{1}{3} \approx 0.3$

   b) Learn some more facts like these. Make a list of all the facts like these that you know. Add
       new facts to your list as you learn them.

REVIEW

Cube

Cuboid

Triangular prism

Cylinder

Hexagonal prism

PRISMS

SECTIONS

Cone

Triangular-based pyramid

Square-based pyramid

PYRAMIDS

Curved edge

Vertices

Plane face

Plane face

Edge

Curved surface

SPHERE
No edges
No vertices

Height

Base

■ How many edges, plane faces, and vertices does a triangular-based prism have?

NETS

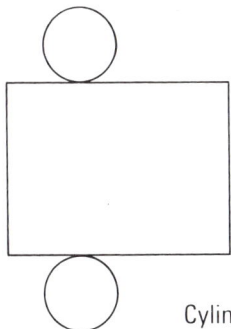

Cylinder

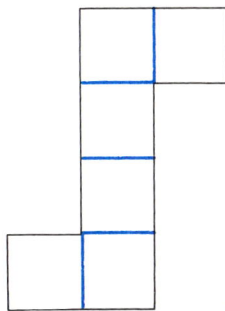

Cube

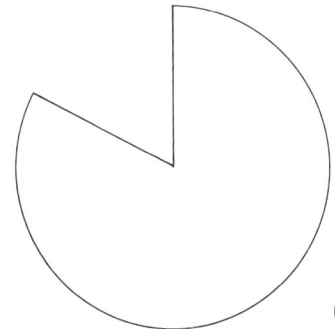

Cone

■ Sketch a different net for a) a cylinder, b) a cube.

A4

CONSOLIDATION

1 a) Which of these solids are prisms, which are pyramids, and which are neither?

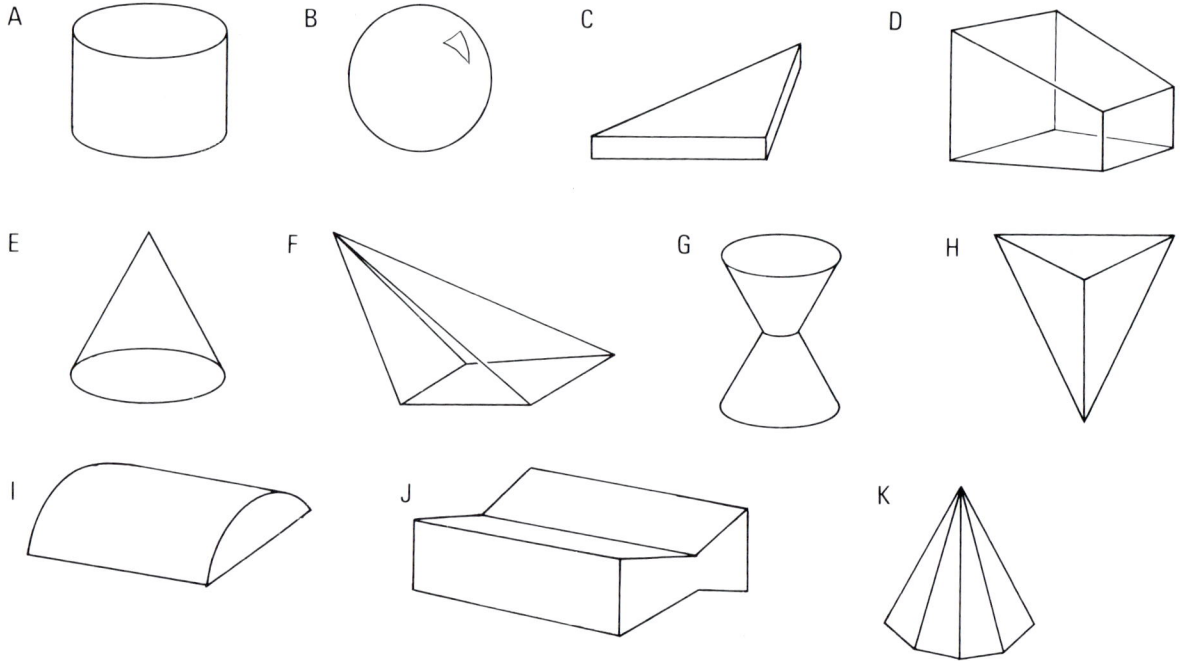

b) Sketch your own example of (i) a prism, (ii) a pyramid.

2 a) Check that this pyramid has: 8 vertices
14 edges
8 plane faces
and 0 curved faces.

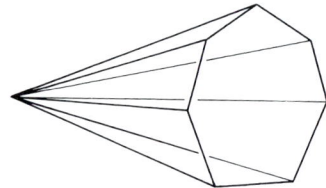

b) Write down how many vertices, edges, plane faces, and curved faces each of the solids in question 1 has.

3 These are the nets of three boxes. (The 'flaps' for glueing have not been included.) The blue lines are fold lines. Make a sketch of each completed box.

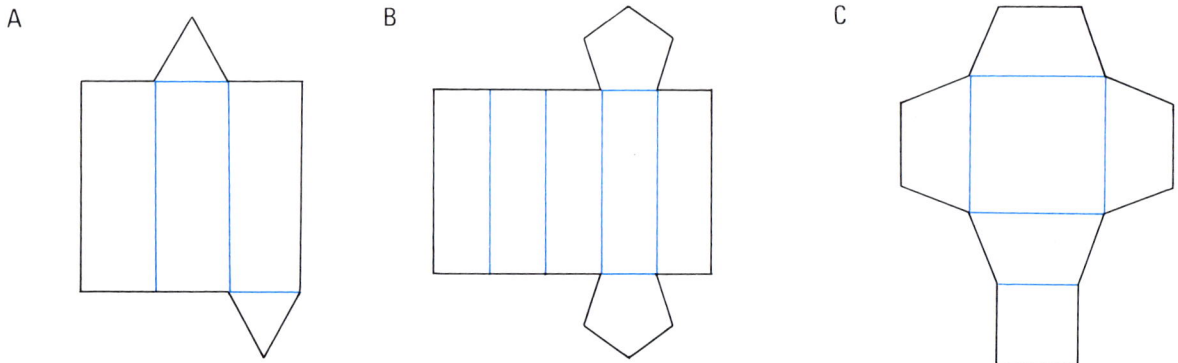

4   a) Draw a solid which has 4 faces, 4 vertices, and 6 edges.

b) Sketch a net for your solid.

5   These are three prism-shaped tubes, made from card.
A and B are nets for some prism-shaped tubes. Sketch *two* possible tubes for each net.

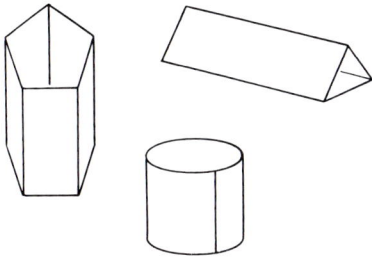

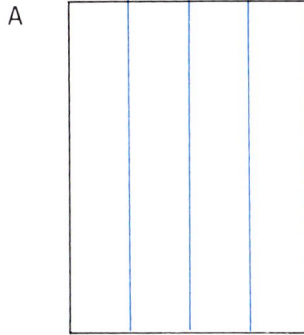

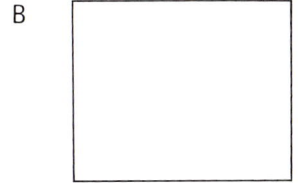

6   Sketch a net for each of these solids.

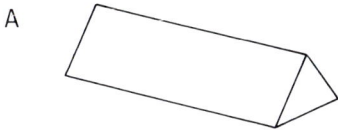

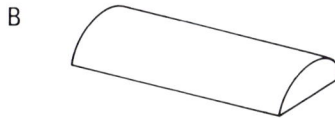

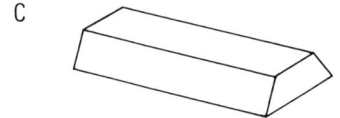

A   B   C

7   Sketch a solid which has: 1 curved surface
                              2 plane surfaces
                              and 2 edges.

======= CHALLENGE =======

8   a) Sketch *three* different tubes which can be made from this net.

b) Here is another net. What does the net for the new solid look like? Sketch it.

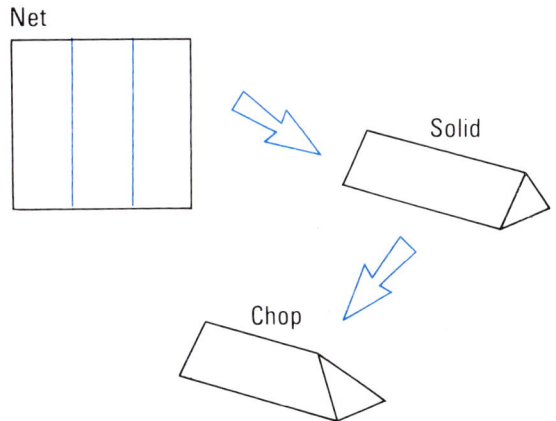

Net

Solid

Chop

A4

ENRICHMENT

Do one of these activities.

1   You need squared paper, glue or tape, and scissors. Make a triangular-based pyramid with height 8 cm and area of base 30 cm².

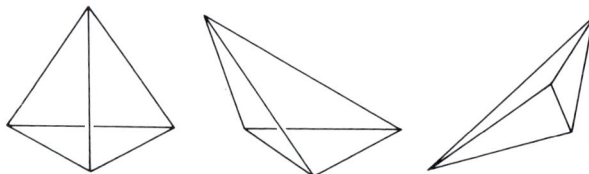

2   You need 1 cm squared paper, thin card, and glue or tape. These are 'chocolate cylinders'. They are 2 cm wide and 1 cm tall.
The manufacturer wants a box which will hold 10 cylinders ... but the box must not be a boring cylinder.

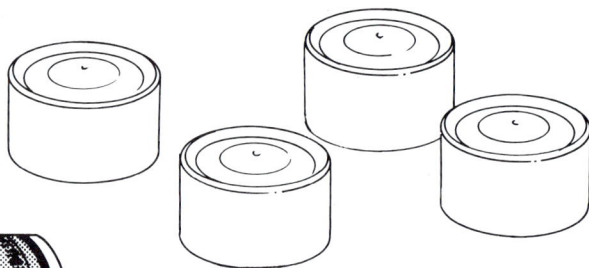

Design your own box. Sketch the net of your box, and make it from thin card.

3   Choose any fragile object which you might want to send by post.

Design and make a box for it. Do your best to make sure that your object will survive the journey safely. (No extra packing such as tissue paper is allowed.)
Write a report to say how you ensured the box protected your object. Include the net of your box in your report.

REVIEW

- We often round measurements or amounts up or down.

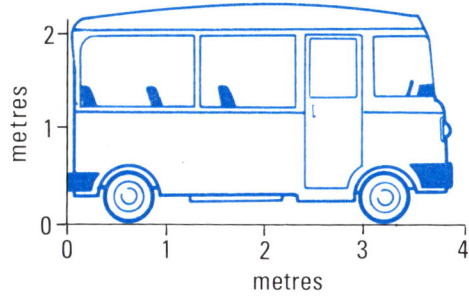

The coach is about 4 m long ... and about 2 m tall.
   (length rounded up)   (height rounded down)

- It can be very important to decide whether to overestimate (*round up*) or underestimate (*round down*).

Lorry and coach drivers should underestimate the height of a bridge.

You should overestimate the time it will take you to cross a road.

■ Should lorry and coach drivers underestimate or overestimate the heights of their lorries and coaches?

■ Should you underestimate or overestimate the speed of the traffic when you are about to cross a road?

1294 is
- 1290 rounded to the nearest 10
- 1300 rounded to the nearest 100
- 1000 rounded to the nearest 1000

£8.47 is
- £8 rounded to the nearest £1
- £8.50 rounded to the nearest 10p
- £8.45 rounded to the nearest 5p

175 is 180 rounded to the nearest 10. (For 'halfway' numbers, we always round up.)

- We use rounded numbers to make estimations. We can round up or down.

About 4 km   2.3 km

5.7 km   3.8 km   3.5 km

About 6 km   About 2 km   About 4 km

Total distance $\approx (6+4+2+4)\,\text{km} = 16\,\text{km}$

■ Estimate the total distance along the road by rounding each amount to the nearest $\frac{1}{2}$ km.

CONSOLIDATION

IN YOUR HEAD

1   a) You have £5. Which of these sets of items can you afford? Write only 'Yes' or 'No' for each one.

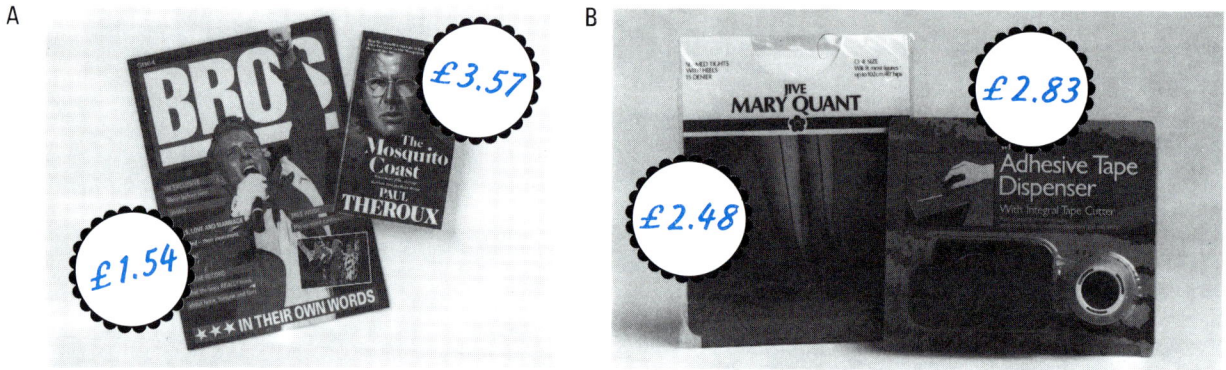

A

£3.57

£1.54

B

£2.83

£2.48

b) You have £10. Which of these sets of items can you afford? Write only 'Yes' or 'No' for each one.

A

£1.78

£4.24

£5.17

B

£3.49

£3.15

£3.12

C

£5.37

28p

£4.53

2   Here is one way of estimating £13.42 + £2.51.

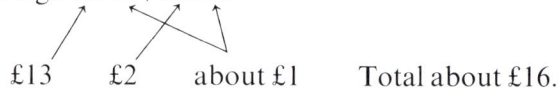

£13   £2   about £1   Total about £16.

Use the same method to do these:

a) £11.67 + £3.27   b) £9.68 + £10.78   c) £3.47 + £3.62 + £5.49

3   Here is one way of estimating £13.42 − £2.51.

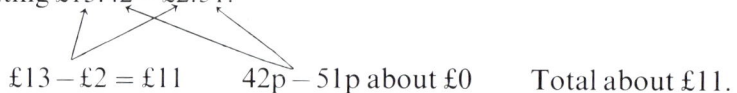

£13 − £2 = £11   42p − 51p about £0   Total about £11.

Use the same method to do these:

a) £17.64 − £3.59   b) £18.09 − £10.93   c) £16.34 − £5.94

4   Here is one way of estimating $9 \times £2.62$. $9 \times £2.62$

$9 \times £2 = £18$ about $9 \times 50\text{p} = £4.50$ Total: about $£22.50$

Here is another way: $9 \times £2.62$ is about $10 \times £2.62$, which is $£26.20$.

Study each method carefully. Use one of the methods to do these:

a) $11 \times £5.36$ b) $4 \times £8.43$ c) $12 \times £5.16$ d) $3 \times £8.89$

5   Here is one way of estimating $£17.57 \div 9$.

$£17.57 \div 9$ About $£18 \div 9$, which is $£2$.

Here is another way:

$£17.57 \div 9$ About $£17.57 \div 10$, which is about $£1.75$.

Study each method carefully. Use one of the methods to do these:

a) $£13.94 \div 9$ b) $£31.48 \div 11$ c) $£17.67 \div 3$ d) $£2.87 \div 6$

6   We can estimate $£39.24 \div 18$ as in this diagram: $£39.24 \div 18$

about $£40$ about 20 $£40 \div 20 = £2$

Estimate each of these. For each one draw a diagram to explain how you made the estimate.

a) $£73.37 \div 11$ b) $4.63\,\text{m} \times 5.8$ c) $19.7 \div 4.2$

A5

7   Maeve wants to estimate the total cost of the four items.

a) Each of you make the estimate in your head. Write down your result.

b) Each of you write down how you made your estimate. (Draw a diagram like that in question 6.)

c) Compare the methods you used: are they the same, or different? Decide between you which method you would prefer to use next time. Write down what you decide, and why.

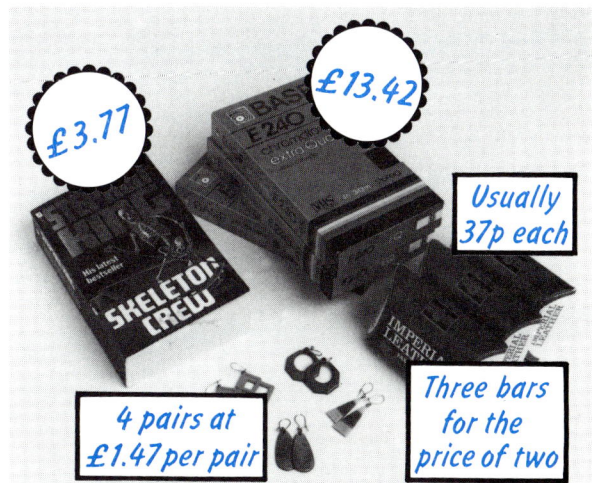

£3.77

£13.42

Usually 37p each

SKELETON CREW

4 pairs at £1.47 per pair

Three bars for the price of two

A5

OBSERVER SOFABEDS

only £149.95, with metal spring action only £199.95

Our modern two seater sofabeds are available with foam pull out action at £149.95 and with a more substantial metal spring action at £199.95. Made in England the sofabeds are constructed on a sturdy wooden frame with base cushions filled with the latest combustion modified heat resistant foam which is less easy to ignite. Additional support is given by the foam bolster at the back and the stitched soft quilting over the seat, back and arms provide extra comfort. The pretty 100% cotton fabric is pink or blue. The sofabeds convert quickly and easily and are very good value for money especially as the price includes delivery by private carrier – please give a daytime telephone number. Pairs of matching 15¾" square foam filled cushions are available at £15.95 and if you would like to make curtains, lengths of material (55" wide) are £6.95 per metre. If you want to give your TV room, study or guest room a new fresh look this sofabed is ideal.

Dimensions: standard sofabed W58 D27 H27ins
Sleeping area 78 x 48ins
Metal spring action W61¾ D29½ H27½ins
Sleeping area 45¼ x 73¼ins
Sofabeds may be viewed at 83/117 Euston Road, NW1

--- please cut ---

Please send:

	Pink	Blue
Sofabed £149.95	✓	
Sofabed £199.95		
Pair of cushions £15.95	✓	✓
Curtain material £6.95 per metre	7 metres	

I enclose crossed cheque with address on reverse value £_____
made payable to Observer Offers.
I wish to pay by Access/Visa/Amex.

Card number: [9 8 3 4 4 6 6 2 2 7 4 5 9]

Signature __R. Harmen Kerr__ Expiry date: 1/11/92
UK mainland only. Subject to availability. Allow 28 days for delivery.
Send to Observer Sofabeds, PO Box 796, London NW1 2RF.

Name __R HARMKEN__
Address __64 STRANTLEY DRIVE__
__WORKSOP NOTTS__
Postcode __NV1 4NZ__ Daytime tel no. __70 9289__

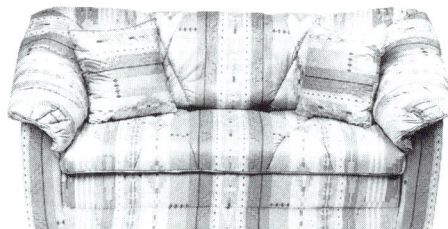

**PHONE
01-387 3313
TO PLACE
ACCESS/VISA
ORDERS.**

8 a) In your head, estimate the total cost of the order. Try to get within about £20 of the true cost.

   b) Does your estimate suggest the total cost is more than £200 or less than £200?

   c) Calculate the exact cost of the order. Were you successful in a)?

9   Estimate how much water you use, on average, each day. In your report explain how you made your estimate, and how accurate you think it is.

A5

A6 AREAS OF SHAPES

REVIEW

● Area of a rectangle = length × width.

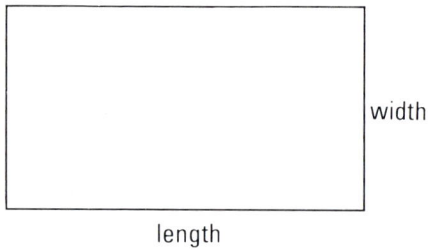

● Area of a parallelogram = base length × height.

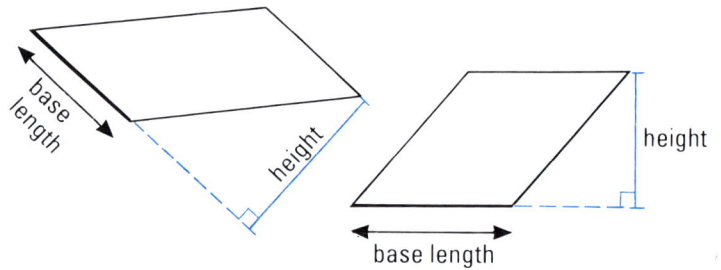

■ How are the areas of parallelograms and rectangles connected?

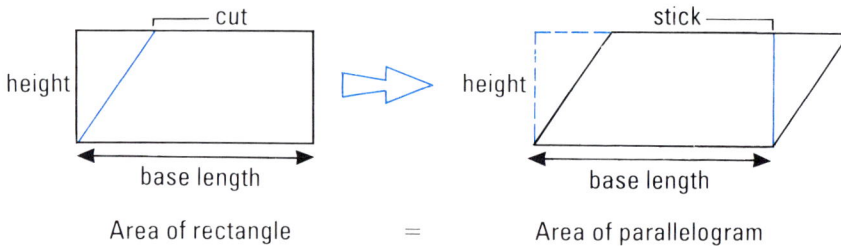

Area of rectangle = Area of parallelogram

● Area of a triangle = $\frac{1}{2}$ × (base length × height).

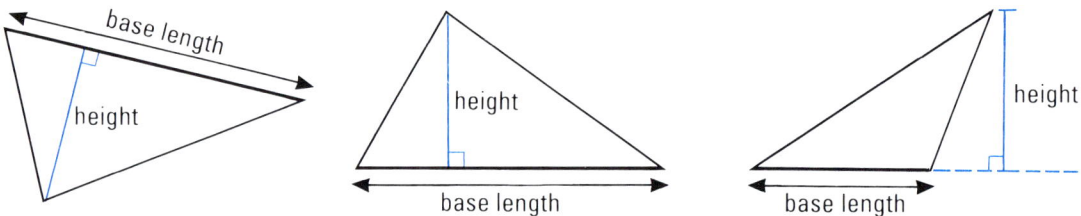

■ How does the area of a triangle relate to the area of ...
a parallelogram ... a rectangle ...?

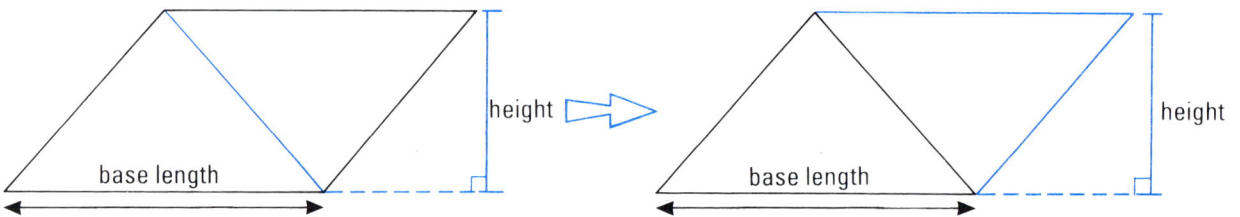

● Area of parallelogram = 2 × area of triangle.

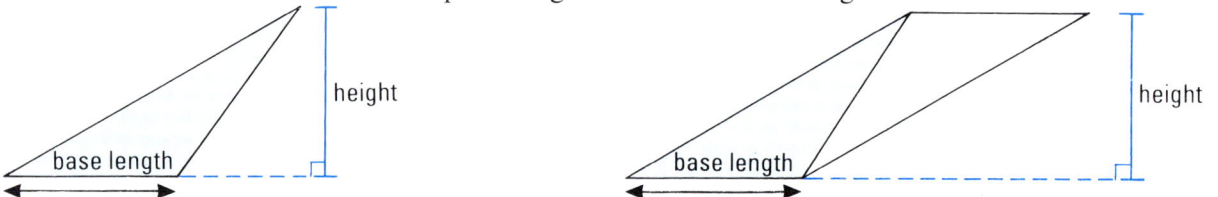

● Area of triangle = $\frac{1}{2}$ × area of parallelogram

CONSOLIDATION

1   The parallelogram has been made into a rectangle.

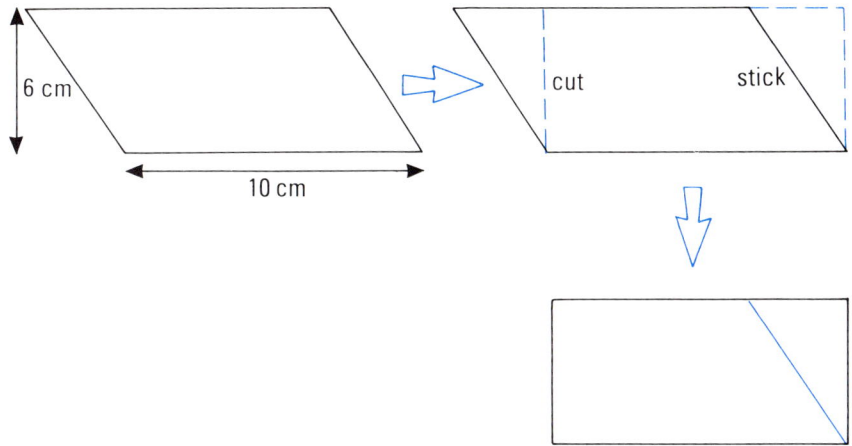

6 cm

10 cm

cut   stick

What is the area of:

a) the rectangle

b) the parallelogram?

2   Use 1 cm squared paper. Draw rectangles which have the same area as these parallelograms.

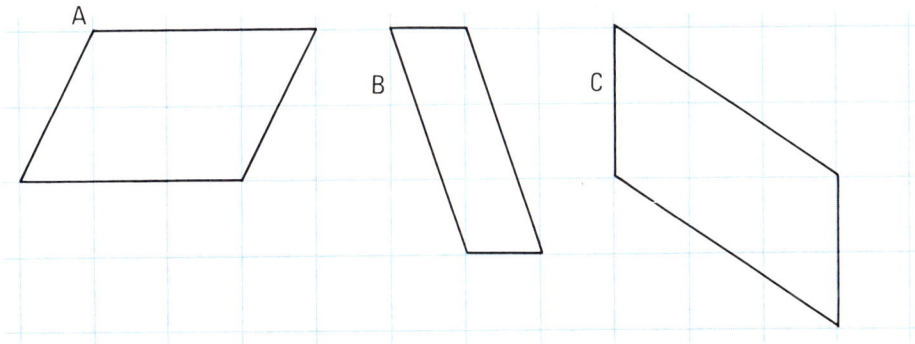

A

B

C

CHALLENGE

3   You need 1 cm squared paper. Make two copies of the parallelogram ABCD. Do not use a set square or protractor.

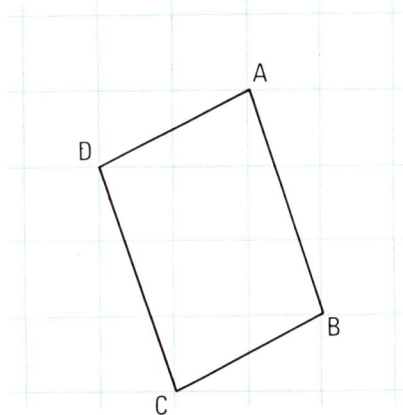

A

D

B

C

a) On your first copy draw a rectangle with the same area. Use AB as one side of your rectangle.

b) On your second copy draw another rectangle with the same area. Use CB as one side of the rectangle.

c) Explain how you know that your rectangles and parallelograms have the same area. (Use sketches to help you if you wish.)

A6

4   All the blue rectangles throughout this question are the same size.

   a) List the parallelograms in order, the one with the smallest
       area first.

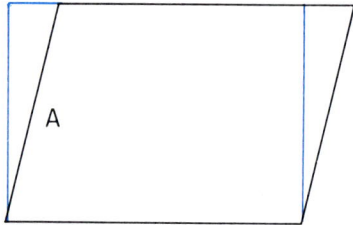

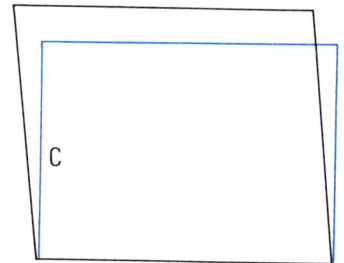

   b) Do the same for these parallelograms,

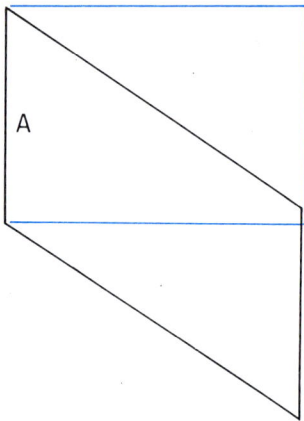

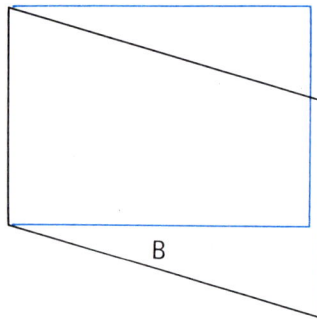

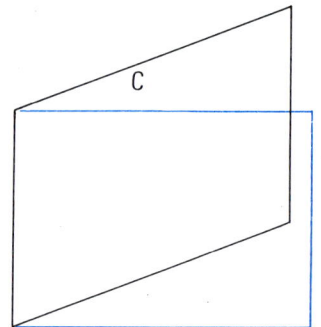

   c) ... and these triangles,

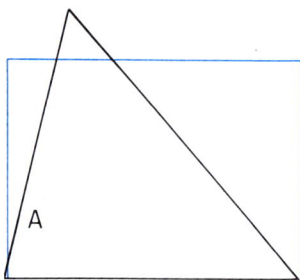

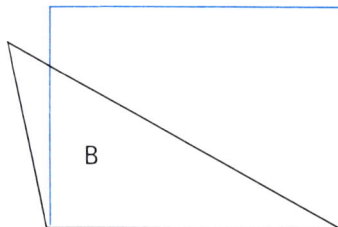

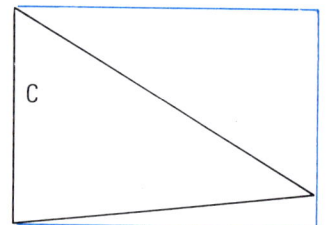

   d) ... and these triangles.

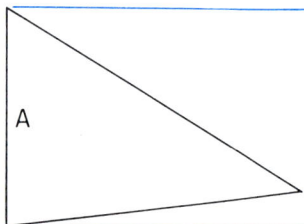

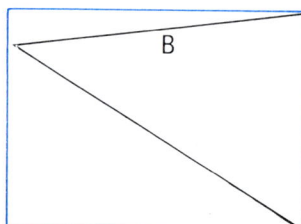

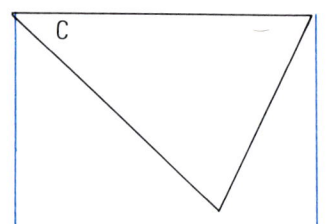

5 a) Count squares to find the area of each shape.

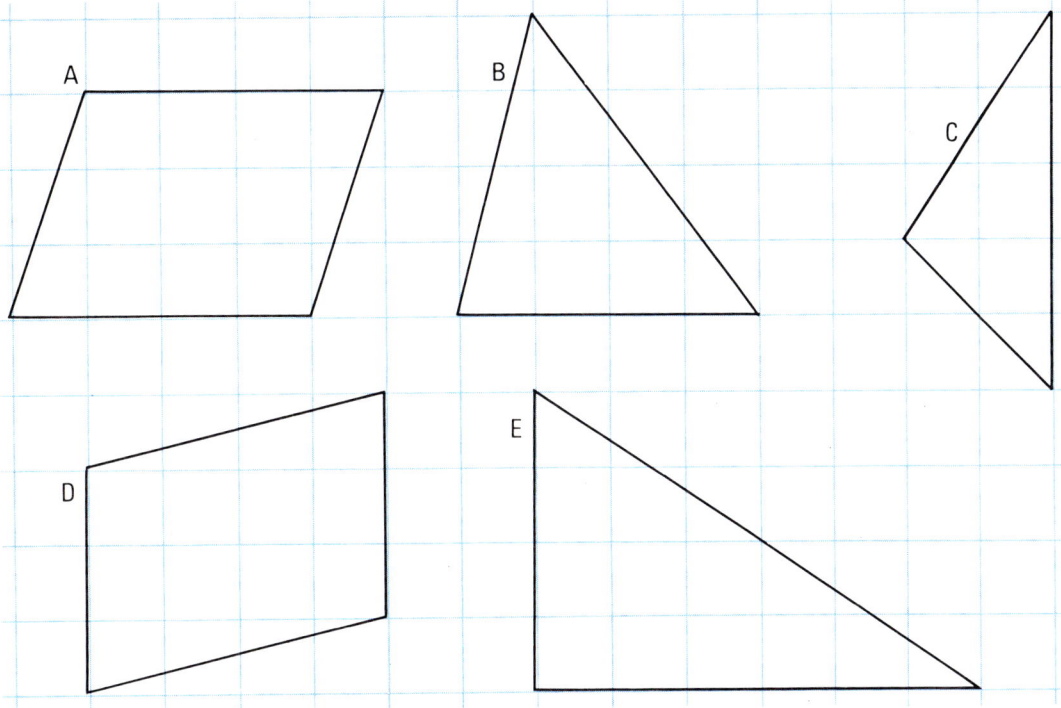

b) Use these formulas to check your results in a).

Area of parallelogram $=$ base length $\times$ height.   Area of triangle $= \frac{1}{2} \times$ base length $\times$ height.

6   Find the area of each shape. Measure any distances you need.

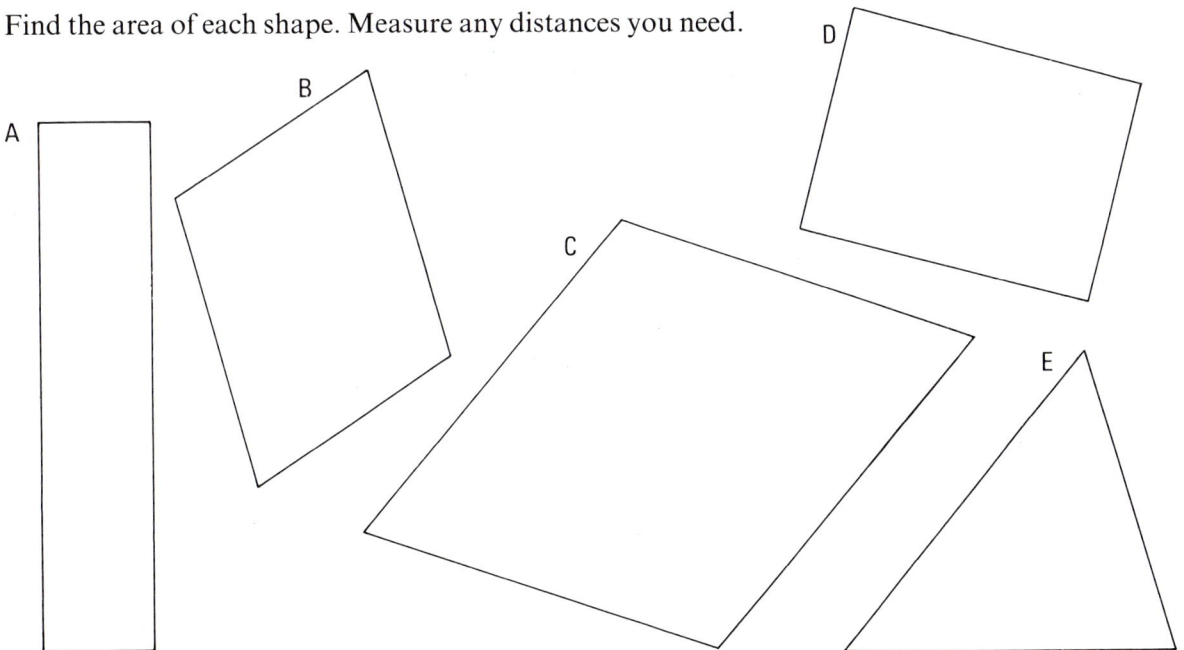

A6

7 a) What is the area of each
      envelope?

   b) Sketch your own
      envelope which has an
      area of 37.5 cm². Write
      down its length and
      width.

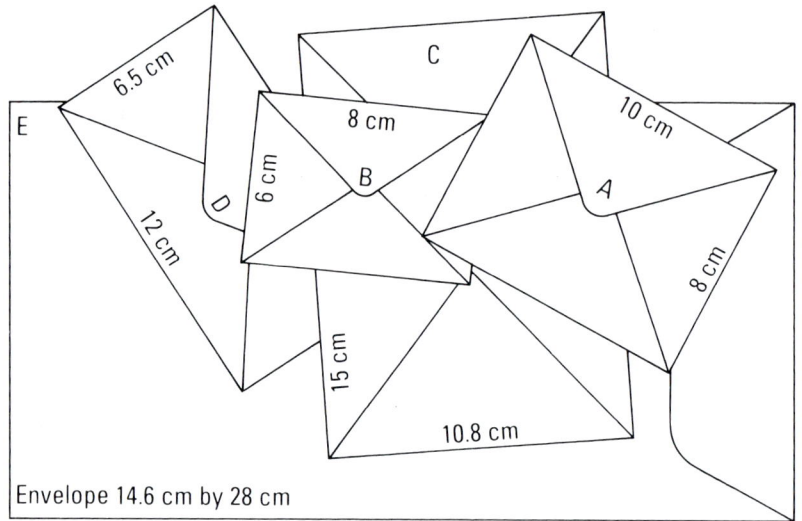

6.5 cm

C

E

8 cm

10 cm

6 cm

B

A

D

12 cm

15 cm

8 cm

10.8 cm

Envelope 14.6 cm by 28 cm

8 These are the backs of
   some designer envelopes.
   The triangular (shaded)
   part is the sticky flap.

   a) What is the area of each
      flap?

   b) Design your own
      envelope. The
      triangular flap must
      have an area of $13\frac{1}{2}$ cm².
      On your design, mark
      the length and width of
      the envelope.

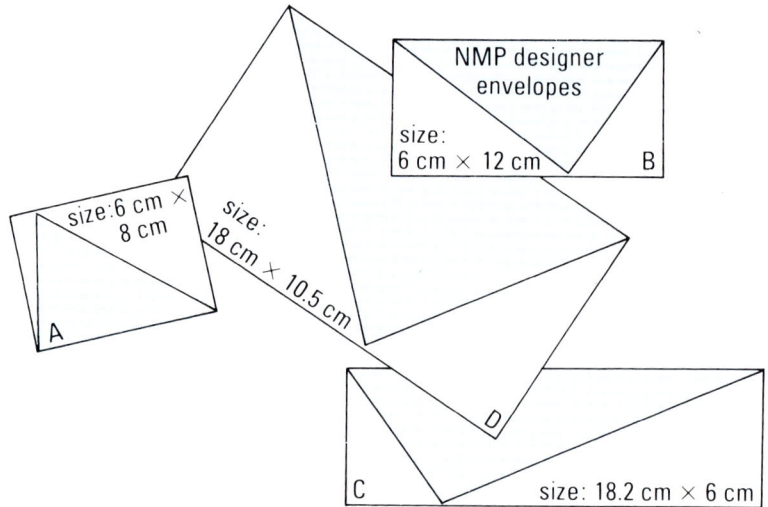

NMP designer
envelopes

size:
6 cm × 12 cm

B

size:6 cm ×
8 cm

size:
18 cm × 10.5 cm

A

D

C

size: 18.2 cm × 6 cm

9 a) Alex needs to find the area of the triangle.
      He says it is half the area of the blue
      rectangle, so it is $7\frac{1}{2}$ cm². Is he correct?
      If you say 'No' explain why, and say
      whether the area is more or less than
      $7\frac{1}{2}$ cm².

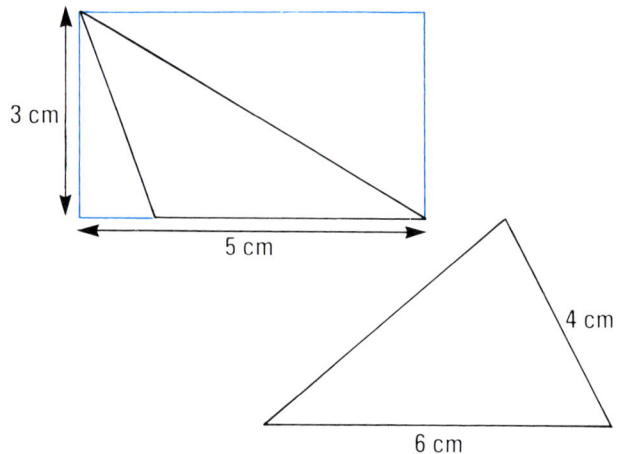

   b) Alex also says that the area of the
      second triangle is 12 cm². Is he correct?
      If you say 'No' explain why, and say
      whether the area is more or less than
      12 cm².

3 cm

5 cm

4 cm

6 cm

A6

ENRICHMENT

1   This is the net of an envelope.

a) What is the length and width of the envelope?

b) Find the area of paper used for the envelope, to the nearest $5\,cm^2$. (You might like to copy the net onto $1\,cm^2$ paper to help you.)

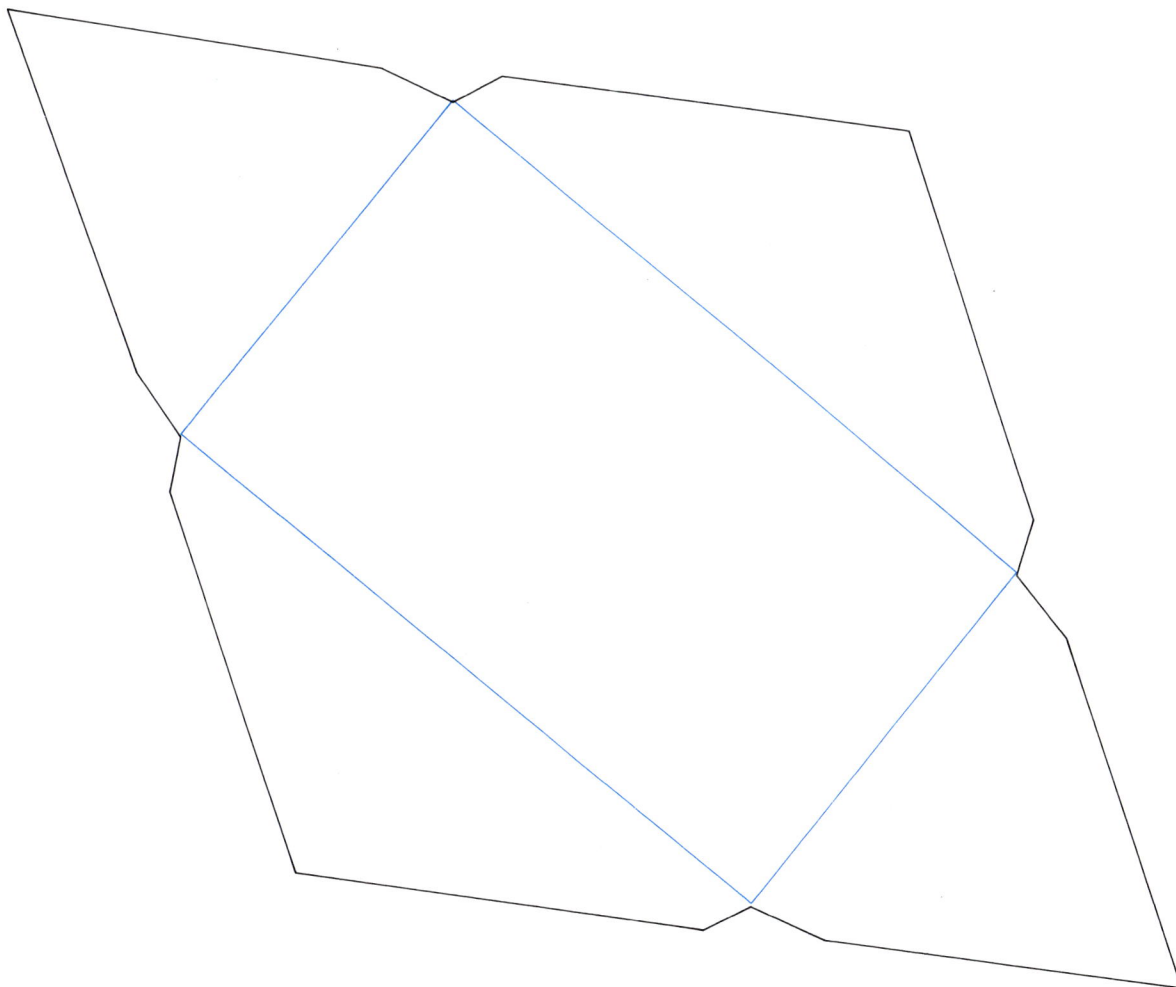

A6

c) Collect some more envelopes. You will find that there are different types of nets. Find as many different types as you can.

d) Decide which design is probably the most economical (that is, uses least paper for the same sized envelope).

e) Design an envelope yourself. Make it, and put your report for this assignment inside it. In your report, include the net for your envelope.

REVIEW

- A shire horse weighs about 1 tonne.
  Guess the weight of the donkey!

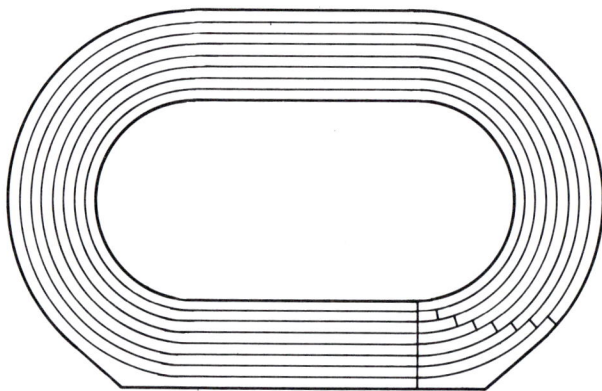

- Sugar is often packed in 1 kg bags.

- A £1 coin weighs about 9.5 g.

- A 1 km race is two-and-a-half circuits of a
  normal running track.

1 km is about $\frac{5}{8}$ of a mile.

- The distance between a man's nose and
  the tips of his fingers is about 1 metre.

● Do this to estimate about 1 cm.

● The distance between the outside of your eye sockets is about 10 cm.

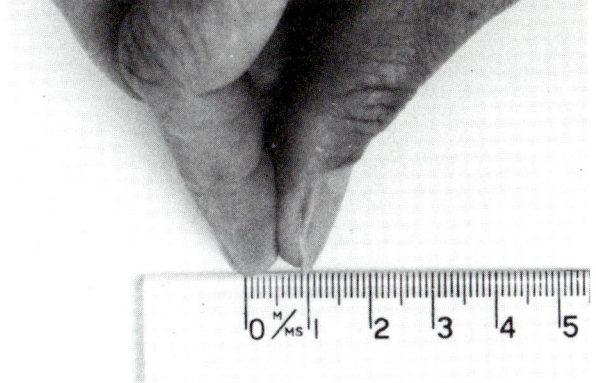

● A 10 cm × 10 cm × 10 cm box holds 1 litre of water.

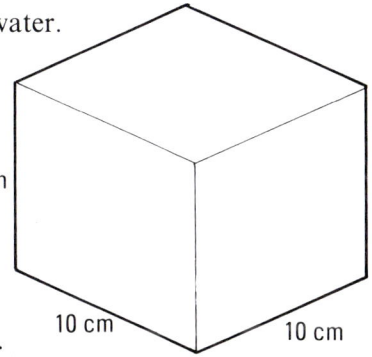

1 litre takes up 1000 cm³ of space.
1 litre of pure water weighs 1 kg.

10 cm
10 cm
10 cm

● Wine bottles normally hold 70 cl or 75 cl of wine (about 4 large or 6 small wine glasses).

BEAUJOLAIS

● A 1 cm × 1 cm × 1 cm box holds 1 ml of water.

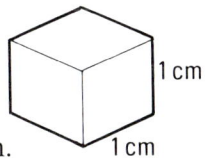

1 ml takes up 1 cm³ of space.
1 ml of pure water weighs 1 gram.

1 cm
1 cm
1 cm

■ In what other ways can we write: 1.54 tonne?   1.54 kg?   15400 g?

■ In what different ways can we write: 1.7 km?   1.7 m?   1.7 cm?

1 tonne = 1000 kg	
1 kg = 1000 g	

1 km = 1000 m
1 m = 1000 mm
1 cm = 10 mm
1 dm = 10 cm

1 cm² ◼ 1 mm²

1 hectare = 10000 m²

1 litre = 1000 ml
1 litre = 100 cl
1 cl   = 10 ml

A7

CONSOLIDATION

WITH A FRIEND

1   For each item, agree on a reasonable estimate for the measurement. Each of you write down what you decide. Sometimes you might want to give a range of possible measurements. (For example, the road bridge 'between … and …'.)

A   The weight of this fully grown pig.

B   The height of this road bridge.

C   The length of the river in this photograph.

D   The mass of this railway engine.

E   The mass of this small can of beans.

F   The amount of milk on this milk float.

G   The amount of wool needed for this sweater.

H   The height of Mount Everest.

I   The diameter of the Moon.

A7

2   Read this carefully.

We can write the height of seedling A in many different ways, for example

$2.3\,\text{cm}, 23\,\text{mm}, 0.023\,\text{m}, 2\,\text{cm}\,3\,\text{mm}, 2\frac{3}{10}\,\text{cm} \ldots$

a) Write the height of seedling **B** in five different ways.

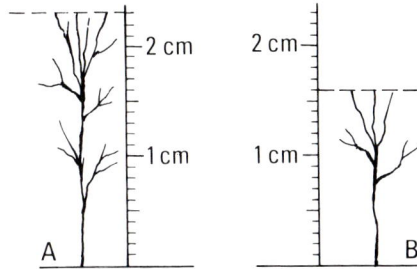

b) Copy and complete each way of writing the length of the boat.

(i) ☐ m ☐ cm   (ii) ☐ . ☐ ☐ m

(iii) ☐ $\frac{☐}{☐}$ m   (iv) ☐ cm

2 m 23 cm

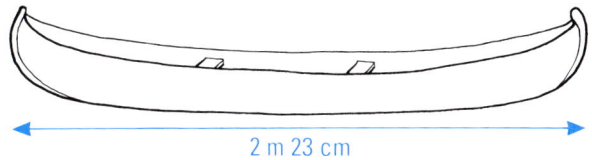

c) The minibus driver fills in the length of the bus on his form like this.

(i) Why is he wrong?
(ii) Write the length correctly in three different ways.

FERRY RATES
complete each part

Car type: Mini Bus VW

Length: 4.9 m

Year of Make: 1985

4 m 9 cm

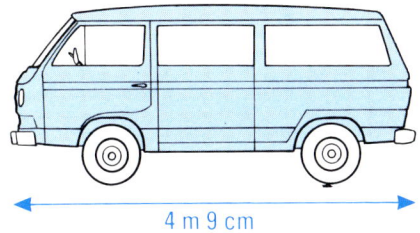

d) Copy and complete these ways of writing the mass of the chimpanzee.

(i) ☐ . ☐ kg

(ii) ☐ kg ☐ g

(iii) ☐ $\frac{☐}{☐}$ kg

e) Write the amount of water in the bottle in three more ways.

75cl
N M P
Sparkling Water

f) Write the distance for this javelin throw in three more ways.

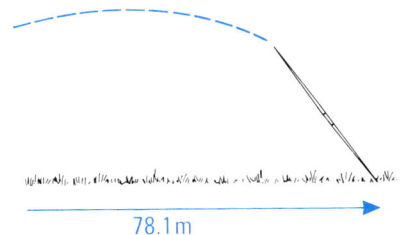

78.1 m

A7

3   a) Write your height:
(i) in centimetres   (ii) in metres   (iii) in kilometres.

b) Write your weight:
(i) in kilograms   (ii) in tonnes.

4 a) A recipe says you need 200 g of butter. Which of these three packs would you buy? Why?

   b) All four orange drinks are the same price. Which is the best buy? Why?

   c) Arrange the signposts in order, the furthest from Land's End first.

   d) Arrange the bookcases in order, longest first.

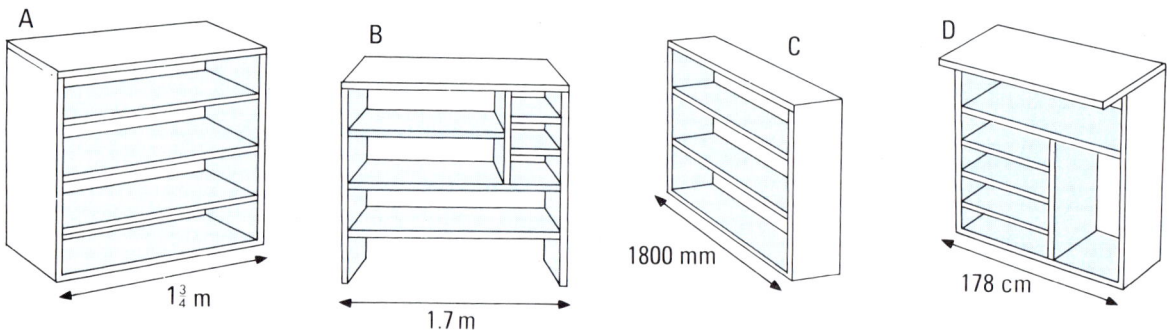

============ CHALLENGE ============

5   1″ (1 inch) is about 2.5 cm.
    Rewrite the dimensions of the cobblestones:

    a) in centimetres
    b) in millimetres.

**We Pave The Way
To Your
Patio - Garden Walls - Pool**

COBBLESTONES
Colour: Buff.
Patterns: Straight, Radius, Tree Planter.
18"x 18"x 1½"......................................£1.73

A7

CORE

━━━━━━━━ CHALLENGE: METRIC TEST ━━━━━━━━

Metric units came into common use in the UK in the late 1970s. At that time many magazines printed quizzes like this one to help people test their knowledge of metric units. (Until the 1970s most people used *imperial units:* pounds, feet, inches, tons, etc.)

14 pounds (lb) = 1 stone (st)
16 ounces (oz) = 1 lb
1 ton = 2240 lb

12 inches (in) = 1 foot (ft)
3 ft = 1 yard (yd)
1 mile = 1760 yd

2 pints (pt) = 1 quart

People didn't find tests like this very easy in the 1970s. Try it yourself.

(Score like this: 14–16 *Very good*; 10–13 *Good*; Below 10 *Could do better!*)

1   Which is heavier?
    1 tonne   1 ton

2   You need 2 litres of water. Would you fetch it in a:
    mug   milk bottle   pail?

3   With this Celsius temperature reading you probably feel:
    normal   feverish   seriously ill?

4   A recipe uses 8 ounces of flour. Which is the nearest weight in grams?
    175 g   200 g   225 g

5   You must wash your new sweater in hand-hot water. In degrees Celsius this would be:
    35°C   110°C   212°C?

6   Jane's hip measurement is 102 cm. She probably wears size:
    12   14   16?

7   You need 6 feet of your chosen fabric for each curtain, say 13 feet for the pair including turnings. So you buy how many metres?
    3   4   5   6.

8   The recommended height for kitchen worktops is 900 mm. This is:
    0.9 m   9 m   9000 cm?

9   Which tin of paint will go further?

10   Your knitting pattern has the instruction Continue in rib until the work measures 12 cm. Is this nearer
    2½in   4 in   5 in?

11   You need your daily dose of medicine. How much does the standard medicine spoon hold?
    5 ml   10 ml   5 cl

12   How many miles away is Little Trudget?
    6   8   10

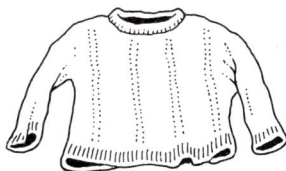

LITTLE TRUDGET
16 KM

13   Your baggage allowance is 20 kg. On the bathroom scales, this is what your baggage weighs. Will you have to pay excess baggage?

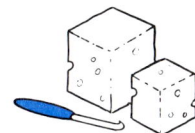

14   The large cube of cheese weighs 445 g. The small cube is half the amount. Is this about ½ lb, 1 lb, or 2 lb?

15   Which jar contains most hand cream? The tall one or the round one?

20 ml

45 cc

16   Which car has the larger engine?

D101 DWW
1500 cc

D211 KLM
1.6 ℓ

1
Litre

1
Quart

A7

Metric and imperial comparisons

In the next three questions you are told how some metric and imperial measures compare. Use the comparisons to do the questions.

METRIC   is roughly   *IMPERIAL*

1   Weight

tonnes
kilograms (kg) $1\,kg \approx 2.2\,lb$ tons
grams (g) $1\,lb \approx 0.45\,kg$ stones (st)
                                                 pounds (lb)
                                                 ounces (oz)

a) You buy 4 kg of potatoes. About how many pounds is this?
b) You buy 8 lb of onions. About how many kilograms is this?
c) Roughly, what is your weight (i) in stones and pounds, (ii) in kilograms?

2   Length/distance

kilometres (km) $1\,km \approx \frac{5}{8}\,miles$ miles
metres (m) $1\,mile \approx 1.6\,km$ yards (yd)
centimetres (cm)   feet (ft or ′)
millimetres (mm)   inches (in or ″)

a) You walk 8 miles. Roughly how many kilometres is this?
b) It is 21 km to the nearest railway station. Roughly how many miles is this?
c) About how tall are you (i) in feet and inches, (ii) in centimetres?

3   Capacity

litres (l) $1\,gallon \approx 4.5\,litres$ gallons (gal)
centilitres (cl) $1\,litre \approx \frac{1}{5}\,gallon$ pints (pt)
millilitres (ml) $1\,litre \approx 1\frac{3}{4}\,pints$
                        $1\,pint \approx 0.6\,litre$

a) The petrol tank of a car holds 12 gallons. About how many litres is this?
b) About how many litres is $\frac{1}{2}$ pt of lemonade?
c) A fish tank holds 300 litres of water. Roughly how many gallons is this?
d) Roughly, how much liquid do you drink each day (i) in pints, (ii) in litres?

4   You go shopping for these things. The shop sells only metric quantities. How much of each do you think you should buy?

a) $\frac{1}{2}$ pint of clotted cream   b) a 2 ft leather thong   c) 3 pints of milk

ENRICHMENT

1   This is a recipe for Lamb in the Pot with Beans. The amounts are given in metric and in imperial units.

This is a recipe from the same book, but the imperial units have been left out. Write down the units missing from brackets A–H.

LAMB IN THE POT WITH BEANS

250 g (8 oz) cannellini beans

500 g (1 lb) lean, boneless lamb

4 large sticks of celery, chopped

1 large onion, sliced thinly

1 tablespoon chopped fresh thyme or 1 teaspoon dried thyme

2 teaspoons chopped fresh rosemary or 1 teaspoon dried rosemary

600 ml (1 pint) chicken stock

300 ml (½ pint) tomato juice

½ teaspoon Tabasco sauce

2 teaspoons paprika

1 clove of garlic, crushed

*Oven temperature:
Gas Mark 4/180°C/350°F*

Put the beans into a saucepan. Cover them with water, bring them to the boil and cook them for 10 minutes. Take them from the heat and leave them to soak for 2 hours. Drain them.

Preheat the oven. Cut the lamb into 2 cm (¾-inch) dice. Layer the beans, lamb, celery, onion and herbs in a deep casserole. Mix the rest of the ingredients together and pour them over the lamb and beans. Cover the casserole and cook it in the oven for 2 hours, or until the beans and lamb are really tender.

VEGETABLE SHEPHERD'S PIE

125 g (A) split red lentils

50 g (B) pot barley or pearl barley

250 g (C) carrots, grated

1 medium-size onion, chopped finely

397 g (D) can of tomatoes

300 ml (E) vegetable stock

750 g (F) potatoes

6 tablespoons milk

75 g (G) Cheddar cheese, grated

*Oven temperature:
Gas Mark 6/200°C/400°F*

Put the lentils, barley, carrots, onion, tomatoes and their juice and stock into a saucepan. Bring them to the boil, cover the pan, and simmer for 40 minutes or until the lentils and barley are soft. Boil the potatoes in their skins. Peel them as soon as they are cool enough to handle and mash them with the milk and cheese.

Preheat the oven. Put the lentil mixture into a 900 ml (H) pie dish. Pile the potatoes on top in an even layer. Make patterns on them with a fork. Put the pie into the oven for 20 minutes or until the ridges on top begin to brown.

ASSIGNMENT: IMPERIAL BATHS

2   Find a large container. It might be a bath, a large tub, a fishtank, … Describe the container using:

a) metric units
b) imperial units.

Give as many measurements as you can. (For example, its mass, its capacity, its length, its mass when full of water …) You might have to estimate some measurements (such as the mass of a bath). Explain how you arrived at each of your measurements or estimates.

A8 SHAPES AND ANGLES

REVIEW

Triangles

Scalene

Isosceles

Equilateral

Right angled

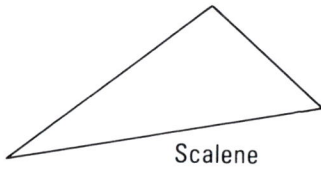

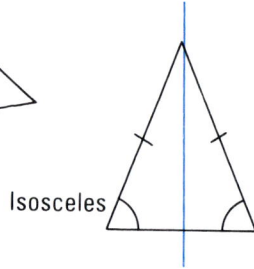

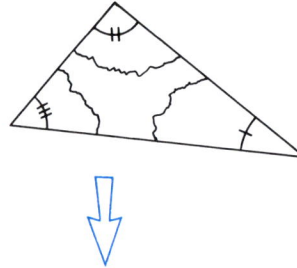

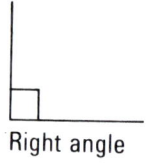

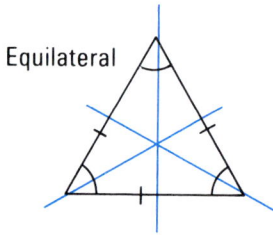

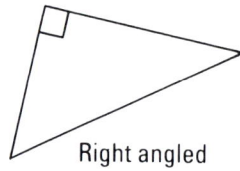

Angles add up to 180°

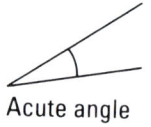

Right angle

Acute angle

Obtuse angle

Reflex angle

■ How many lines of symmetry does each triangle have?

■ What is the order of turn symmetry of each triangle?

Quadrilaterals

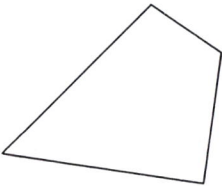

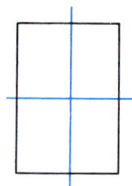

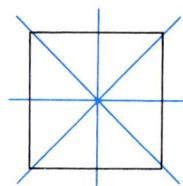

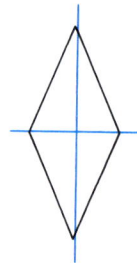

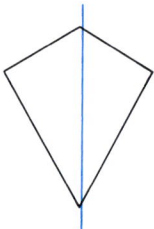

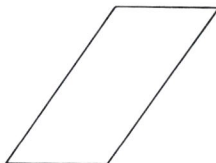

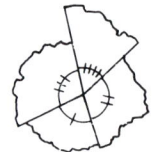

Rectangle

Square

Rhombus

Kite

Parallelogram

Trapezium

Angles add up to 360°

■ How many lines of symmetry does each quadrilateral have?

■ What is the order of turn symmetry of each quadrilateral?

A8

CONSOLIDATION

████████ ACTIVITY ████████

1   You may need dotted isometric and
    dotted squared paper, and scissors.

    a) Using right-angled triangles like A
        we can make other shapes like these.

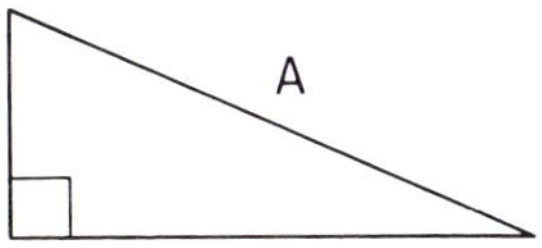

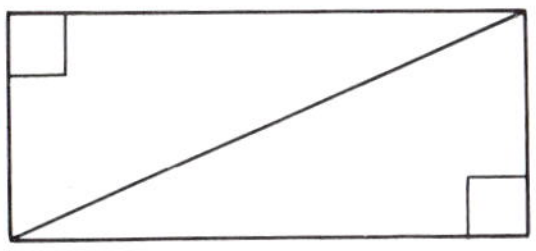

    A rectangle 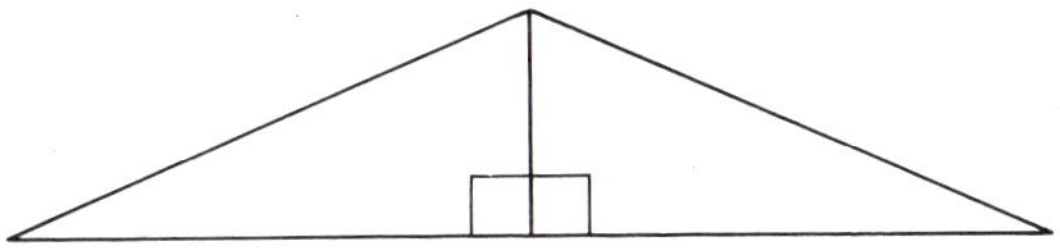   An isosceles triangle 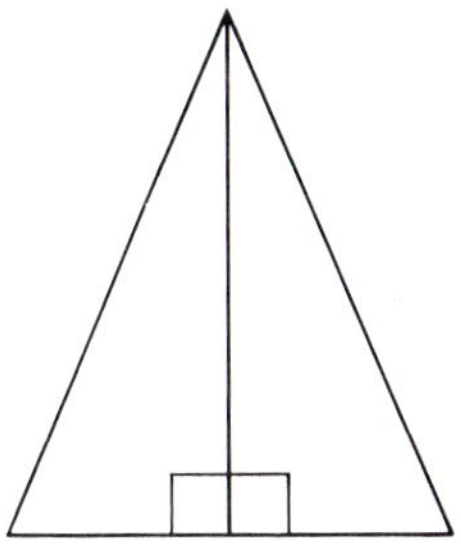   A different isosceles triangle

    Make sketches to show how to make:

    (i) a parallelogram   (ii) a different parallelogram
    (iii) a rhombus (this uses 4 triangles)   (iv) a kite.

    (You might like to draw and cut out some triangles to help you.)

    b) Sketch and name all the different types of
        triangles and quadrilaterals which can
        be made from isosceles triangles
        identical to triangle B.

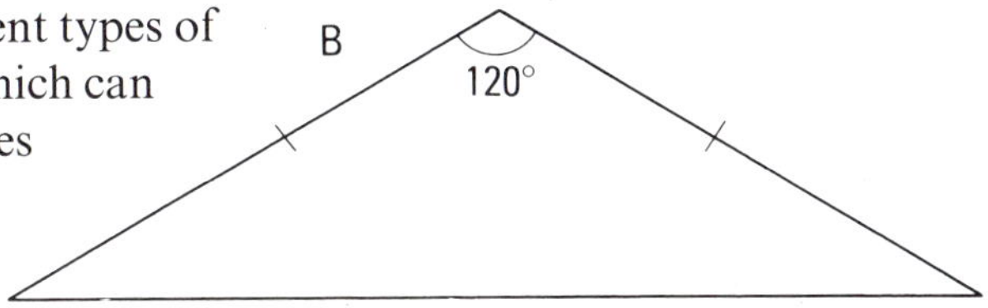

    B   120°

    c) Repeat part b)
        for this isosceles
        triangle.

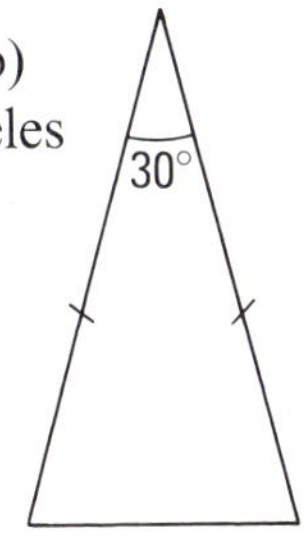

    30°

    d) Repeat part b)
        for an
        equilateral
        triangle.

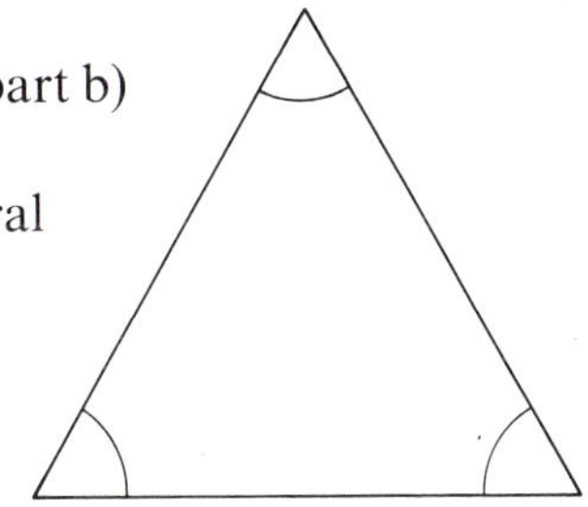

    e) Repeat part b)
        for any scalene triangle.

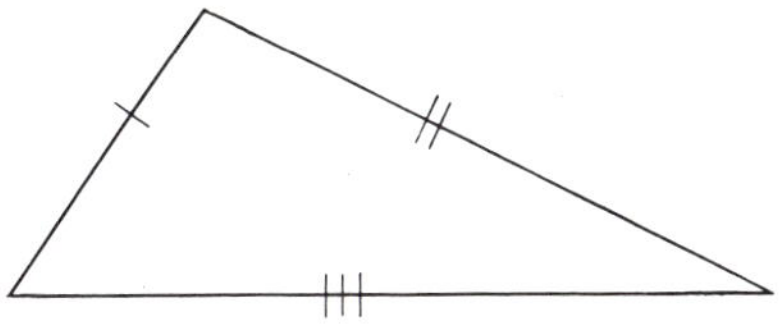

    f) Repeat part b)
        for an isosceles
        right-angled
        triangle.

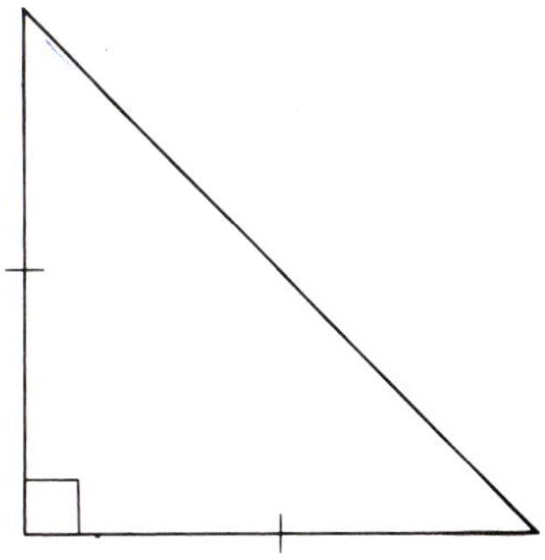

A8

WITH A FRIEND

2   We can see an isosceles triangle in a pair of stepladders. Between you, think of everyday objects where you might see:

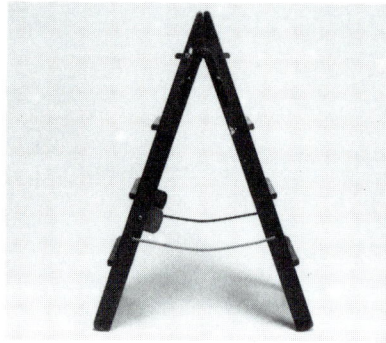

a) a right-angled triangle

b) a rhombus

c) a kite

d) a trapezium.

Each of you write down what you think of.

3   You need a protractor.

a) Try to draw a triangle with angles this size. Explain what happens and why.

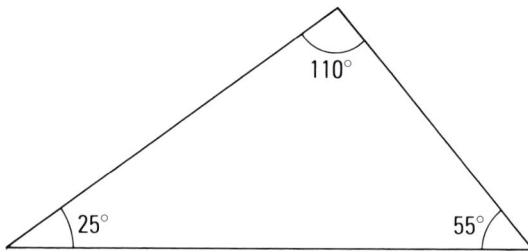

b) Try to draw a quadrilateral with angles this size. Explain what happens and why.

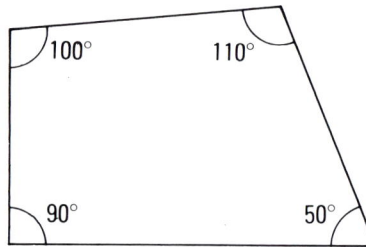

4   Two of these cannot really be triangles. Which are they and why not?

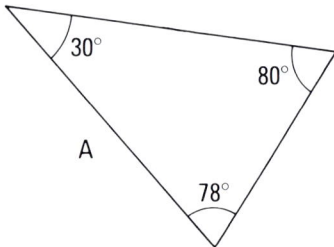

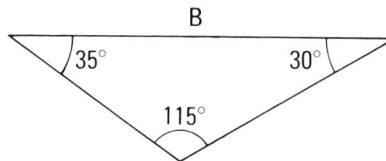

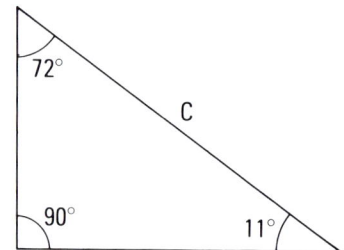

5   Two of these cannot really be quadrilaterals. Which are they and why not?

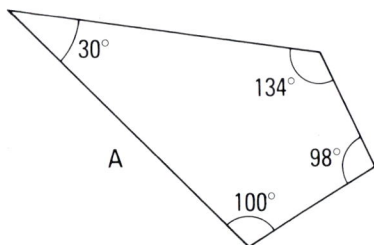

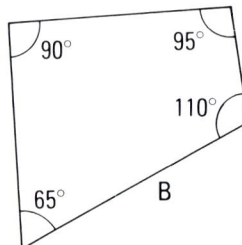

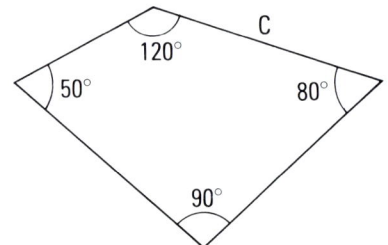

A8

6   a) What is the size of each angle marked with a question mark?

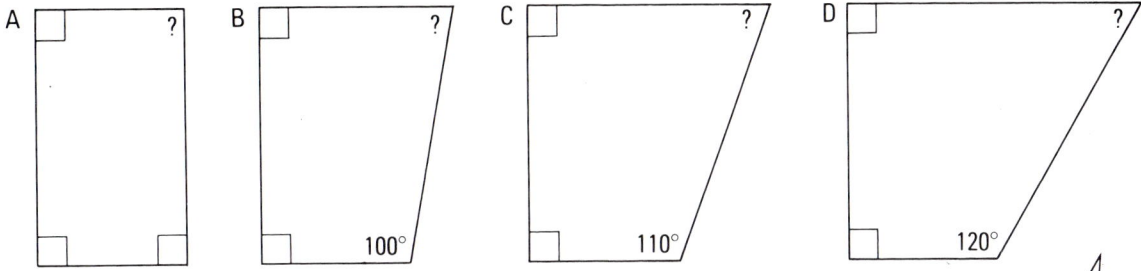

b) What is the size of each angle marked with a question mark?

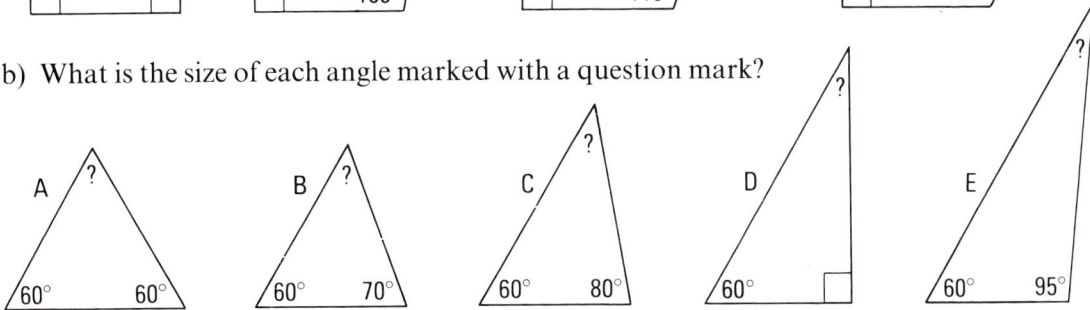

7   These are four pieces of card: a circle, two triangles, and a quadrilateral. What is the size of each angle you cannot see?

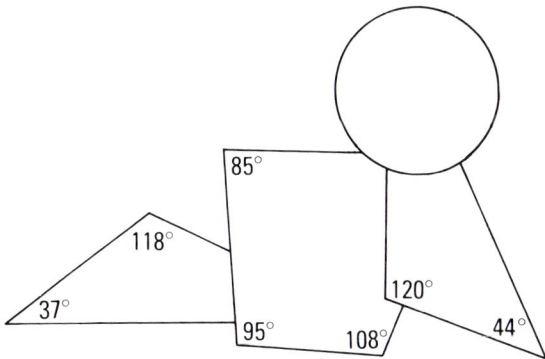

8   The angle at the apex of this barn roof is 126°. The shape you can see is an isosceles triangle. What is the size of the angles marked with a question mark?

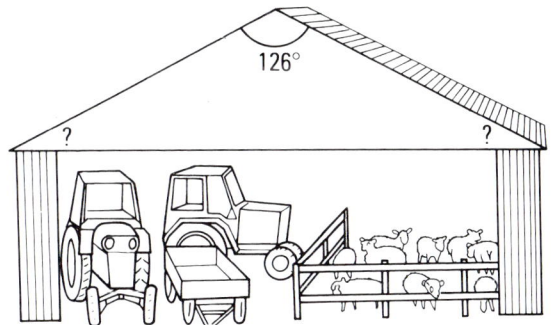

CHALLENGE

9   In this drawing **AB = AC** and **AD = BD**. How many degrees is the angle marked:

a) ■   b) ▲

c) ●   d) *

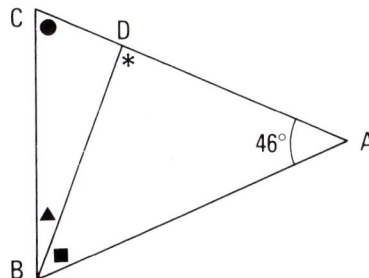

A8

ENRICHMENT

EXPLORATION: INSIDE A CIRCLE

1   You need compasses, a ruler, a very sharp
    pencil, and a protractor. You will need to
    draw and measure very accurately.

a) Draw a circle, any radius. Inside the circle
   draw some triangles. One side of each
   triangle must be a diameter of the circle.

   Measure the angles of your triangles.
   There is something special about one of
   the angles of every triangle you draw.
   What is it?

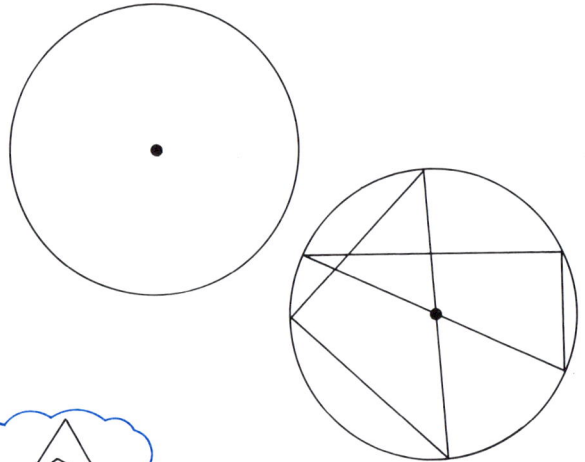

b) Draw some circles. Inside each one, draw a concave quadrilateral, with one vertex at the centre
   of the circle. Measure the angles marked ✳ and ●. What is special about them?

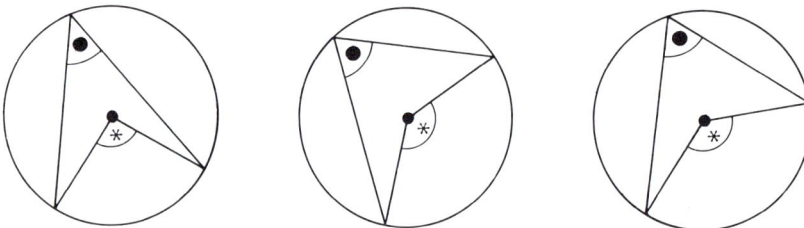

c) Draw some circles. Inside each one draw a quadrilateral. Measure the angles of the
   quadrilaterals. Make a table of your results.

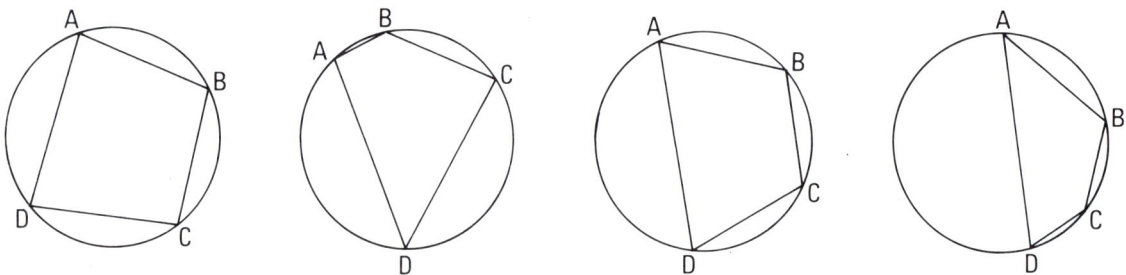

Is there anything special about the angles
which is not special about other
quadrilaterals (that is, those which don't
fit inside circles)?

Angle at	A	B	C	D
	46°	110°		

REVIEW

Eight-ninths ($\frac{8}{9}$) of the area is shaded.

$\frac{8}{9}$ ——— Numerator
——— Denominator

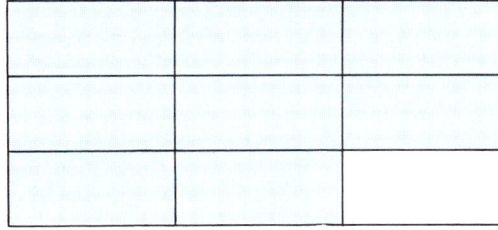

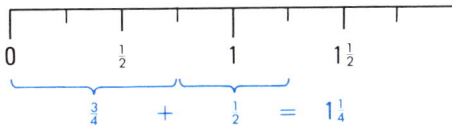

$\frac{3}{4}$ + $\frac{1}{2}$ = $1\frac{1}{4}$

$\frac{3}{4}\ell$ + $\frac{1}{2}\ell$ = $1\frac{1}{4}\ell$

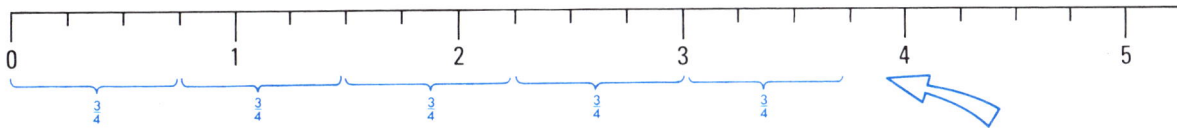

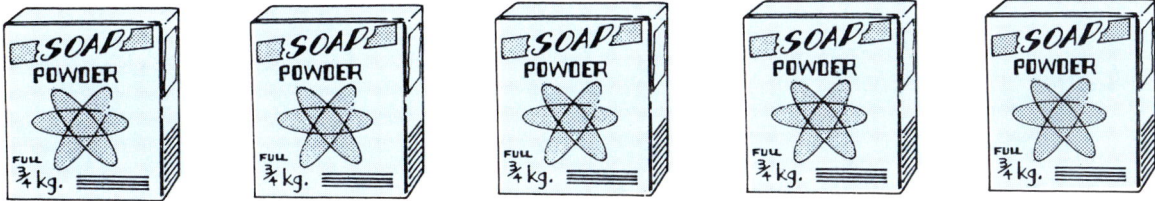

$\frac{3}{4}$ kg × 5 = $3\frac{3}{4}$ kg

■ What is $3\frac{3}{4}$ kg ÷ 5?   ■ What is $3\frac{3}{4} \div \frac{3}{4}$?

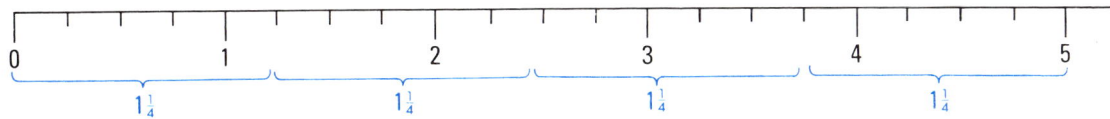

$1\frac{1}{4}$ m $1\frac{1}{4}$ m $1\frac{1}{4}$ m $1\frac{1}{4}$ m

■ What is:

a) $1\frac{1}{4} \times 4$
b) $5 \div 4$
c) $5 \div 1\frac{1}{4}$?

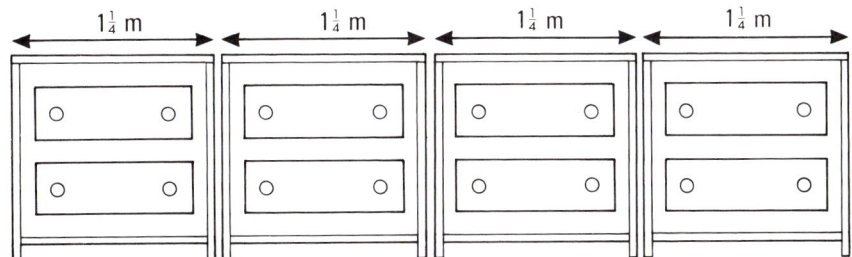

CONSOLIDATION

A9

1   These designs are traditional patchwork patterns used for quilts. Some patterns use three fabrics, some use two.

Flock of geese

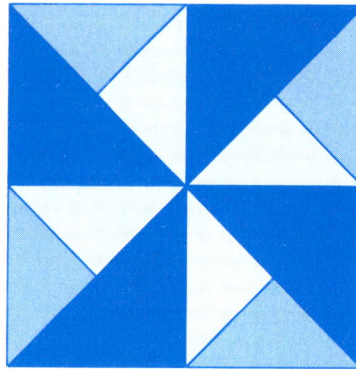

Windmill

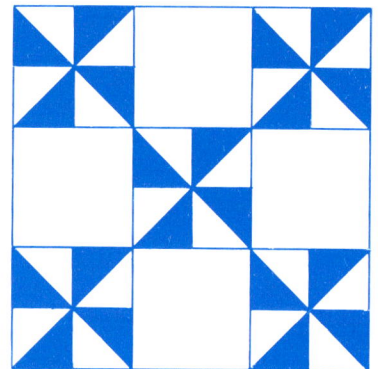

Pin wheels

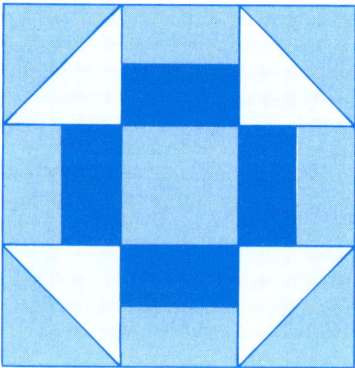

Churn dash

Beggar's blocks

Old Maid's puzzle

a) Choose one of the patterns with two fabrics. Find what fraction of the quilt is covered by each fabric.

b) Choose one of the patterns with three fabrics. Find what fraction of the quilt is covered by each fabric.

c) Use 1 cm squared paper. Design a patchwork pattern which uses two fabrics. One fabric should cover $\frac{9}{25}$ of the pattern.

d) Design another patchwork which uses three fabrics. Write down what fraction of the pattern is covered by each fabric.

═══ WITH A FRIEND ═══

2 a) Here is a 'never-ending' squares pattern for a patchwork quilt. Suppose the pattern could really be continued forever … the squares getting smaller and smaller. Decide between you what fraction of the quilt would be covered by each type of fabric. Each of you write down what you decide, and why.

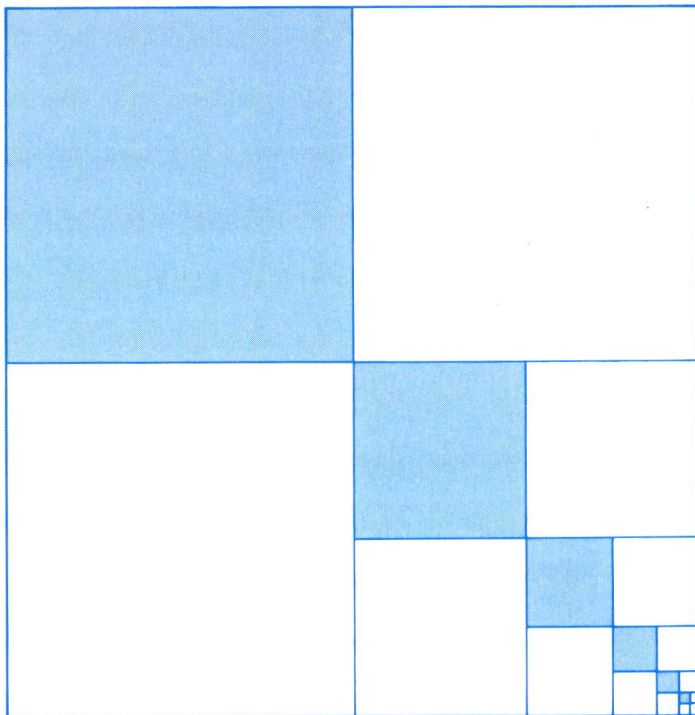

   b) Here is another 'never-ending' pattern for a quilt. Decide between you what fraction of the quilt would be covered by each type of fabric. Each of you write down what you decide, and why.

   c) Each of you design your own 'never-ending' patterns for a quilt. Try to decide what fraction of each of the quilts would be covered by each type of fabric.

Mixed numbers

1   Check from the diagram in the *Take note* that $2\frac{3}{4}$m is 'eleven $\times \frac{1}{4}$ metres'.
    We can write this as $\frac{11}{4}$m. (We say 'Eleven over four metres'.)

    Write each of these in the same way: a) $1\frac{1}{4}$m   b) $2\frac{1}{4}$m   c) $3\frac{3}{4}$m

═══════════════ TAKE NOTE ═══════════════

To help us with fractions we can draw line diagrams.

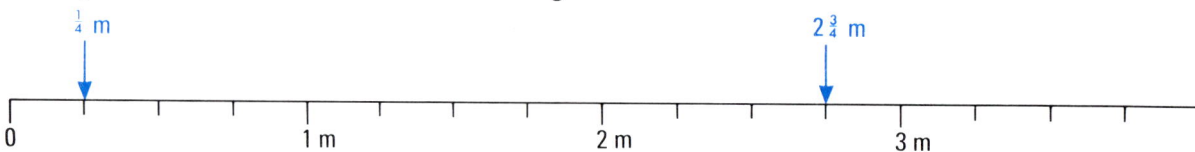

2   Use the diagrams to help you
    to complete these:

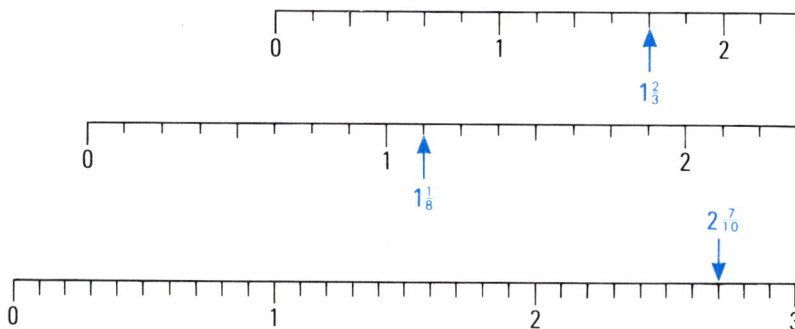

a) $1\frac{2}{3} = \frac{\square}{3}$ b) $1\frac{5}{6} = \frac{\square}{6}$

c) $1\frac{3}{4} = \frac{\square}{4}$ d) $2\frac{1}{2} = \frac{\square}{2}$

e) $1\frac{1}{8} = \frac{\square}{8}$ f) $2\frac{3}{8} = \frac{\square}{8}$

g) $1\frac{4}{5} = \frac{\square}{5}$ h) $2\frac{7}{10} = \frac{\square}{10}$

3   Write each of these as mixed numbers
    (the diagrams in question 2 will help you):

    Whole numbers and vulgar fractions;
    for example, $2\frac{1}{8}$, $3\frac{1}{4}$ .....

    a) $\frac{7}{3}$ b) $\frac{7}{2}$ c) $\frac{9}{4}$ d) $\frac{8}{6}$ e) $\frac{11}{5}$ f) $\frac{14}{5}$ g) $\frac{21}{10}$ h) $\frac{33}{10}$

4   a) Copy and complete these. (It may help to draw line diagrams.)

      (i) $1\frac{1}{6} = \frac{\square}{6}$ (ii) $2\frac{3}{8} = \frac{\square}{8}$ (iii) $2\frac{5}{9} = \frac{\square}{9}$ (iv) $3\frac{7}{12} = \frac{\square}{12}$

    b) Copy and complete:

      (i) $\frac{11}{2} = \square\frac{\square}{2}$ (ii) $\frac{17}{3} = \square\frac{\square}{3}$ (iii) $\frac{23}{8} = \square\frac{\square}{8}$ (iv) $\frac{32}{9} = \square\frac{\square}{9}$

═══════════════ CHALLENGE ═══════════════

5   a) Try to find a quick way of changing from a mixed number ($2\frac{1}{2}$, $3\frac{7}{9}$ ...) to a 'top heavy fraction'
      ($\frac{11}{4}$, $\frac{17}{8}$ ...).
      Write down any method that you find.

    b) Now try to find a quick method for doing the opposite. Write down any method that you find.

Fraction calculations

1   The 'Spice of Life' shop buys its wholemeal flour in 3 kg bags. It sells the flour in $\frac{1}{2}$ kg bags.

a) How many $\frac{1}{2}$ kg bags can be filled from a 3 kg bag?

b) Alan packed twelve $\frac{1}{2}$ kg bags. How many 3 kg bags did he use?

c) Copy and complete the following – your results in a) and b) will help:

(i) $\frac{1}{2} \times 6 = \square$ (ii) $\frac{1}{2}+\frac{1}{2}+\frac{1}{2}+\frac{1}{2}+\frac{1}{2}+\frac{1}{2} = \square$ (iii) $3 \div \frac{1}{2} = \square$ (iv) $12 \times \frac{1}{2} = \square$

2   The yoghurt maker holds sixteen $\frac{1}{4}$ pint cartons.

a) How many cartons can be filled from the jug?

b) How much more milk is needed to fill the remaining cartons for the yoghurt maker?

c) Copy and complete the following – your results in a) and b) will help:

(i) $2\frac{1}{2} \div \frac{1}{4} = \square$ (ii) $16 \times \frac{1}{4} = \square$

(iii) $4 - 2\frac{1}{2} = \square$ (iv) $1\frac{1}{2} \div \frac{1}{4} = \square$

3   Suzie and Alan need to fix the two blocks of wood together.

a) Which of these screws do you think would be best? Why?

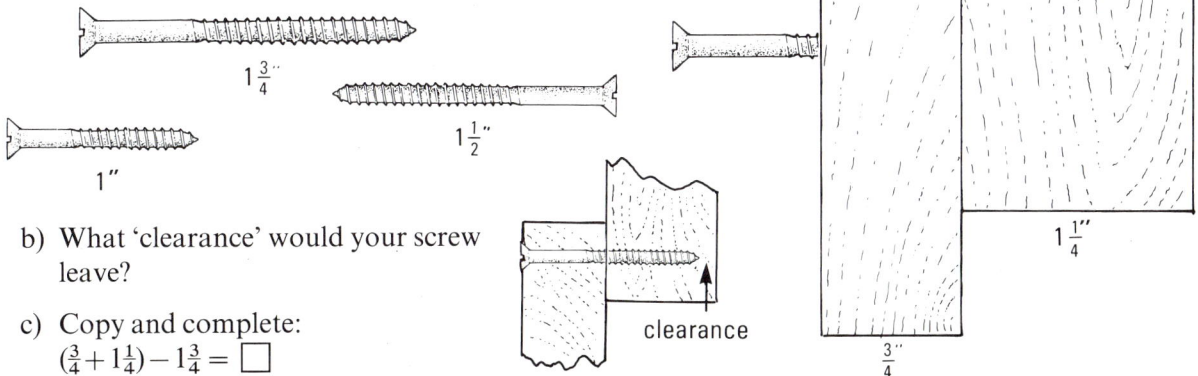

$1\frac{3}{4}''$

$1\frac{1}{2}''$

$1''$

$1\frac{1}{4}''$

$\frac{3}{4}''$

clearance

b) What 'clearance' would your screw leave?

c) Copy and complete:
$(\frac{3}{4}+1\frac{1}{4})-1\frac{3}{4} = \square$

A9

4   The weighing scale is marked in eighths and quarters of a pound (lb).

a)   What is the weight of potatoes in the pan?

b)   Another $\frac{3}{4}$ lb is added. What does the scale read now?

c) $\frac{3}{8}$ lb is removed. What is the final weight of potatoes in the pan?

d)   Copy and complete the following – your results in a), b), and c) will help:

   (i) $1\frac{1}{8} + \frac{3}{4} = \square$ (ii) $1\frac{7}{8} - \frac{3}{8} = \square$

e)   Use the scales to help you to do these:

   (i) $\frac{1}{8} + \frac{1}{4} = \square$ (ii) $\frac{1}{2} + \frac{3}{8} = \square$ (iii) $\frac{3}{4} + \frac{3}{8} = \square$

   (iv) $2 - \frac{5}{8} = \square$ (v) $1\frac{1}{2} - \frac{3}{8} = \square$ (vi) $2\frac{1}{2} - 1\frac{7}{8} = \square$

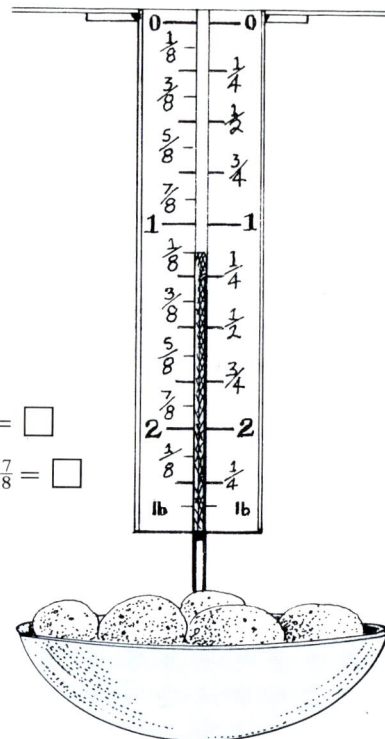

═══════ CHALLENGE ═══════

f)   Use the scales to help you to do these:

   (i) $\frac{7}{8} \times 2 = \square$ (ii) $\frac{5}{8} \times 3 = \square$ (iii) $2 \div \frac{1}{8} = \square$

   (iv) $1\frac{1}{8} \div \frac{1}{8} = \square$ (v) $2 \times 1\frac{3}{8} = \square$ (vi) $1\frac{7}{8} \div \frac{3}{8} = \square$

5   This is a design for a stained-glass window.

a)   What fraction of the glass will be stained:

   (i) blue   (ii) green   (iii) red?

b)   Copy and complete (use the window to help you):

   (i) $\frac{1}{16} + \frac{1}{16} = \square$

   (ii) $\frac{1}{16} + \frac{1}{16} + \frac{1}{16} + \frac{1}{16} = \frac{1}{\square}$

   (iii) $2 \times \frac{1}{16} = \frac{1}{\square}$

   (iv) $4 \times \frac{1}{16} = \frac{1}{\square}$

   (v) $8 \times \frac{1}{16} = \frac{1}{\square}$

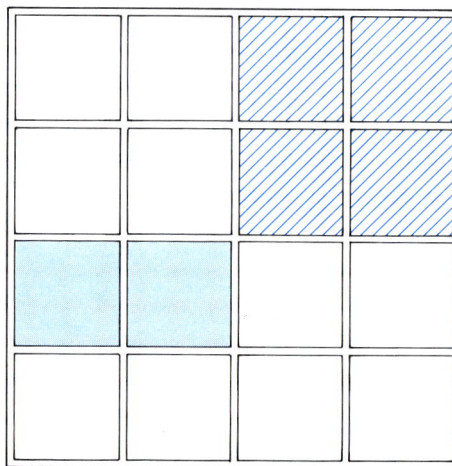

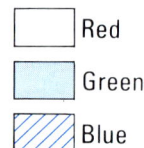

Red

Green

Blue

CHALLENGE: ADVERTISING SPACE

6 Advertising space for a newspaper is sold like this.

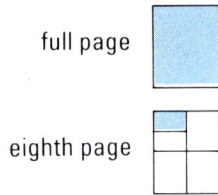

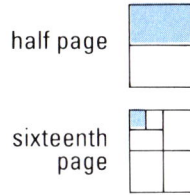

full page

half page

quarter page

eighth page

sixteenth page

a) This amount of space is sold on page 4 for Tuesday. What fraction of the page has been sold?

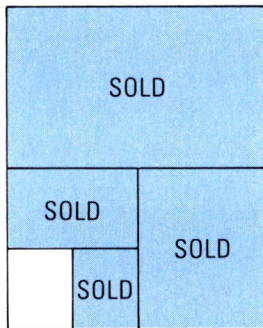

SOLD

SOLD

SOLD

SOLD

Write your result:

(i) using three $+$s
(ii) as a number of sixteenths: $\frac{\square}{16}$.

b) What fraction remains unsold?

c) $\frac{7}{16}$ of page 5 has been sold for Tuesday. Sketch page 5. Shade in the amount sold.

d) Here is a 'grid' for each page. Use it to help you to calculate these total amounts sold.

(i) $\frac{1}{2}+\frac{1}{4}+\frac{1}{8}=\frac{\square}{8}$ (ii) $\frac{1}{4}+\frac{1}{16}=\frac{\square}{16}$

(iii) $\frac{1}{2}+\frac{1}{8}+\frac{1}{16}=\frac{\square}{16}$ (iv) $2\times\frac{3}{16}=\frac{\square}{8}$

(v) $4\times\frac{2}{16}=\frac{1}{\square}$ (vi) $\frac{5}{16}\times 3=\frac{\square}{\square}$

(vii) $\frac{3}{16}+\frac{3}{8}=\frac{\square}{\square}$ (viii) $\frac{1}{4}+\frac{5}{16}+\frac{3}{8}=\frac{\square}{\square}$

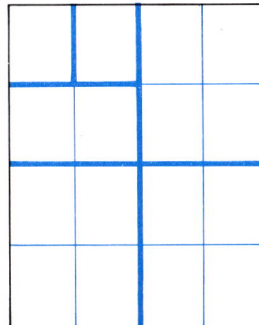

e) $\frac{15}{16}$ of page 5 has been sold, but then $\frac{3}{8}$ is cancelled.

(i) What fraction is still sold?
(ii) What fraction is unsold?

REVIEW

- Scale drawings are used by:

  interior decorators
  architects
  surveyors
  model makers
  craftspeople
  engineers ...
  and many other people.

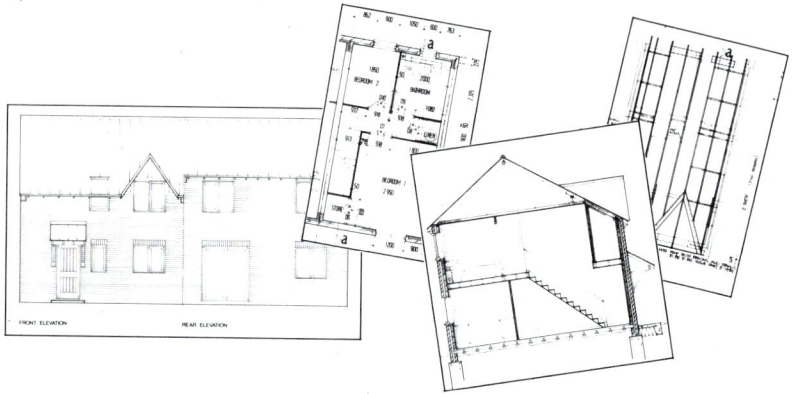

- A map is a scale drawing of a piece of land.

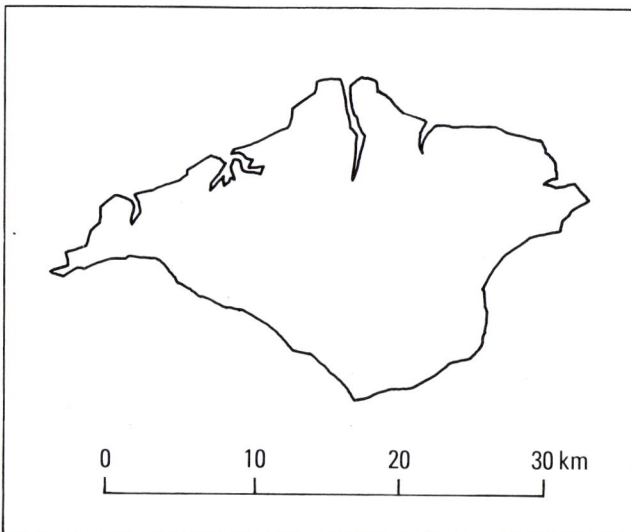

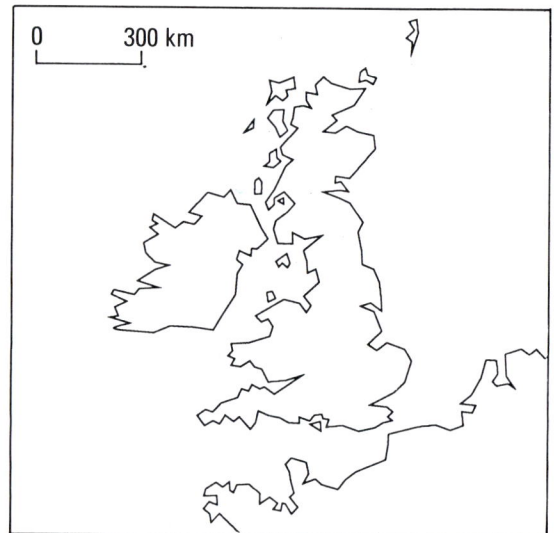

This is a scale drawing of a tower block.

Scale:　1 cm to 10 m

which is 1 cm to 1000 cm

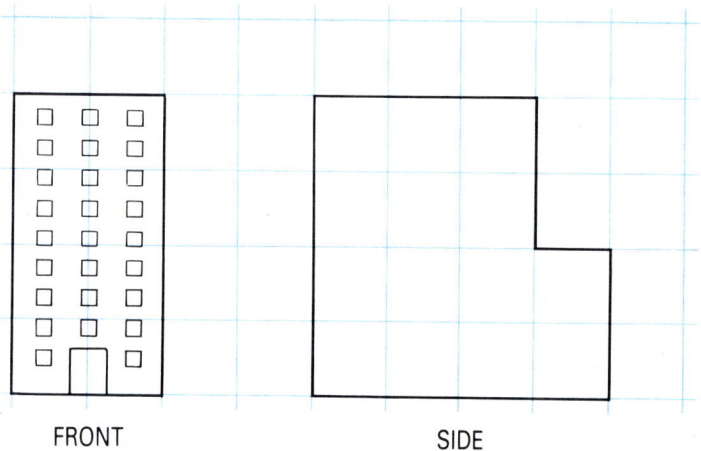

which is 1 : 1000.

FRONT　　　　　SIDE

■ The tower block is drawn to a new scale of 1 : 500. Is the scale drawing larger or smaller than the one here? How many times larger or smaller?

CONSOLIDATION

1   These are all scale drawings of a car park barrier when closed. A is drawn to a scale of 1:50 (1 cm to 50 cm).

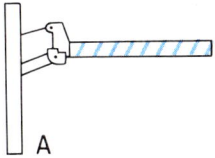

A

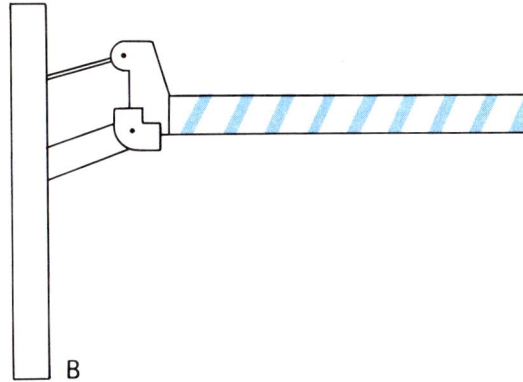

B

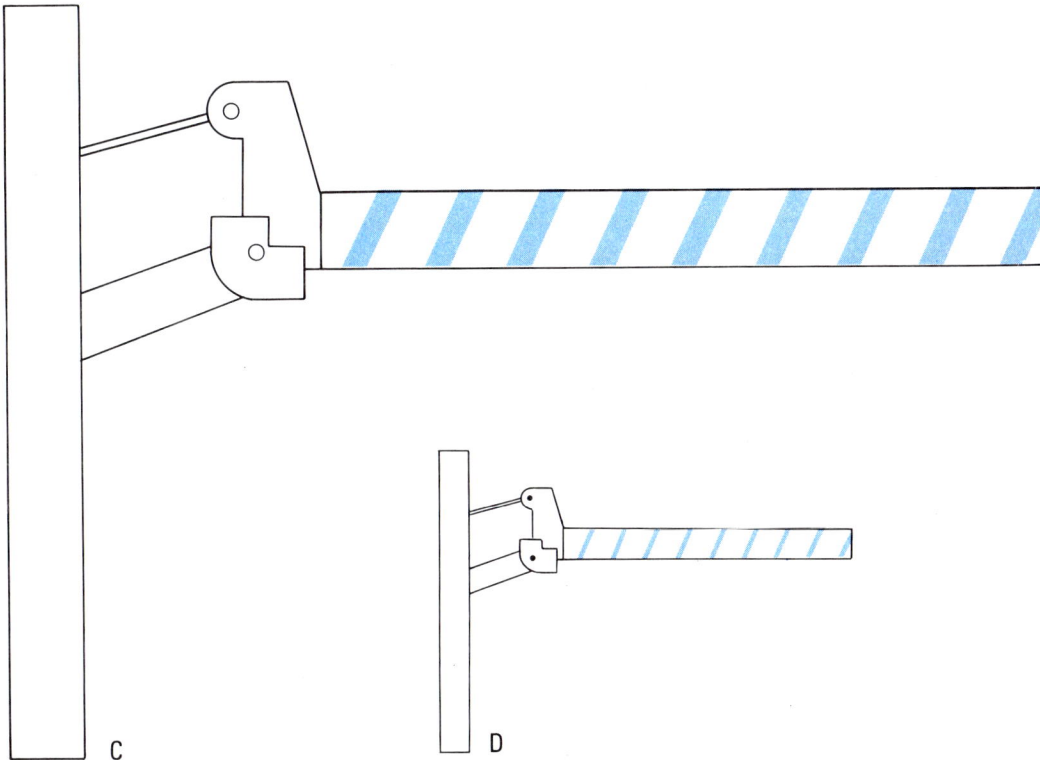

C

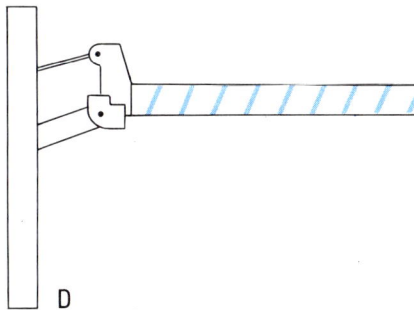

D

E

A10

a) Roughly, how high is the striped barrier above the ground when the barrier is closed?

b) Roughly, how wide do you think the entrance is to the car park?

c) What do you think is the scale of each of the other drawings, B, C, D, and E?

d) Use 1 cm squared paper. Make your own scale drawing of the barrier. Use a scale of 1:40.

2 These are some of the main household pests. The drawings in the circles are life size.

a) Roughly what scale is used for each of the larger drawings?
Write each one like this ☐:1.

b) Make your own drawing of the
death watch beetle.
Use a scale of 10:1.

House Longhorn Beetle

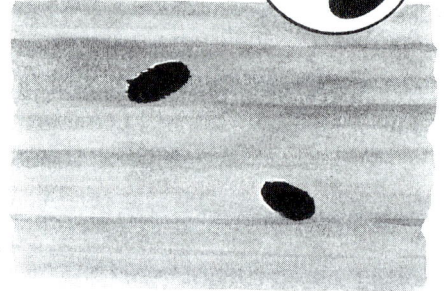

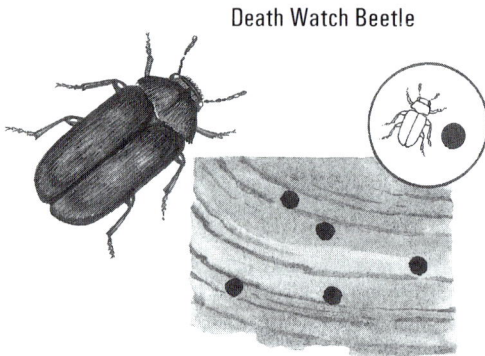

Powder Post
Beetle

Death Watch Beetle

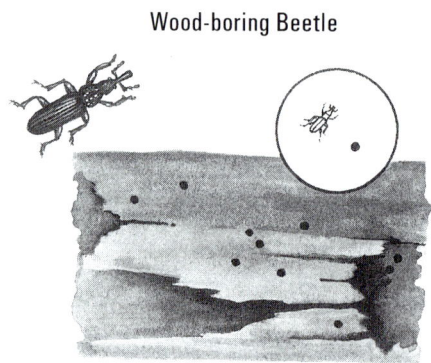

Wood-boring Beetle

c) This is an enlarged
photograph of a pond skater
beetle. The scale is 5:1.
Make your own life-size
drawing of the beetle.

3   You need 5 mm squared paper.
Shelley is arranging furniture in her new kitchen. This is her plan.

a) Copy and complete this table:

Kitchen length	400 cm
Kitchen width	
Hall door width	
Back door width	
Window width	
Depth of sink cupboards	

Window

Sink unit with cupboards

A10

Scale 1:40

Back
door

Door to hallway

b) Draw a plan of the kitchen on 5 mm
squared paper. Use a scale of 1:20.

c) This is a list of the kitchen furniture.

		Real distance
Cooker	length	55 cm
	depth	55 cm
Refrigerator	length	60 cm
	depth	60 cm
Washing machine	length	75 cm
	depth	40 cm
Table	length	85 cm
	depth	45 cm
Cupboard work surface	length	100 cm
	depth	55 cm

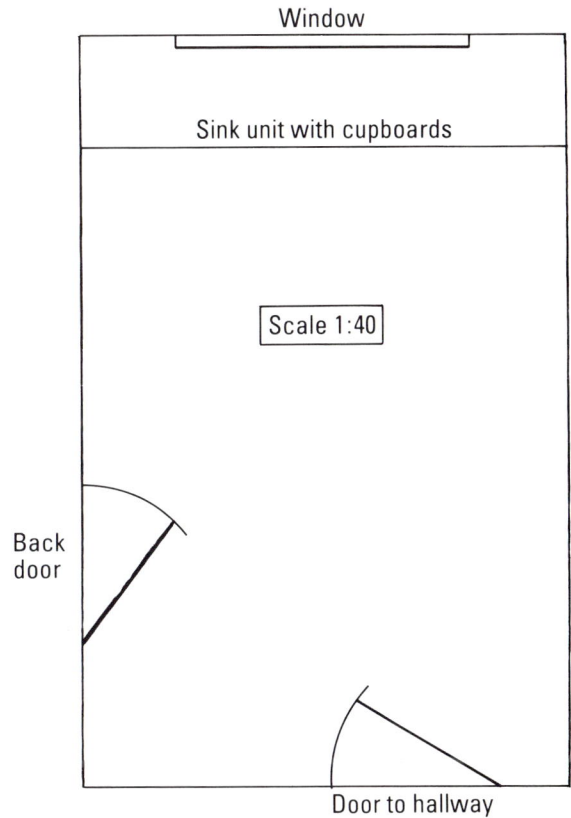

Make scale drawings of the furniture on
5 mm squared paper. Use a scale of 1:20.
Name each piece of furniture and cut it out.
All the pieces of furniture stand on
the floor. Arrange them in the way you
think best. Draw their final positions on
your plan.

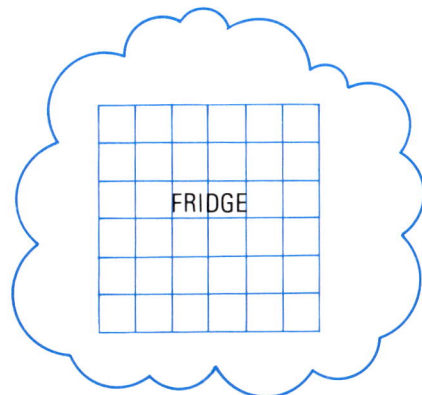

FRIDGE

Map scales

1 It is 3 km by road from Jim's house to Carol's house. Carol drew this map.

   a) What is the map distance from Jim's house to Carol's house (to the nearest $\frac{1}{2}$ cm)?

   b) Roughly what scale did Carol use?

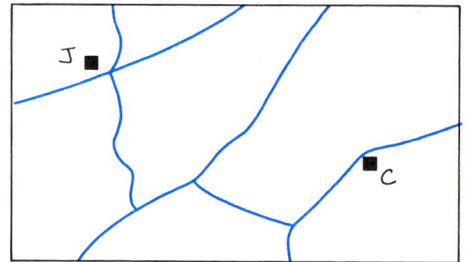

A10

2 a) Guess how far it is from Manchester to Liverpool:

   A   450 km   B   45 km   C   450 m

   b) On a motorist's map this distance is shown as 18 cm. The scale of the map is 1:250 000. Check to see if your guess was right.

3 a) Guess how far it is from Big Ben to Nelson's Column in Trafalgar Square:

   A   8.5 km   B   85 m   C   850 m

   b) On a town plan of London this distance is shown as 8.5 cm. The scale of the map is 1:10 000. Check to see if your guess was right.

4 a) One Saturday between breakfast and lunch Jim and Carol walked from Grassington to Malham Tarn. Guess how far they walked:

   A   1200 m   B   120 km   C   12 km

   b) On a walker's map the distance from Grassington to Malham Tarn is 48 cm. The scale of the map is 1:25 000.
   Check to see if your guess was right.

ENRICHMENT

CHALLENGES

1 These are three
   photographs of the same
   lamp filament. The one on
   the left is a $\times 1000$
   enlargement of the real
   filament. We can write this
   as $1000:1$.
   Estimate the scale of the
   other two photographs.
   Write each one like this:
   $\square:1$.

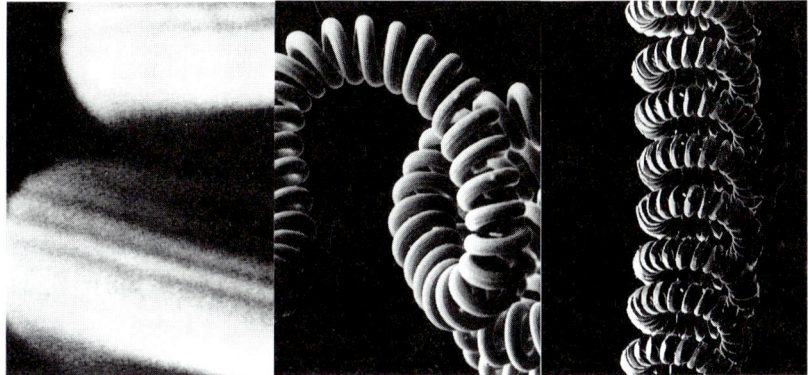

2 The photograph shows a
   tiny microprocessor
   on a 1 penny coin. The
   actual diameter of a
   1 penny coin is 20 mm.
   Estimate the length and
   width of the
   microprocessor. Draw it
   full size on 1 mm paper.

3 Roughly, how far is it by road from Cow House Farm to Ivy House Farm? (The scale of the map
   is $1:25000$.)

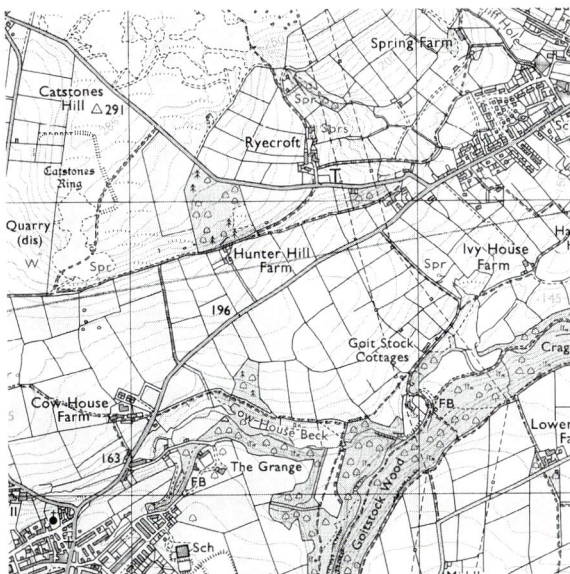

ASSIGNMENTS

4   Find out how far it is from your home to the school. Decide what scale you would use to show this on a plan in your book. Draw a sketch plan to show the route from your home to school. Write down the scale you have chosen to use.

5   a) Find a street plan of the town where your school is. What is the scale of this plan? What does 1 cm on the plan represent?

    b) Find a map that shows your school, and a town at least 5 km away. What is the scale of this map? What does 1 cm on the map represent?

    c) Find a map that shows the town nearest to where you live, and a town at least 50 km away. What is the scale of this map? What does 1 cm on the map represent?

A10

REVIEW

- Thermometers use numbers such as ⁻1, ⁻1.5, ⁻5, ..., to represent temperatures below zero.

■ Which is warmer: 1°C or ⁻2°C?   ⁻3°C or ⁻10°C?

- Banks use amounts such as ⁻10.00, ⁻12.65, ..., to represent money which is owed (that is, overdrawn amounts).

Deposit(£)	Withdrawn(£)	Balance(£)
		10.00
	15	⁻5.00
12		7.00
	18	?

■ What amount is missing from the account sheet?

- Changes in the price of shares on the stock market are shown using positive and negative numbers:
⁺2 means a rise of 2p
⁻3 means a fall of 3p.

OIL		
Ampol	⁺2	222
Aran Energy	⁻3	11
Atlantic Reserve	⁻4	8
BP	⁺6	72
IC Gas	⁺5	485

■ The price shown for IC Gas is the price at the end of the day.
What was the price at the beginning of the day?
What was the price of Aran Energy shares at the beginning of the day?

In NMP we have used numbers written in colour to represent changes. For example:

$$7 \quad + \quad {}^-3 \quad = \quad 4$$

Starting temperature　　A fall of 3°C　Final temperature

$$^-4 \quad + \quad {}^+5 \quad = \quad 1$$

Starting temperature　　A rise of 5°C　Final temperature

■ What number is missing from each sentence?

$$10 + {}^-4 = \square \qquad \square + {}^+3 = 6$$

$$\square + {}^+2 = {}^-1 \qquad {}^-4 + \square = {}^-6$$

A11

CONSOLIDATION

1 The following numbers are temperatures. Arrange them in order, coldest first.

$^-$4, 5, 0, $^-$1, 3, $^-$2.1, 1.5, $^-$5.1, 1.2

2 Copy this line accurately.

```
        -4        -3                        0       1              3
```

On your line mark positions for these numbers: 2, $^-$2, $^-$1, $^-$3.5, $^-$0.2, 1.7, $^-$1.7, 2.8, $^-$2.8

3 a) The temperature in Calgary, Canada, changes from $^-$10°C to $^-$3°C. Write the change as a number. (Use coloured pencil, or circle the number, to show that it represents a change.)

  b) Write these temperature changes as numbers:

   (i) from 0°C to 10°C
   (ii) from 0°C to $^-$10°C
   (iii) from 10°C to $^-$10°C
   (iv) from $^-$10°C to 10°C.

  c) It is 5°C in Glasgow. The temperature changes by $^-$7°C. What is the new temperature?

  d) The temperature in Colwyn Bay falls by 9°C to $^-$3°C. What was the original temperature?

  e) Copy and complete each of these sentences. Your results in c) and d) will help.

   (i) $5 + {}^-7 = \square$ (ii) $\square + {}^-9 = {}^-3$

4 Write a question like the ones in question 3c) and d) for each of these sentences. Then answer your question.

a) $4 + {}^-5 = \square$ b) $^-2 + {}^+7 = \square$ c) $\square + {}^-2 = 0$ d) $3 + \square = {}^-2$

5 This is part of a bank account record sheet. Copy and complete it.

Deposit(£)	Withdrawn(£)	Balance(£)
		50.00
	30.00	?
	40.00	?
	?	$^-$50.00
?		120.00

CORE

Adding and subtracting

1 A firm of architects has six bank accounts.

   a) Which account has the most money in it?

   b) Which account is most overdrawn?

   c) Which account has the better balance
      (has more money in it, or is less overdrawn):

      (i) A or B   (ii) A or C   (iii) B or E?

	Balance(£)
Account A	$^-90$
Account B	$^-30$
Account C	$^+40$
Account D	$^-2670$
Account E	$^+8$
Account F	$^+1990$

we use '$-$' to show the account is overdrawn

we use '$+$' to show the account is in credit

   d) The firm decides to combine accounts B and C. How much money is there in B and C altogether?

   e) This number sentence can be used to work out how much there is altogether in accounts A and E. Copy and complete it.

$$^-90 \quad + \quad ^+8 \quad = \quad ?$$
Overdrawn by £90   £8 in credit   Overdrawn by £? altogether

A11

   f) Write a number sentence to work out how much there is altogether in accounts D and F.

2 a) Tessa's bank account is overdrawn by £10. She deposits £32 into the account. Use this number sentence to work out the new balance for her account:

   $^-10 + {}^+32 =$

   b) Lloyd has £20 in his bank account. He withdraws £32. Copy and complete this number sentence to work out the new balance for his account:

   $^+20 + {}^?? =$

   c) Tessa and Lloyd decide to combine their accounts. Write a number sentence to work out the balance of the joint account.

3 A firm of dress manufacturers has these bank accounts.

   a) How much more is there in account A than account B?

   b) This number sentence can be used to work out a). Copy and complete it.

   $^+18 - {}^-102 =$

	Balance (£)
Account A	$^+18$
Account B	$^-102$
Account C	$^-8$
Account D	$^+52$

   c) Write a number sentence to work out:

      (i) how much better account D is than account A
      (ii) how much better account C is than account B
      (iii) how much worse account A is than account D.

4   Copy and complete each sentence in each group. (Think of *comparing* bank accounts.)

   a) (i) $^+50 - {}^+30 = ?$ b) (i) $^+50 - {}^+5 = ?$ c) (i) $^+50 - {}^+40 = ?$
       (ii) $^+50 - {}^+45 = ?$ (ii) $^+50 - {}^+1 = ?$ (ii) $^+30 - {}^+40 = ?$
       (iii) $^+50 - {}^+52 = ?$ (iii) $^+50 - {}^-2 = ?$ (iii) $^+5 - {}^+40 = ?$
       (iv) $^+50 - {}^+90 = ?$ (iv) $^+50 - {}^-30 = ?$ (iv) $^-5 - {}^+40 = ?$

5   Copy and complete each sentence in each group. (Think of *combining* bank accounts.)

   a) (i) $^+100 + {}^+10 = ?$ b) (i) $^-10 + {}^+100 = ?$
       (ii) $^+100 + {}^-2 = ?$ (ii) $^-10 + {}^+2 = ?$
       (iii) $^+100 + {}^-90 = ?$ (iii) $^-10 + {}^-2 = ?$
       (iv) $^+100 + {}^-110 = ?$ (iv) $^-10 + {}^-100 = ?$

6   Copy and complete each sentence. (Think of comparing or combining bank accounts.)

   a) $^-10 - {}^-2 = ?$ b) $^-10 + {}^-2 = ?$ c) $^+50 - {}^-2 = ?$ d) $^+3 + {}^-20 = ?$

A11

7   These stories all go with this number sentence.

   $^+10 + {}^-12 = {}^-2$

   Copy and complete each story.

   | |
   | Yusef gets to the cinema 10 minutes after the film starts. Zelda arrives 12 minutes ... Yusef, which is 2 minutes ... the film starts. |

   | Erskine works in an office block. He is on the tenth floor above ground level. He then travels down 12 floors in the lift. He ends up on the ... floor ... ground level. |

   | At 12 noon the temperature in the greenhouse was 10 °C. By midnight it had ... by 12 °C, to ... °C. |

   | Ten people get on a bus and 12 people get .... There are now 2 ... people on the bus than before. |

8   Write two different stories for this number sentence: $^-2 + {}^-12 = {}^-14$.

9   The stories on the right go with this number sentence.

   $^+4 - {}^-10 = {}^+14$

   Copy and complete the stories.

   | Gladys eats 4 more cherries than Louis. Glenda eats 10 fewer cherries than Louis. Gladys eats 14 ... cherries than .... |

   | The temperature in Berlin is $^-10$ °C. In Lisbon it is $^+4$ °C, which is 14 °C ... than in Berlin. |

10 Write two different stories for this number sentence: $^-10 - {}^-2 = -8$.

━━━━━━━━━━ TAKE NOTE ━━━━━━━━━━

We often write 4 for $^+4$, 3 for $^+3$, etc., and -4 for $- {}^+4$, -3 for $- {}^+3$, etc.

$0 - {}^-4$ is equivalent to $^+4$. (We say 'zero subtract negative four'.)

$0 - {}^+4$ (zero subtract positive four)⎫
$0 + {}^-4$ (zero add negative four)⎬ are equivalent to $^-4$.

================ THINK IT THROUGH ================

11 a) Table A shows average weekly earnings for various occupations.
Table B compares the earnings of a train driver with some other occupations. Copy and complete it.

Table A

	£ per week
Medical practitioner	425
Finance, insurance specialist	403
University teacher	321
Office manager	302
Scientist	280
Production manager	276
Accountant	268
Engineer (civil or structural)	264
Policeman (below Sergeant)	250
Primary teacher	227
Electrician	210
Telephone fitter	207
Maintenance fitter	205
Train driver	203
Heavy goods driver	188
Plumber	186
Postman	178
Bricklayer	158
General clerk	151
Painter, decorator	150
Dustman	147
Unskilled building worker	140
Butcher	129
Hospital porter	125
General farm worker	121
Total workers	207
of which: Manual	147
Non-manual	244

Table B

Occupation	Weekly earnings compared with a Train driver £
Bricklayer	$^-45$
Electrician	$^+7$
Hospital porter	$^-78$
Medical practitioner	–
Postman	–
Scientist	

This $-$ means a Bricklayer earns £45 less than a Train driver

Table C

Occupation	Weekly earnings compared with an Engineer £
Bricklayer	$^-106$
Electrician	$^-54$
Hospital porter	–
Medical practitioner	–
Postman	–
Scientist	–

A11

b) Copy and complete Table C.

c) Which occupations are being used as the comparison in Table D and Table E?

Table D

Occupation	Weekly earnings compared with a [ ] £
Bricklayer	$^-69$
Electrician	$^-17$
Hospital porter	$^-102$
Medical Practitioner	$^+198$
Postman	$^-49$
Scientist	$^+53$

Table E

Occupation	Weekly earnings compared with a [ ] £
Bricklayer	$^+37$
Electrician	$^+89$
Hospital porter	$^+4$
Medical Practitioner	$^+304$
Postman	$^+57$
Scientist	$^+159$

d) This sum uses numbers from Table B: $^+7 - {}^-25 = \square$
   (i) Find the answer.
   (ii) What does the answer represent?
   (iii) This sum uses the corresponding numbers in Table C. $^-54 - {}^-86 = \square$
       Check that it produces the same answer.
   (iv) Write down the corresponding sums for Tables D and E.

REVIEW

● Shapes fold onto themselves along a line of symmetry.

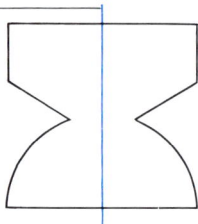

line of symmetry

One line of symmetry

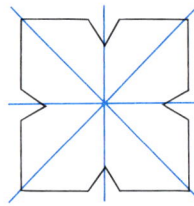

Four lines of symmetry

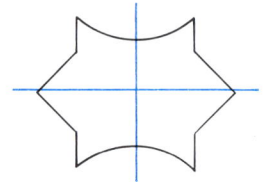

Two lines of symmetry

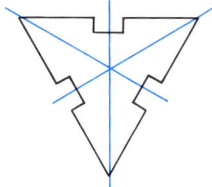

Three lines of symmetry

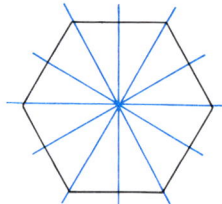

Six lines of symmetry

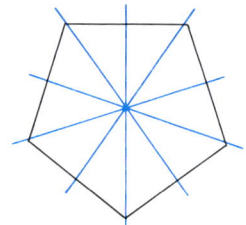

Five lines of symmetry

Turn the S. It fits this position twice in a full turn.

Turn centre

180°

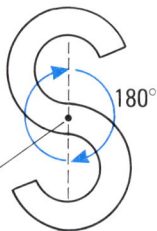

Turn symmetry of order 2, or 180° turn symmetry

Turn the shape. It fits this position three times in a full turn.

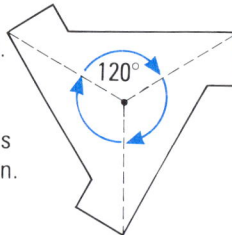

120°

Turn symmetry of order 3, or 120° turn symmetry

Turn the shape. It fits this position four times in a full turn.

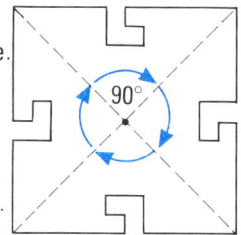

90°

Turn symmetry of order 4, or 90° turn symmetry

Regular pentagon (5 sides)

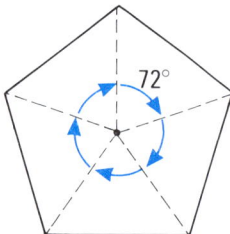

72°

Turn symmetry of order 5, or 72° turn symmetry

Regular hexagon (6 sides)

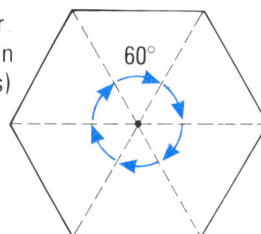

60°

Turn symmetry of order 6, or 60° turn symmetry

Regular octagon (8 sides)

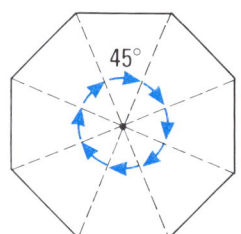

45°

Turn symmetry of order 8, or 45° turn symmetry

■ Do shapes with turn symmetry always have a line of symmetry?

■ Do shapes with line symmetry always have turn symmetry?

CONSOLIDATION

1   Here is a square tile. The pattern has
    one line of symmetry.

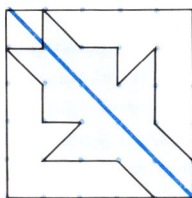

You need dotted squared paper.
Here are some broken square tiles.
Each pattern has two lines of
symmetry. Copy and complete the
tiles.

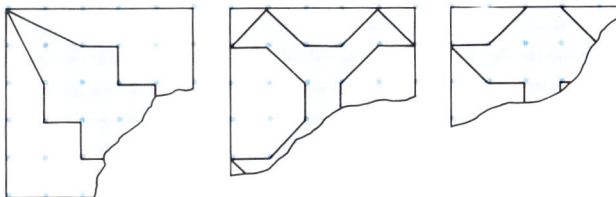

2   Here is another square tile. The pattern
    has turn symmetry of order 4.

A12

You need dotted squared paper.
Here are some broken square tiles.
Each pattern has turn symmetry of
order 4. Copy and complete the tiles.

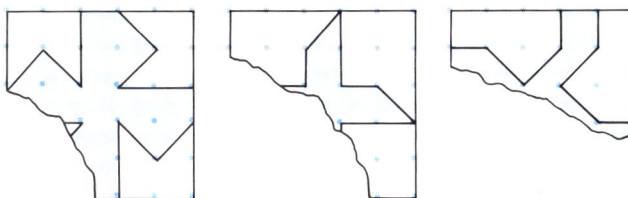

CHALLENGE

3   You need dotted isometric paper.
    Copy this hexagon tile. Draw a pattern on it
    which has turn symmetry of order 3.

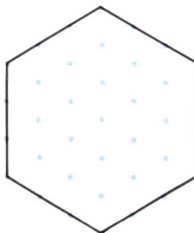

4   Check that this drawing has two lines of symmetry and 180°
turn symmetry.

   a) Copy the drawing and mark in the lines of symmetry.

   b) In this activity there is no need to use a protractor or a
   ruler. Sketches will do.

      (i) Copy the "pendulum". Add another pendulum to it so
      that the final figure has just one line of symmetry.
      (ii) Add another pendulum so that the figure still has only
      one line of symmetry.

   c) Start again with a pendulum. Add two pendulums so that the final figure has 120° turn
   symmetry. What is its order of symmetry?

   d) Add two more pendulums to this figure so that the final figure has:

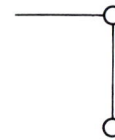

      (i) turn symmetry of order 4
      (ii) turn symmetry of order 2
      (iii) turn symmetry of order 1.

   e) Write down how many lines of symmetry each of your drawings has in part d).

5   a) What size is the smallest angle you can turn each of these shapes before they fit back onto
   themselves?

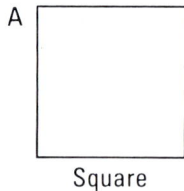

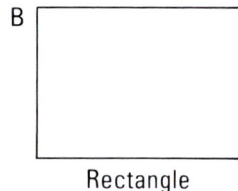

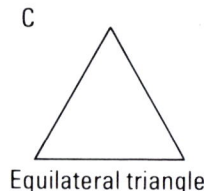

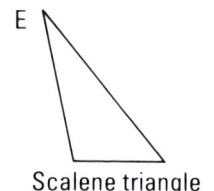

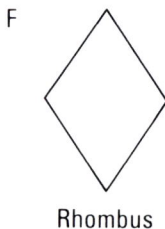

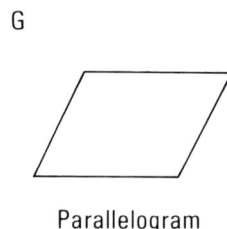

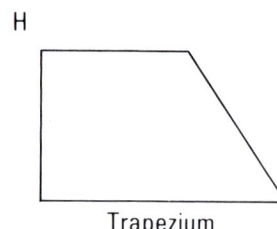

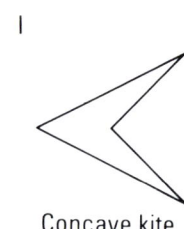

   A   Square   B   Rectangle   C   Equilateral triangle   D   Isosceles triangle   E   Scalene triangle

   F   Rhombus   G   Parallelogram   H   Trapezium   I   Concave kite

   b) How many lines of symmetry does each shape in part a) have?

6   Sketch and name these quadrilaterals.

   a) Turn symmetry of order 2 but no lines of symmetry.

   b) Turn symmetry of order 2 and exactly two lines of symmetry (name two possibilities).

   c) Turn symmetry of order 1 and one line of symmetry (name two possibilities).

ACTIVITIES: MAKING SYMMETRICAL SHAPES

You need dotted squared paper and scissors, and isometric dotted paper.

7   Cut out these two shapes. Fit the shapes together to make a shape with:

   a) one line of symmetry

   b) turn symmetry of order 2.

Sketch each shape you make. On your sketches mark any lines of symmetry and any centres of symmetry.

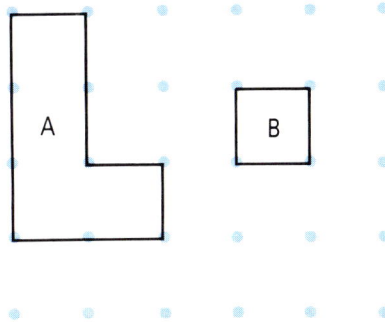

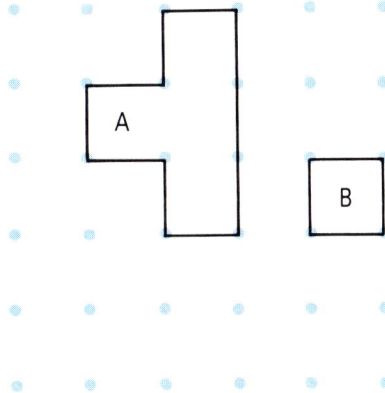

8   Cut out these two shapes. Fit the shapes together to make a shape with:

   a) turn symmetry of order 4

   b) one line of symmetry.

Sketch each shape you make. On your sketches mark any lines of symmetry and any centres of symmetry.

A12

9   a) Cut out the two shapes. Fit them together to make a shape with line symmetry. (You may turn the shapes over if you wish.) Sketch your symmetrical shape, and mark in its line of symmetry.

   b) Design a third shape which will fit together with the other two to make a shape with turn symmetry. Sketch the 3-piece symmetrical shape, mark in its centre of symmetry, and shade in your new piece.

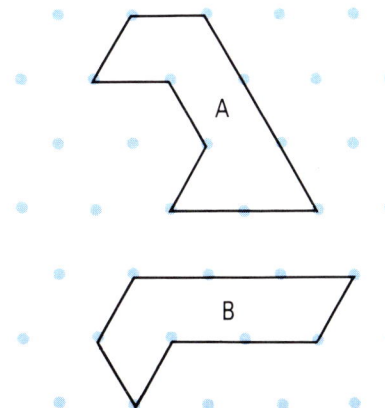

ENRICHMENT

ASSIGNMENT: RANGOLI PATTERNS

1   Hindus and Sikhs sometimes use Rangoli patterns to decorate their houses on special occasions. Here is one method for producing Rangoli patterns. Read through the steps first.

Divide a square of dots into 4 equal parts

Draw some lines in one quarter

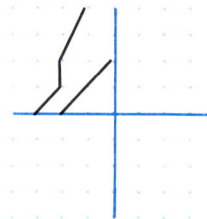

Reflect in the two blue lines

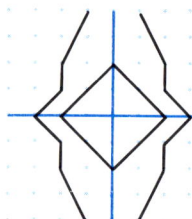

Reflect in the two blue diagonal lines

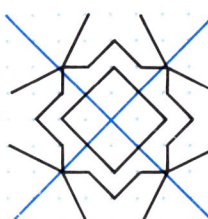

Repeat the pattern

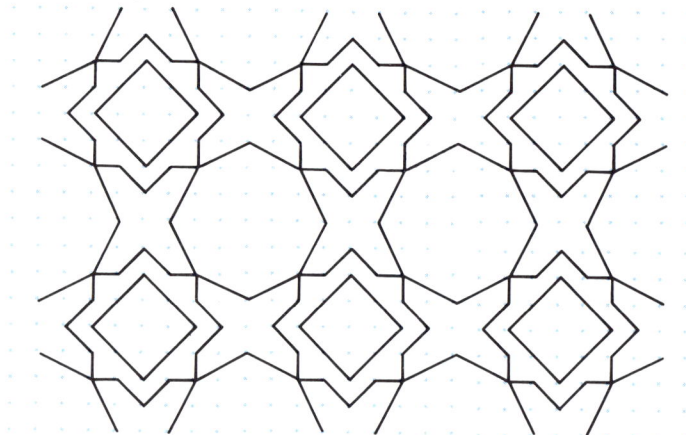

a) Draw your own Rangoli patterns.

b) Pick out parts of one of your patterns, like the parts below. Write about their symmetry (the number of lines of symmetry, the order of turn symmetry).

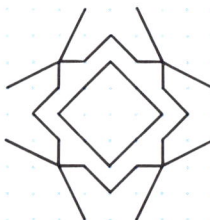

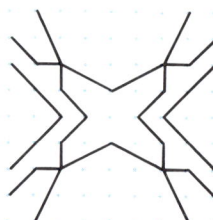

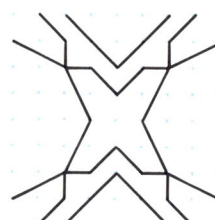

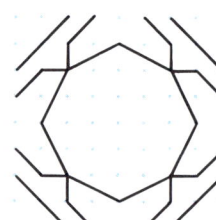

c) Start a Rangoli pattern from a triangle or hexagon instead of a square.

CONSOLIDATION

LIFE EXPECTANCY AT BIRTH

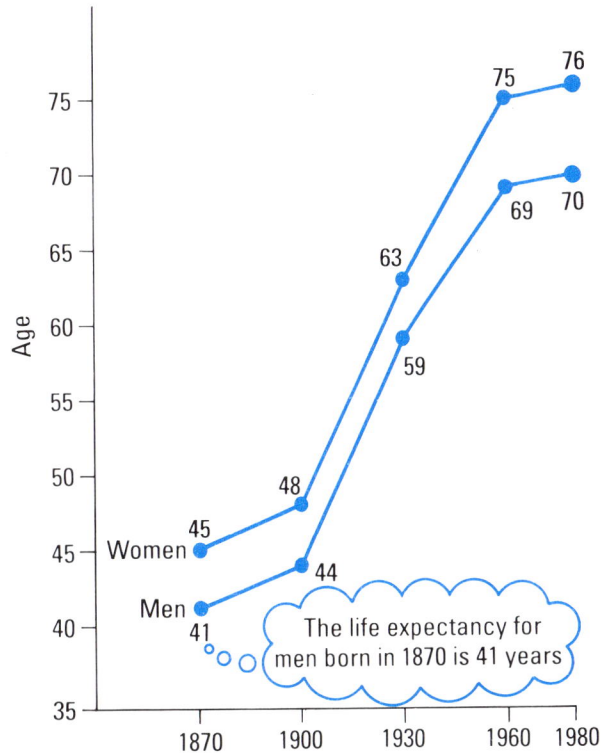

The life expectancy for men born in 1870 is 41 years

1 a) The graph shows that men born in 1870 were expected to live, on average, to an age of only 41 years. (This is called their 'life expectancy'.) What was the life expectancy of women in 1870?

   b) What was the life expectancy of men in 1930?

   c) What was the life expectancy of women in 1930?

   d) Men born in 1900 were expected, on average, to live 3 years longer than men born in 1870. What can you say about women born in 1930 and 1900?

   e) During the past 100 years has life expectancy increased or decreased?

   f) Write one or two sentences to explain your answer to part e). Try to give two reasons.

   g) From the graph estimate the life expectancy for:
      (i) men born in 1950   (ii) women born in 1950.

─── CHALLENGE ───

   h) What do you think the life expectancy will be for (i) men, (ii) women born in the year 2000? Explain how you arrived at your decisions.

A13

2 Graphs A to C give information about the number of people in the world.
You can see from graph A that in 1950
there were about 50 million people in
England, Wales, and Scotland.

a) Roughly, how many were there in 1850?

b) Check that in 1800 there were about
10 million people ... and in 1840 there
were about 20 million people. So the
population doubled in 40 years.
Roughly, how long did it take the
population to double again, that is, from
20 million to 40 million?

c) From the graph estimate:

(i) the number of people in 1900
(ii) the population of England, Wales, and Scotland in the year 2000
(iii) how many more people there will be in 2000 than there were in 1900.

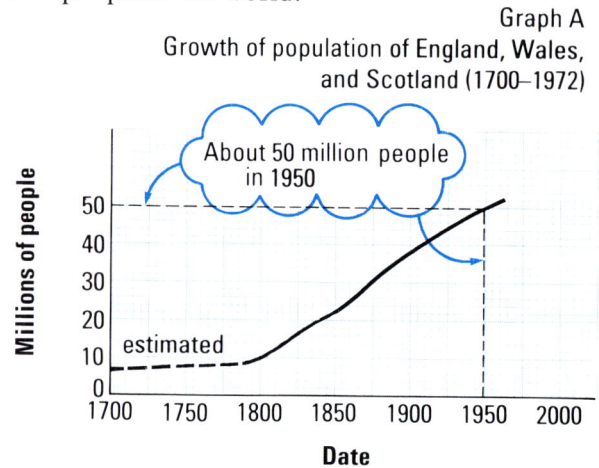

Graph A
Growth of population of England, Wales,
and Scotland (1700–1972)

About 50 million people in 1950

d) You can see from graph B that the world
population in 1950 was about 2300
million. What was the world population
in 1850?

e) Give two dates between which the world
population doubled. Say how many
people there were in the world at each
date.

f) Is the world population increasing more
rapidly now than in 1800, or less rapidly
now than in 1800? Why do you think this
is? Try to give two reasons.

g) Use graph C to estimate how many
people there will be in India in the year
2000.

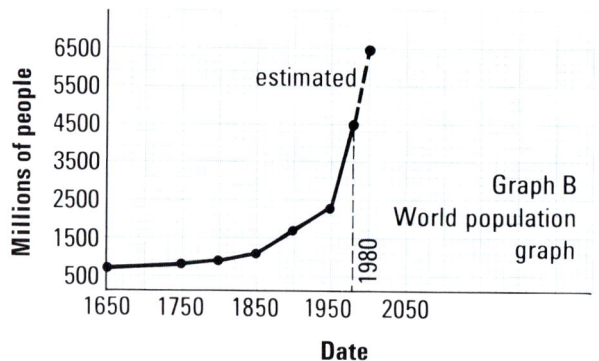

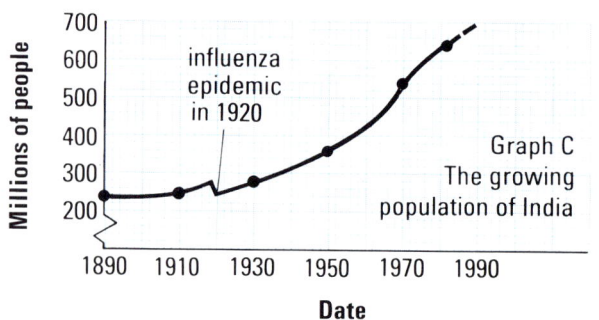

Graph B
World population
graph

estimated

Graph C
The growing
population of India

influenza
epidemic
in 1920

A13

CHALLENGE

h) Between 1900 and 1950 which grew more rapidly – the population of India or the population
of England, Wales, and Scotland? Write one or two sentences to explain how you decided.

CORE

Conversion graphs

1   1 pint is about 568 ml.

a) Copy and complete this conversion
table for pints and millilitres.

Pints	$\frac{1}{4}$	$\frac{1}{2}$	$\frac{3}{4}$	1	2	3	4	5
ml		284		568				

b) Use your table to do these conversions:
(i)   This recipe needs $2\frac{1}{2}$ pints of milk. How many
millilitres is this?
(ii)   A coffee percolator holds $1\frac{3}{4}$ pints of water. How
many millilitres is this?
(iii) Approximately, how many pints is 1 litre (1000 ml)?

9

PANCAKE MIX

$2\frac{1}{2}$ pints of milk
6 eggs

c) The graph shows the
information in the table.
Sometimes it may be
better to use the graph
for conversions. Or it
may be better to use the
table. Give an example
for each case.

d) A recipe uses $3\frac{1}{4}$ pints of
milk. Use the graph to
estimate how many
millilitres this is. Check
the result in your table.

e) Use the graph to
estimate how many
pints of wine there are
in each of these bottles.

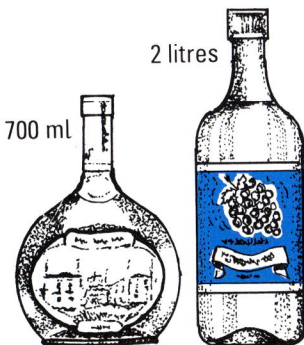

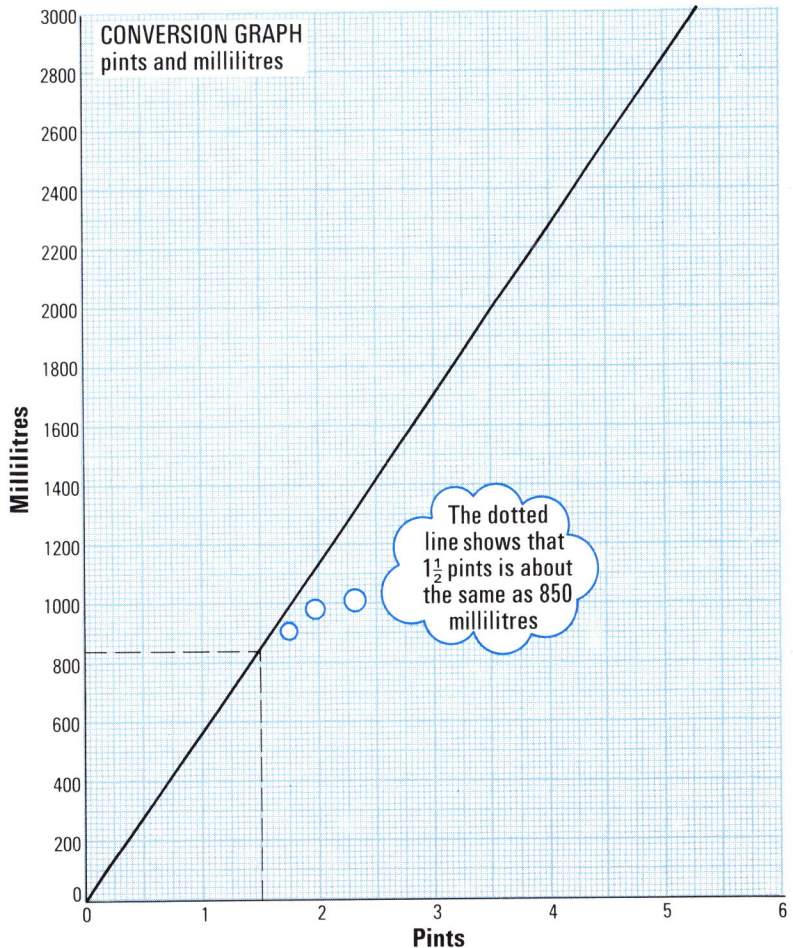

CONVERSION GRAPH
pints and millilitres

The dotted
line shows that
$1\frac{1}{2}$ pints is about
the same as 850
millilitres

2 litres

700 ml

2 In some countries distances are shown in miles. The signpost on the left used to be at Limerick, Eire. It was changed to kilometres.

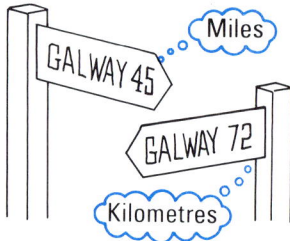

GALWAY 45   Miles

GALWAY 72   Kilometres

a) How many kilometres is each mile?

b) Here is a conversion graph for miles and kilometres. The distances below are in miles. Rewrite them in kilometres.

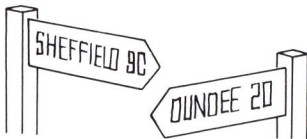

SHEFFIELD 90

DUNDEE 20

c) These distances are in kilometres. Rewrite them in miles.

PARIS 120

BRUSSELS 50

d) Write the distance from your home to a town or village between 10 and 30 miles away:

(i) in miles
(ii) in kilometres.

How far would you travel, in kilometres, in 10 return journeys?

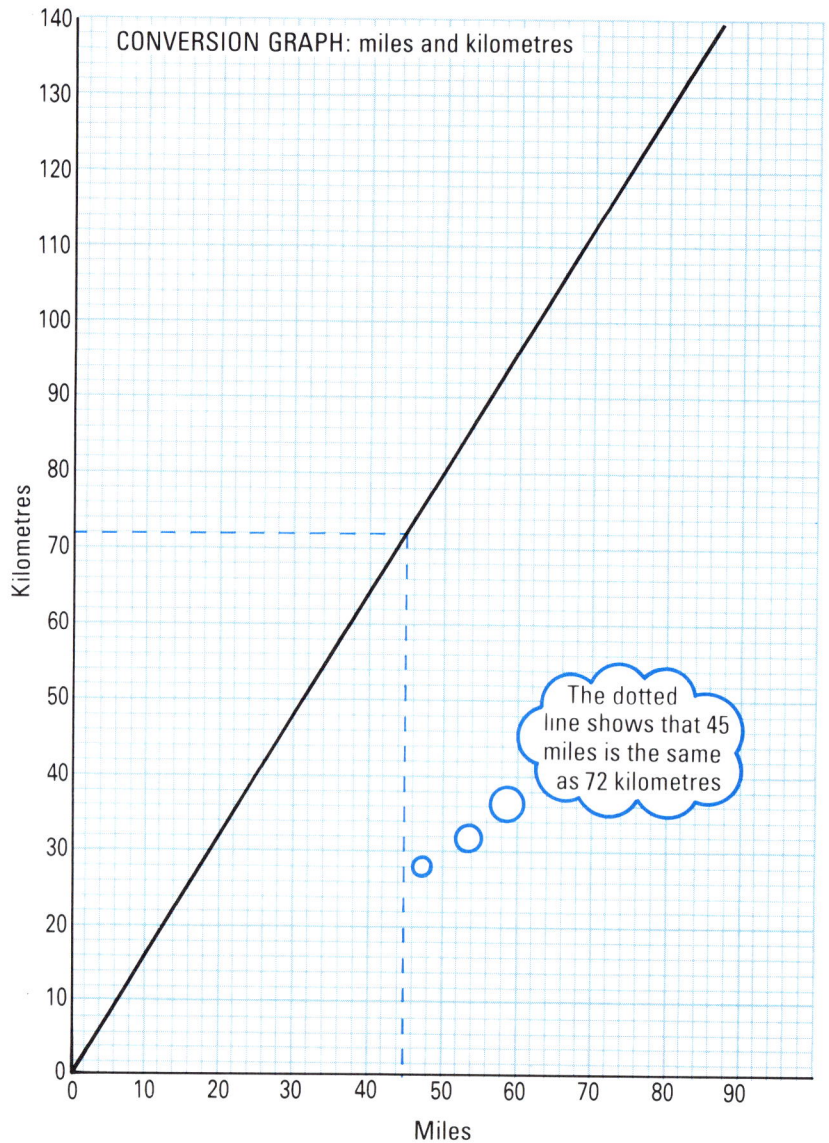

CONVERSION GRAPH: miles and kilometres

The dotted line shows that 45 miles is the same as 72 kilometres

Kilometres

Miles

A14 ANGLES AND DISTANCES

REVIEW

- Compass points help us to give simple directions.

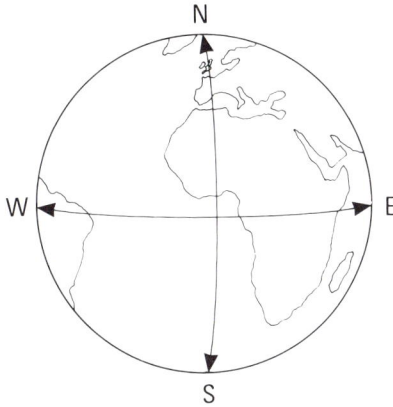

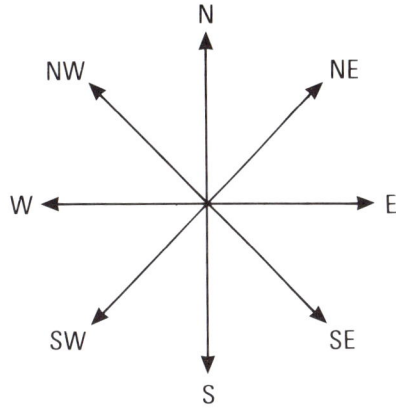

▪ Is Japan East of the USA or West of the USA?

▪ What do we mean by 'Middle East' and 'Far East'?

▪ Which coast in the USA is known as the 'West Coast'?

To give directions from a place we can imagine a compass at that place, as in the diagram.

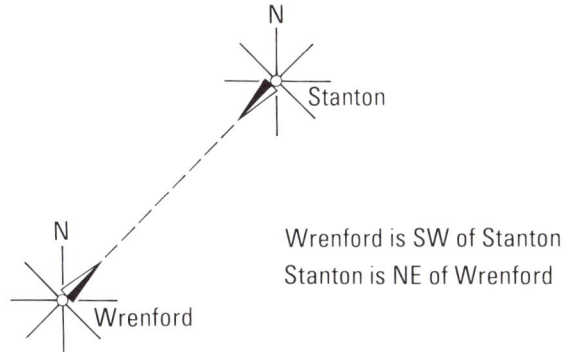

Wrenford is SW of Stanton
Stanton is NE of Wrenford

Bearings give directions more accurately than points of a compass do. We write bearings using three digits (004°, not 4°). 000° is due North (N).

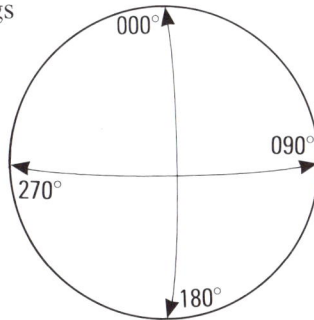

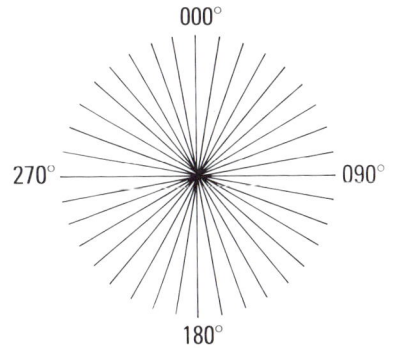

We measure a bearing from the North, in a clockwise direction.
To give a bearing from a place, we can imagine a circular protractor at that place.
The bearing of Swinton from Rawmarsh is 087°.
The bearing of Rawmarsh from Swinton is 267°.

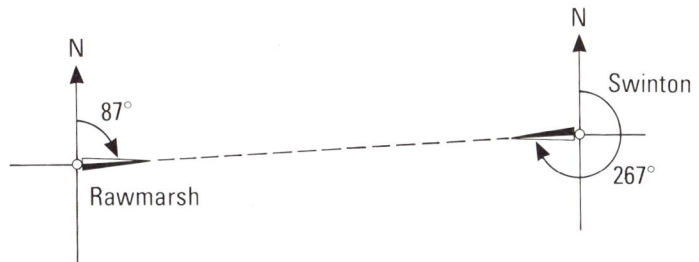

A14

CONSOLIDATION

1 The seven villages are snowbound. The helicopter is visiting them all with supplies, starting from NMP Base.

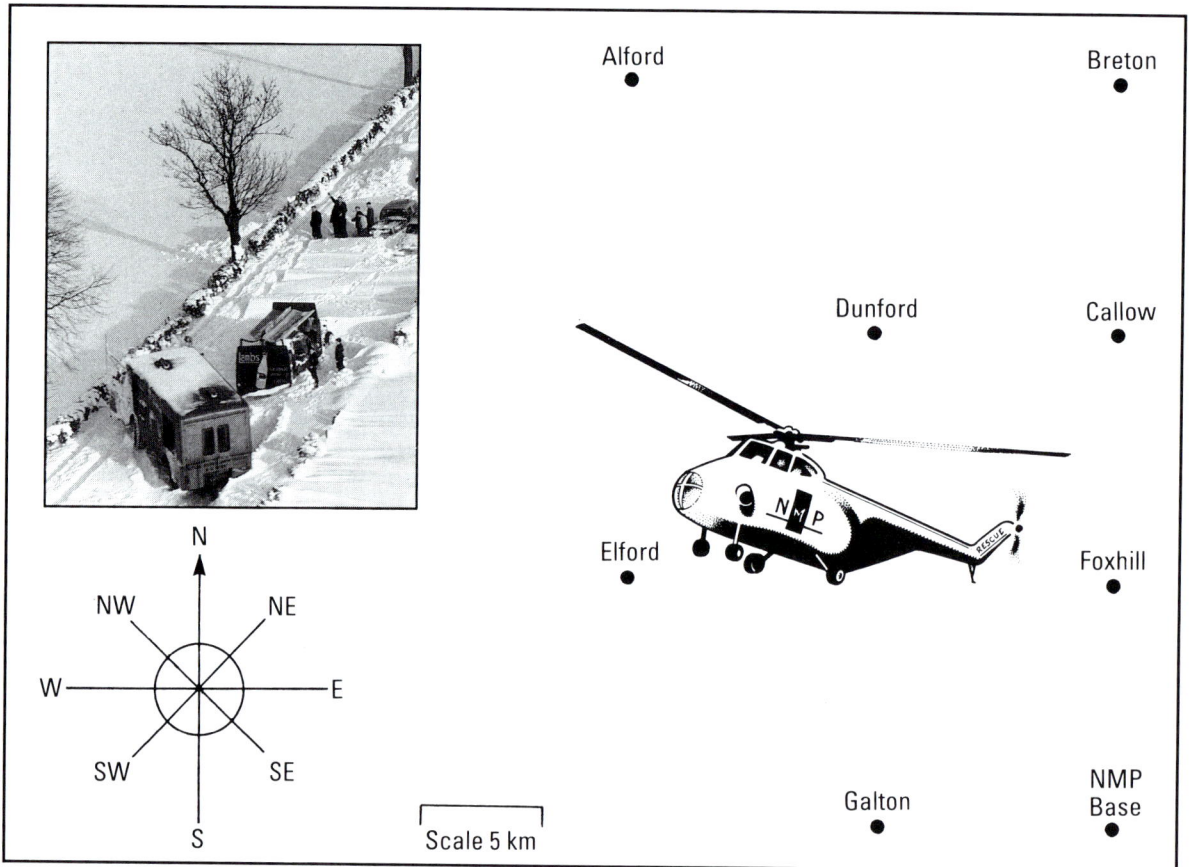

a) Plot a route it can take to visit all the villages and return to base. Sketch your route.

b) Write down the compass direction the helicopter flies on each section of your route. For example:

NMP Base	$\longrightarrow$	Foxhill	N
Foxhill	$\longrightarrow$	Galton	SW
Galton	$\longrightarrow$	...	

c) Approximately, how far does the helicopter fly on your route?

d) Find a shortest route around the villages (there is more than one possibility). Write down the villages in the order in which the helicopter visits them.

e) Approximately how many kilometres does the helicopter fly on a shortest route?

Do you remember . . . ?

We say that the *bearing* of HMS Survey from the lighthouse is 120°.

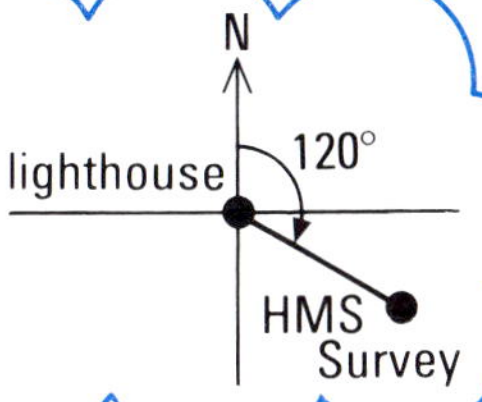

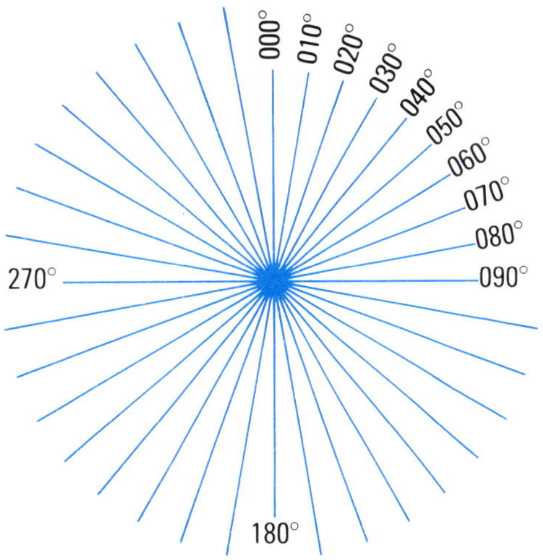

2 The map shows the major cities of European countries. Use the circular protractor to estimate these bearings:

a) the bearing of Paris from London
b) the bearing of London from Stockholm
c) the bearing of Copenhagen from Prague
d) the bearing of Stockholm from Warsaw.

3 Approximately, what is the distance in kilometres from:

a) London to Paris
b) London to Warsaw
c) Madrid to Vienna?

The State Opera House, Vienna

Prague Castle

Scale
0 100 200 300 400 miles
0 100 200 300 400 500 600 km

N

4 Write down only the answers.

a) Cardiff is approximately West of Bristol. Roughly, what is the direction of Bristol from Cardiff?

b) Roughly, Grantham is on a bearing of 180° from Lincoln. Approximately, what is the bearing of Lincoln from Grantham?

c) Kelso is North West of Newcastle. What is the direction of Newcastle from Kelso?

d) Ely is on a bearing of 090° from Coventry. What is the bearing of Coventry from Ely?

e) Horace walks along Grey Street, travelling in a North Easterly direction. In which direction does he walk on his return journey?

f) The sun rises in the East. In which direction do shadows point:

  (i) early in the morning
  (ii) late in the evening?

g) The wind is blowing *from* the South-East. Suddenly, it changes through 180°.

  (i) *From* which direction is it blowing now?
  (ii) *In* which direction is it blowing now?

h) 'The North wind (wind *from* the North) doth blow and we shall have snow, And what will the robin do then, poor thing?' On what bearing is the wind blowing?

i) 'Westward Ho!' What is this direction as a bearing?

j) The shadows of these bushes are pointing to the North West. In which direction is the sun from the bushes? Write your result:

  (i) using compass directions
  (ii) as a bearing.

k) The bearing of Aberdeen from Dundee is 030°. What is the bearing of Dundee from Aberdeen?

ENRICHMENT

CHALLENGES

1   You need a protractor and a ruler. Melton is 6 km from
Walsley on a bearing of 300°.
Here are some more towns:

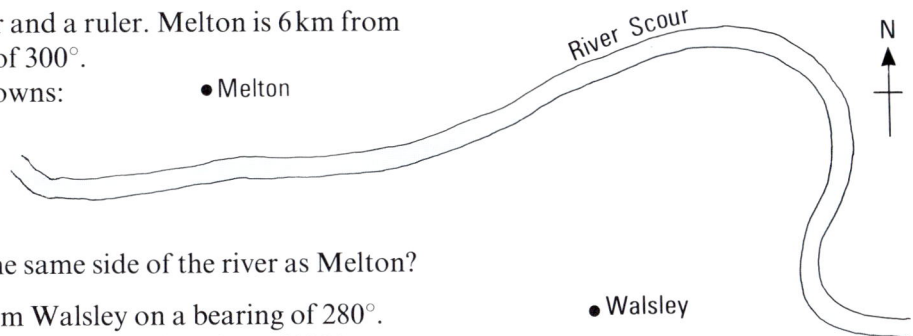

River Scour

• Melton

N

Which of these
do you think are on the same side of the river as Melton?

• Walsley

Wandswall: 10 km from Walsley on a bearing of 280°.
Westleigh: 7 km from Walsley on a bearing of 030°.
Streetham: 4 km from Melton on a bearing of 100°.
Elephant: 1 km from Melton on a bearing of 220°.
Dulton: North West of Walsley and East of Melton.

2   a)   Draw your own plan of an imaginary island, and give it a name. Show the North direction
on your map. Show the scale of your map.

   b)   Show these places on your map:
   • a place called North Headland
   • a place called South Harbour

   • a farm slightly to the East of the centre
     of the island
   • a place called West Bay.

   c)   A climber has fallen on the cliffs at North Headland. Write down clear instructions
for a rescue party to find her. The rescue party must start from the Farmhouse.

   d)   A helicopter meets the rescue party at North Headland. Give clear instructions to get the
helicopter to the West Bay rescue centre.

REVIEW

The coffee in the machine below was made with 3 teaspoonfuls of coffee and 2 mugs of water. We say the *ratio* of coffee to water is 3 to 2. We write 3:2.

Coffee to water ratio 3:2

A

These will give the same strength of coffee as A: 6:4, $1\frac{1}{2}$:1, 12:8, ...

Coffee to water ratio 4:3

B

These will give the same strength of coffee as B: 8:6, 2:$1\frac{1}{2}$, 12:9, ...

■ Write down an example of your own which gives the same strength coffee as A.

■ Write down an example of your own which gives the same strength coffee as B.

■ Which is the stronger coffee, A or B? How can you tell?

Green paint A is mixed from 2 cans of yellow and 3 of blue; 2:3.

Paint B is a different shade of green from A.

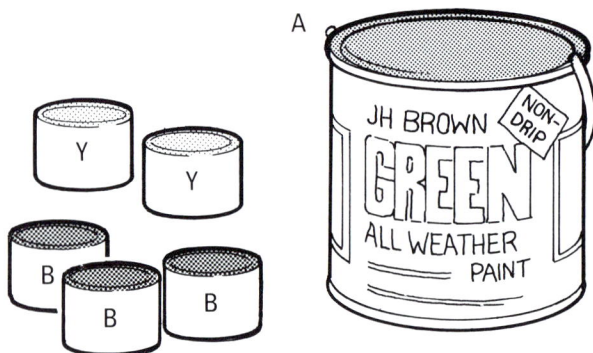

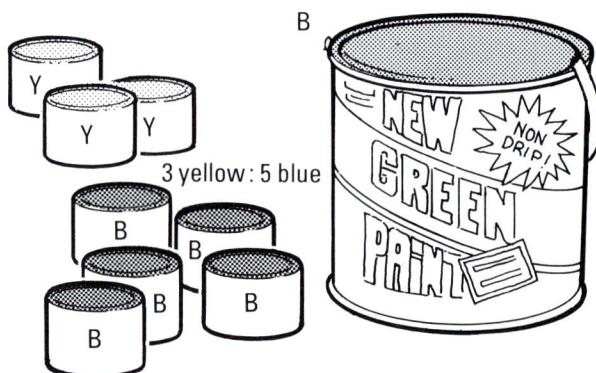

A

JH BROWN NON-DRIP GREEN ALL WEATHER PAINT

Y Y
B B B

B

3 yellow : 5 blue

NEW GREEN PAINT NON-DRIP!

Y Y Y
B B B B B

These will give the same shade of green as A: 2:3, 4:6, 6:9, ...

These will give the same shade of green as B: 3:5, 6:10, 9:15, ...

■ Which is the lighter green, A or B? How can you tell?

CONSOLIDATION

1   The ratio of small buttons to large buttons on this card is 3:2.

a)   Draw a card with the same ratio of small buttons to large buttons. Put 9 small buttons on your card.

b)   A larger card also has buttons in the ratio 3:2. There are 10 large buttons. How many small buttons are there?

c)   Six of the large buttons are removed from the card. Some small buttons are also removed. Tariq says the ratio is still 3:2. Can he be correct? Explain your answer.

2   Estimate these ratios.

a)   The speed of a top class sprinter: your own speed.
b)   Your age now: your teacher's age now.
c)   The number of hours you spend sleeping: the number of hours a newborn baby spends sleeping.
d)   Your age ten years ago: your teacher's age ten years ago.

3   You need 1 cm squared paper.

a)   Draw examples of rectangles whose length:width ratio is:

(i)   2:3   (ii)   1:1   (ii)   1.5:1.

4 cm

2 cm

length:width ratio is 4:2 or 2:1

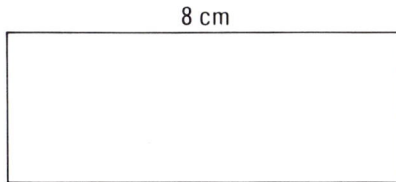

b)   What special kind of rectangle are shapes with length:width ratio 1:1?

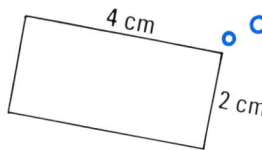

8 cm

3 cm

length:width ratio is 8:3

4   The information on the map says the scale is 1:1 000 000.

a)   Is this
A   1 cm to 1 km
B   1 cm to 10 km
C   1 cm to 100 km?

b)   Do you think there is a misprint on the map? Why?

c)   Write down what you think the scale is likely to be for the map.

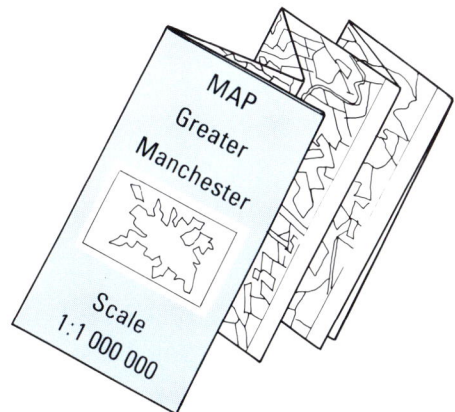

MAP
Greater
Manchester

Scale
1:1 000 000

A15

EXPLORATION: TEARING SHEETS

5   The length:width ratio of this sheet of paper is 3:2 (which is the same as 30:20).

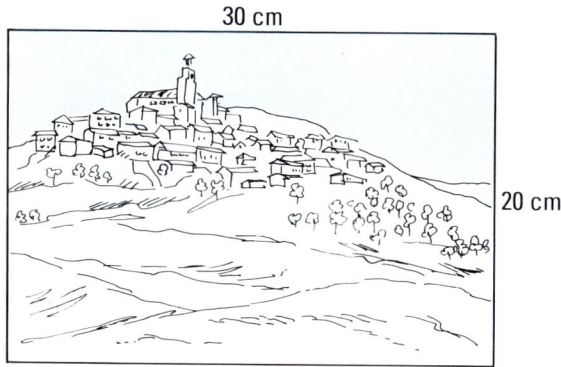

30 cm

20 cm

a) It is torn in half, across its length. Explain why the length:width ratio of each half is 4:3.

b) One of the half-sheets is torn in half again, across its length.

Copy and complete:
The length:width ratio of each half is 3:☐ .

c) One of the halves is torn again. What is the length:width ratio of each half?

d) The tearing continues. Explore what happens to the length:width ratios. Write down what you discover.(Always write the ratios as simply as you can, using whole numbers. For example, 2:3 rather than 20:30; 1:4 rather than $\frac{1}{2}$:2.)

e) Explore what happens for these starting sheets.

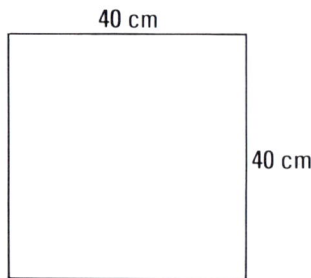

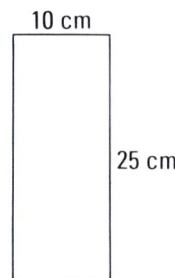

40 cm

40 cm

10 cm

25 cm

f) What happens for other starting sheets? Are there any regular patterns? Write a report about what you discover.

A15

A16 DEALING WITH FRACTIONS

CORE

░░░░░░░░░░ IN YOUR HEAD ░░░░░░░░░░

1 Write down only the answers.

$\frac{1}{4}$ kg $\frac{1}{2}$ kg

a) How much flour altogether?

b) Fill the three mugs. How many litres are left in the bottle?

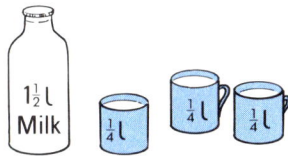

$1\frac{1}{2}$ l Milk $\frac{1}{4}$ l $\frac{1}{4}$ l $\frac{1}{4}$ l

c) What fraction of the tiles is blue: $\frac{1}{4}, \frac{1}{5}$ or $\frac{5}{15}$?

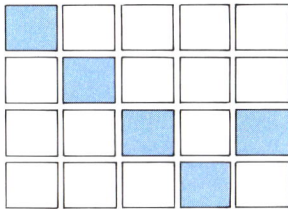

d) The lane is $\frac{3}{8}$ km long. $\frac{1}{8}$ km has been 'tarmaced'. What fraction of a kilometre has yet to be done?

e) What fraction of the chocolate has been eaten?

f) How much milk is there altogether?

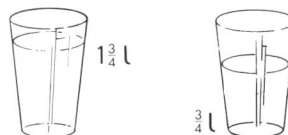

$1\frac{3}{4}$ l $\frac{3}{4}$ l

g) What is the total length of copper tubing: $1\frac{1}{4}$ m, $1\frac{1}{2}$ m or $1\frac{3}{64}$ m?

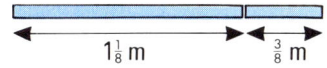

$1\frac{1}{8}$ m $\frac{3}{8}$ m

h) What is the total mass of the lorry and its load?

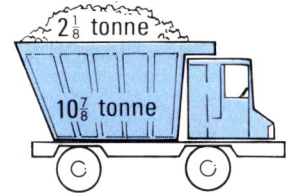

$2\frac{1}{8}$ tonne $10\frac{7}{8}$ tonne

i) How many $\frac{1}{4}$ litre mugs can be filled from the bottle?

$2\frac{3}{4}$ l

j) What is the perimeter of the square: $1\frac{1}{2}$ m, $1\frac{1}{8}$ m or $\frac{9}{64}$ m?

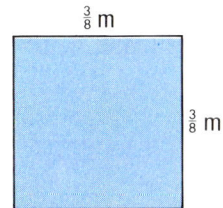

$\frac{3}{8}$ m $\frac{3}{8}$ m

k) What is the total amount of cider?

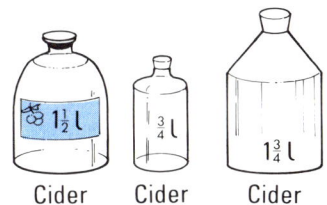

$1\frac{1}{2}$ l $\frac{3}{4}$ l $1\frac{3}{4}$ l
Cider   Cider   Cider

l) The four tennis balls weigh $\frac{1}{4}$ kg altogether. What does each one weigh?

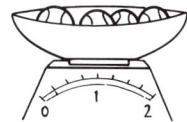

A16

2   a) Check from the line diagram that: $\frac{2}{3}m + \frac{2}{3}m + \frac{2}{3}m + \frac{2}{3}m = 2\frac{2}{3}m = \frac{8}{3}m$

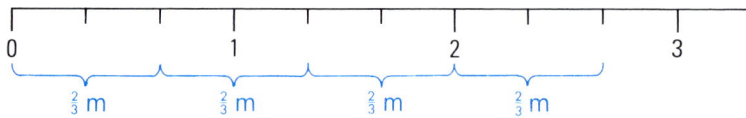

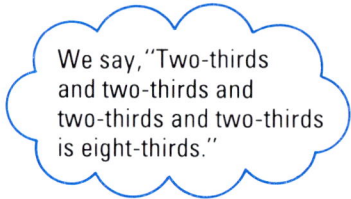

> We say, "Two-thirds and two-thirds and two-thirds and two-thirds is eight-thirds."

   b) Copy and complete:

   (i) $\frac{2}{3}m + \frac{2}{3}m = \frac{\square}{3}m$ (ii) $\frac{2}{3}m + \frac{2}{3}m + \frac{2}{3}m = \square\,m$

   c) Use the line diagrams to help you to do these additions. Write your results as mixed numbers.

   (i) $\frac{3}{5}m + \frac{4}{5}m$ (ii) $1\frac{1}{5}m + \frac{3}{5}m$
   (iii) $\frac{7}{10}m + \frac{6}{10}m$ (iv) $\frac{2}{3}m + 1\frac{2}{3}m$
   (v) $\frac{5}{6}m + \frac{2}{6}m$ (vi) $1\frac{1}{6}m + \frac{7}{6}m$
   (vii) $1\frac{5}{6}m + \frac{1}{3}m$

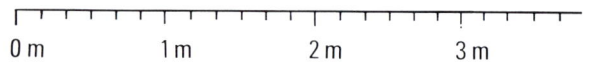

3   Complete this sentence in three different ways: $\frac{\square}{5} + \frac{\square}{5} = \frac{4}{5}$

4   Complete this sentence in three different ways: $\frac{\square}{3} + \frac{\square}{3} = 1\frac{2}{3}$

THINK IT THROUGH

5   Do these additions. Draw line diagrams to help you if you wish.

   a) $\frac{7}{12} + \frac{4}{12}$ b) $2\frac{1}{8} + \frac{7}{8}$ c) $3\frac{2}{3} + \frac{2}{3}$ d) $\frac{5}{6} + \frac{7}{6}$

   e) $\frac{13}{14} + \frac{4}{14}$ f) $\frac{3}{25} + \frac{20}{25}$ g) $\frac{5}{7} + \frac{8}{7}$ h) $2\frac{5}{9} + 3\frac{5}{9}$

6   a) Check from the line diagram that: $1\frac{2}{3} - 1\frac{1}{3} = \frac{1}{3}$

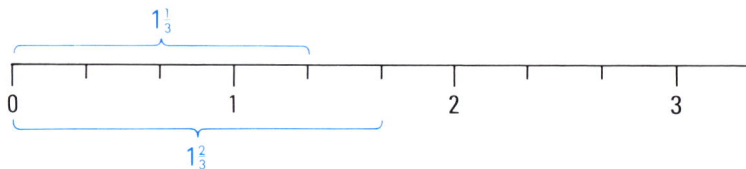

   b) Copy and complete (use the line diagram to help you):

   (i) $2 - 1\frac{1}{3} = \frac{\square}{3}$ (ii) $2\frac{2}{3} - 1\frac{1}{3} = \frac{\square}{3}$ (iii) $3 - 1\frac{1}{3} = \frac{\square}{3}$

   (iv) $2 - 1\frac{2}{3} = \frac{\square}{3}$ (v) $2\frac{1}{3} - 1\frac{2}{3} = \frac{\square}{3}$ (vi) $3 - 1\frac{2}{3} = \square\frac{\square}{3}$

7   Use the line diagrams to help you to do these subtractions:

   a) $\frac{4}{5} - \frac{1}{5}$ b) $1\frac{1}{5} - \frac{2}{5}$
   c) $2\frac{3}{5} - 1\frac{2}{5}$ d) $\frac{7}{8} - \frac{3}{8}$
   e) $1\frac{1}{4} - \frac{3}{4}$ f) $\frac{3}{4} - \frac{3}{8}$
   g) $2\frac{7}{8} - 1\frac{5}{8}$ h) $2\frac{1}{4} - \frac{7}{8}$.

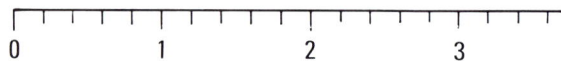

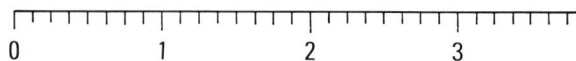

A16

8 Copy and complete this sentence in three different ways: $1 - \frac{\square}{7} = \frac{\square}{7}$

9 Copy and complete this sentence in three different ways: $1\frac{1}{12} - \frac{\square}{12} = \frac{\square}{12}$

10 Do these subtractions. Draw line diagrams to help you if you wish.

a) $\frac{9}{14} - \frac{3}{14}$ b) $3\frac{1}{5} - 1\frac{2}{5}$ c) $4\frac{5}{6} - 3\frac{1}{6}$ d) $1\frac{1}{10} - \frac{7}{10}$

e) $1\frac{4}{15} - \frac{8}{15}$ f) $\frac{24}{25} - \frac{9}{25}$ g) $10\frac{1}{10} - 3\frac{7}{10}$ h) $5\frac{1}{2} - \frac{3}{8}$

CHALLENGE

11 The internal diameter of this pipe is $7\frac{3}{8}$ inches.
The external diameter is $10\frac{7}{16}$ inches.
How thick is the wall of the pipe?

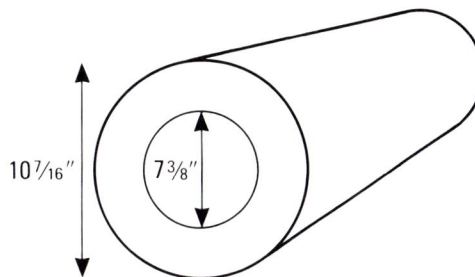

$10\frac{7}{16}''$ $7\frac{3}{8}''$

CHALLENGE

12 Grandma Higgins spends many hours knitting scarves. She knits each one 1 m long.
This is a drawing of one of Grandma Higgins scarves. It shows how much she knitted each day.
What fraction of the whole scarf had she knitted after:

a) two days b) three days?

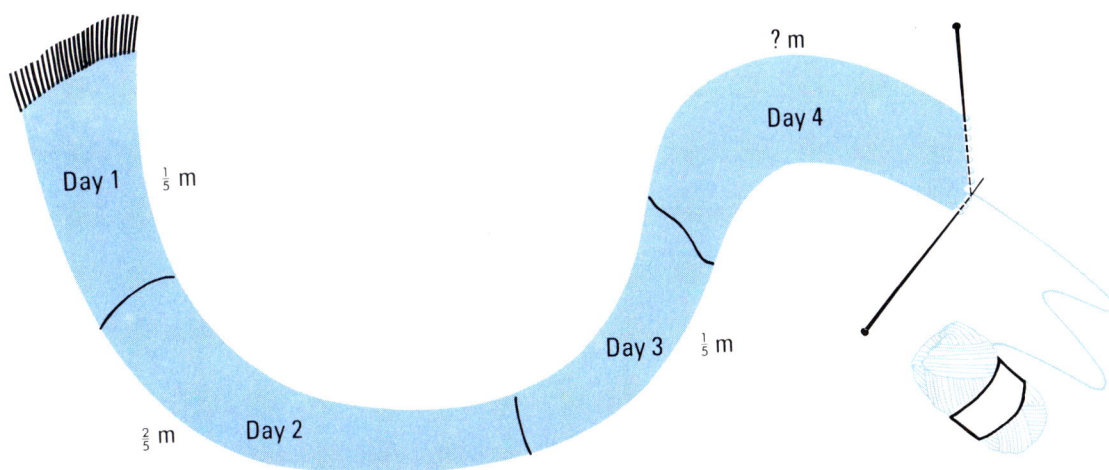

Day 1 $\frac{1}{5}$ m

$\frac{2}{5}$ m   Day 2

Day 3 $\frac{1}{5}$ m

? m

Day 4

What fraction was left to be knitted after:

c) one day d) two days?

e) Grandma Higgins finished the scarf on the fourth day. What fraction of a metre did she knit
on Day 4?

A16

f) This is another of Grandma Higgins' scarves. What fraction of it is white?

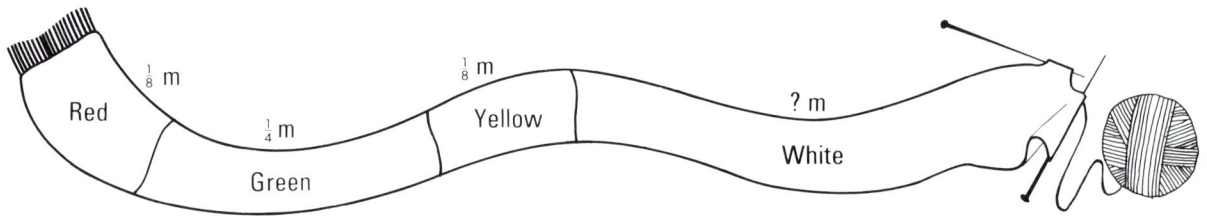

g) Here is another scarf. ...What fraction is red?

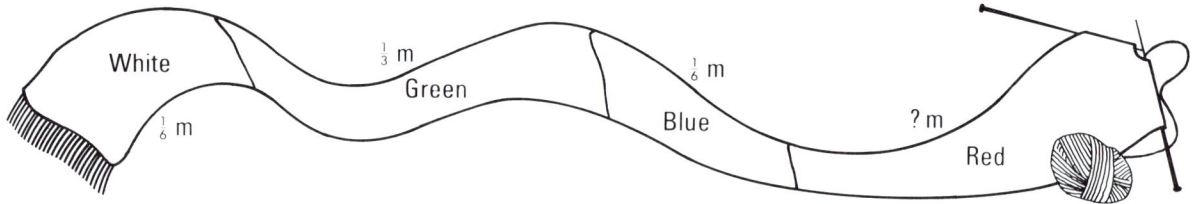

h) ... and a fourth. What fraction is yellow?

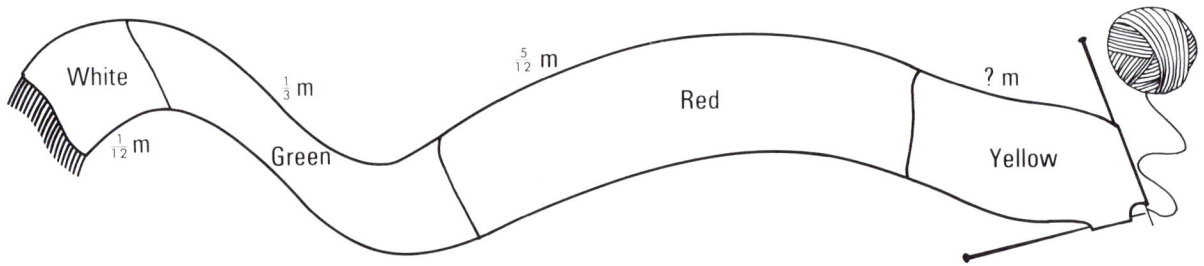

REVIEW

Border patterns are often made by:

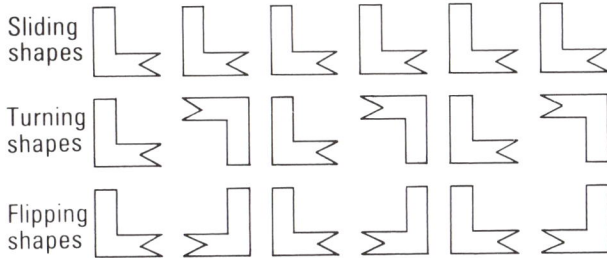

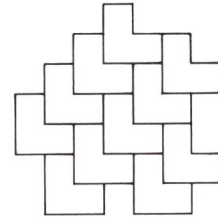

Sliding shapes

Turning shapes

Flipping shapes

This is a *tessellation* of letter Ls.

■ Draw a tessellation of Ts.

We can slide (or, *translate*) a shape like this:

$AA' = BB'$

Slide

■ Draw this shape and a translation of it on squared paper. Show the slide with an arrow.

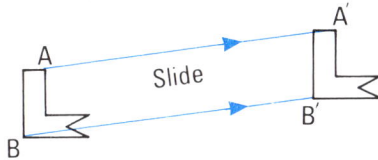

Rotate a shape.

$AO = OA'$

O — Centre of rotation

TURN

Angle of rotation

■ Draw this shape and a rotation of it on squared paper. Mark in the centre and angle of rotation.

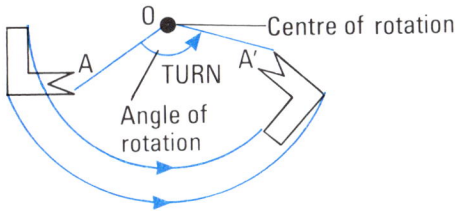

Reflect a shape.

$AX = XA'$
$BY = YB'$

Line of reflection

■ Draw this shape and a reflection of it, on squared paper. Mark in the line of reflection.

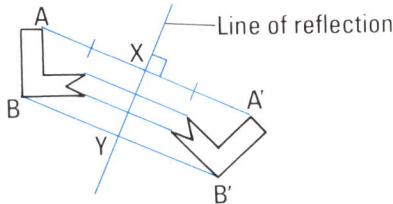

Enlarge a shape.

$OA' = 2OA$, etc.
$A'B' = 2AB$, etc.

Centre of enlargement

■ Draw this shape and an enlargement of it on squared paper. Mark the centre of enlargement. Write down the scale factor.

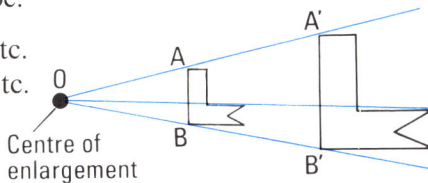

The scale factor for enlargement is $\times 2$.

Reduce a shape.

$OA' = \frac{1}{2}OA$, etc.
$A'B' = \frac{1}{2}AB$, etc.

The scale factor for enlargement is $\times \frac{1}{2}$ or $\div 2$.

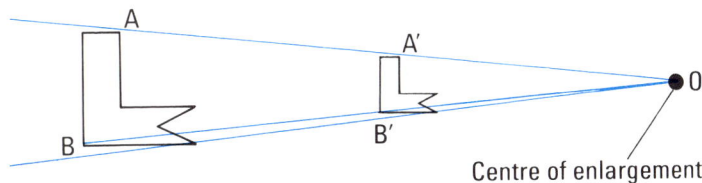

Centre of enlargement

CONSOLIDATION

A17

1   You need 1 cm dotted squared paper.

a) Copy the letter J and the line AB. Complete the reflection of the J in the line.

b) Which of these are true and which are false?

   A   The new letter J can be translated back onto the original J.
   B   The new letter J can be rotated back onto the original J.
   C   The new letter J can be reflected back onto the original J.

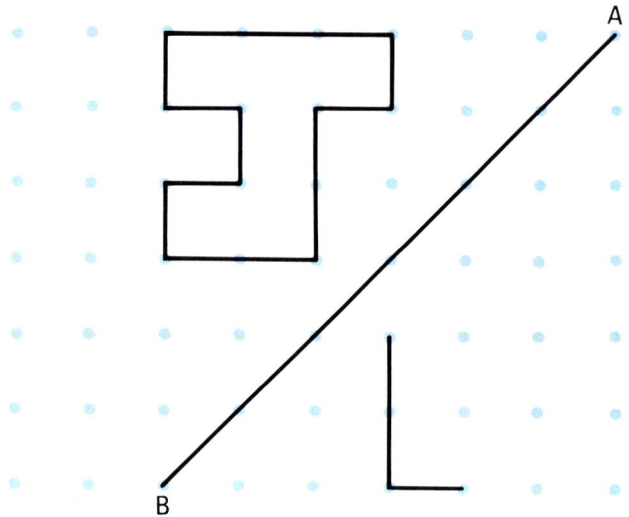

c) The letter F is translated so that the point P finishes up at P′. Draw the new position of the F.

d) Which of these are true and which are false?

   A   The new letter F can be reflected back onto the original F.
   B   The new letter F can be rotated back onto the original F.
   C   The new letter F can be translated back onto the original F.

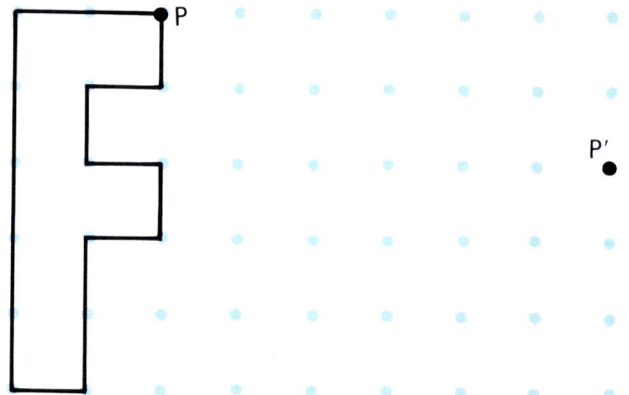

e) The letter L is rotated through 90° clockwise so that the point M finishes up at M′.

   (i)   Draw the new position of the L.
   (ii) On your drawing, mark in the centre of rotation.

f) Which of these are true and which are false?

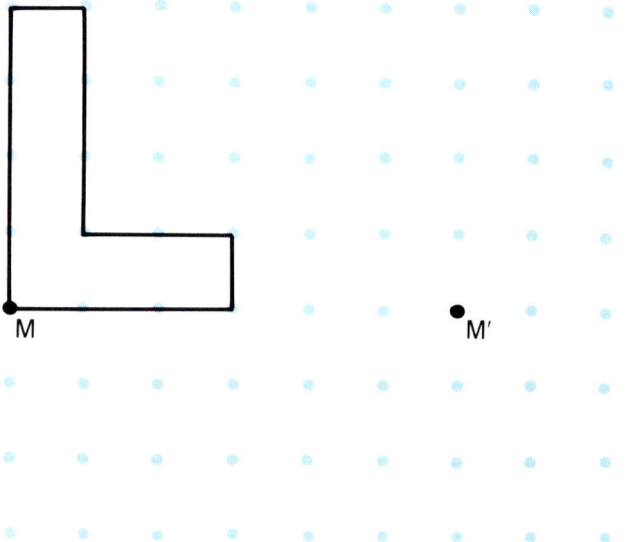

   A   The new letter L can be reflected back onto the original L.
   B   The new L can be rotated back onto the original L.
   C   The new L can be translated back onto the original L.

A17

ACTIVITY

2   You need 1 cm dotted squared paper and tracing paper.

a) Try to answer questions (i)–(iv) by looking and thinking. (You may find them difficult!) If you are not sure of your answers:

- copy the triangles onto dotted squared paper
- trace your drawing
- use your drawing and your tracing to check your answers.

  (i)   Triangle A can be reflected onto *two* of the triangles 1 to 8. Which two?
 (ii)   Triangle A can be translated onto *two* of the triangles 1 to 8. Which two?
(iii)   Triangle A can be rotated onto *two* of the triangles 1 to 8. Which two?
(iv)   Triangle A cannot be rotated, translated, or reflected onto two of the triangles 1 to 8. Which two?

b) Copy triangles A, 1, 3, 4, and 6 onto dotted squared paper. Mark in:

  (i)   two lines of reflection for the triangles in a)(i)
 (ii)   two centres of rotation for the triangles in a)(iii).

c) Draw another triangle like the ones you found in a)(iv). Give it the number 9.

A page of griffins

3   Here are six griffins.

a) Which griffin is a translation of griffin 2?

b) Which griffin is a reflection of griffin 1, in line RS?

c) Griffin 3 is a rotation of griffin 1. Which is the centre of rotation, A, B, or C?

d) There is a rotation with centre D that rotates one griffin onto another.

   (i) Which two griffins are involved?
   (ii) Roughly, what is the angle of rotation?

e) Griffin 2 is a reflection of griffin 1 … in which line?

f) Griffin 5 is an enlargement of griffin 4.
   (i) Which is the centre of enlargement, G, H, or I?
   (ii) What is the scale factor for enlargement?

g) Griffin 6 can be rotated onto one other griffin. Which one?

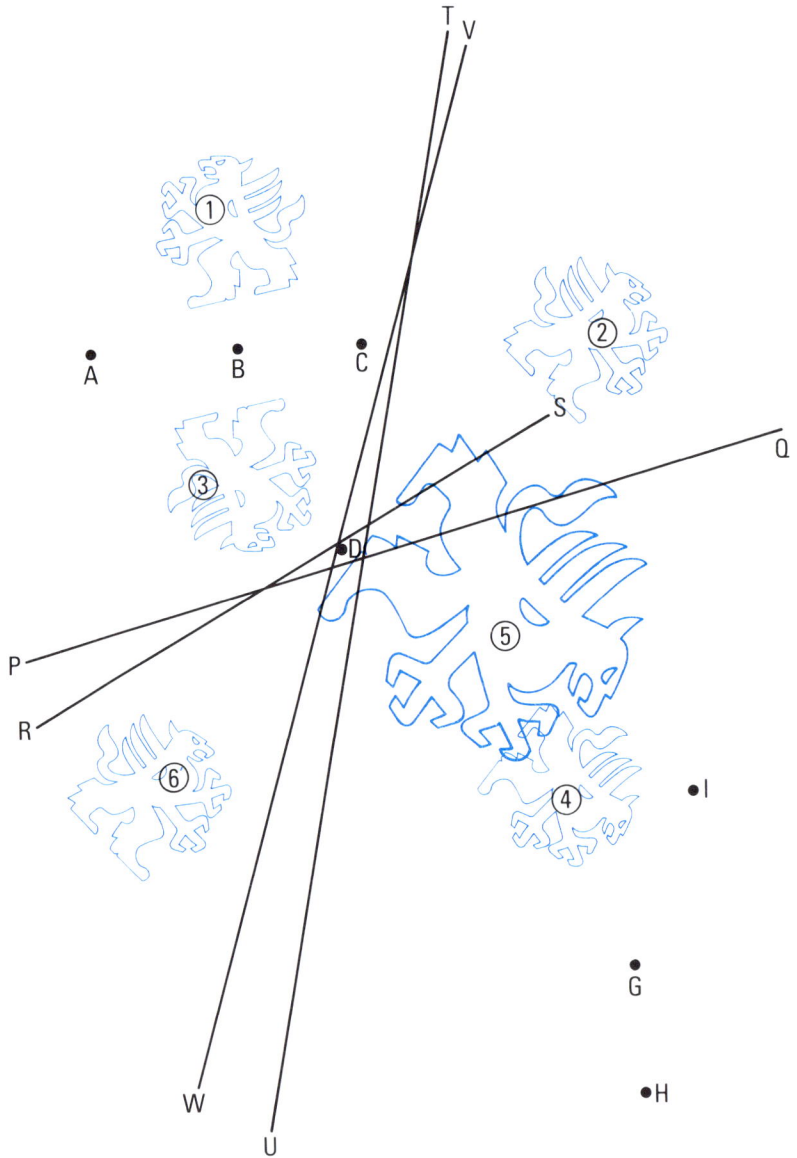

REVIEW

These piles of nuts and bolts obey the rule:

$$n + 1 = b$$

↑ number of nuts in a pile   ↑ number of bolts in that pile

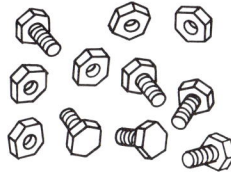

■ Draw a pile in which there are 4 nuts. How many bolts are there?

■ In a large pile there are 60 bolts. How many nuts are there?

The rule $n + 1 = b$ can be written in these different ways: $n = b - 1$
(Check by looking at the piles of nuts and bolts that each one is correct.) $b - n = 1$

■ Draw three different piles of drawing pins and matchsticks which agree with this rule:

$$2 \times d = m$$

↑ number of drawing pins in a pile   ↑ number of matchsticks in that pile

■ Rewrite the rule like this: $d = \ldots$ (Use your drawings to help you.)

We can represent rules on a graph.

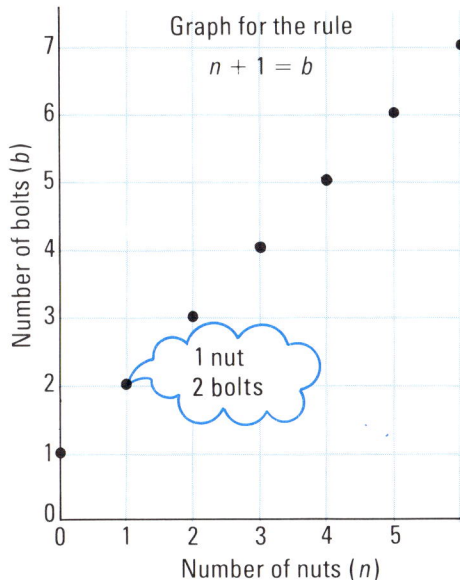

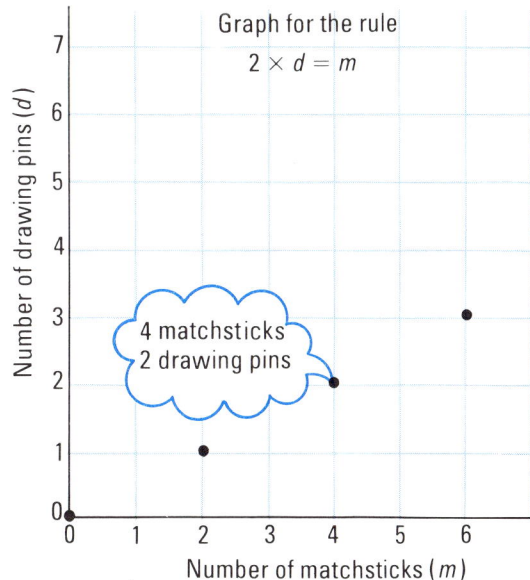

Graph for the rule $n + 1 = b$

1 nut
2 bolts

Number of bolts (b) vs Number of nuts (n)

Graph for the rule $2 \times d = m$

4 matchsticks
2 drawing pins

Number of drawing pins (d) vs Number of matchsticks (m)

■ Draw a graph for this rule:

$$p = b - 2$$

↑ number of pins   ↑ number of buttons

CONSOLIDATION

A18

1 'Axis' sells gift sets of pencils and pens. This is a rule about the number of pencils and ballpoint pens in different gift sets:

$$p \quad = \quad b+3$$

↑ number of pencils

↑ number of ballpoint pens

a) Sketch two different piles of pencils and ballpoint pens which agree with the rule.

b) In one gift set there are 5 ballpoint pens. How many pencils are there?

c) In another gift set there are 10 pencils. How many ballpoint pens are there?

d) Here are some rules written in words. Which ones are correct for the gift sets?

A There are three times as many ballpoint pens as pencils in each set.
B There are three more ballpoint pens than pencils in each set.
C There are three more pencils than ballpoint pens in each set.
D There are three fewer ballpoint pens than pencils in each set.

e) This graph represents the sets of pens and pencils. Copy and complete it.

f) These are different ways of writing the same rule. Copy and complete each one:

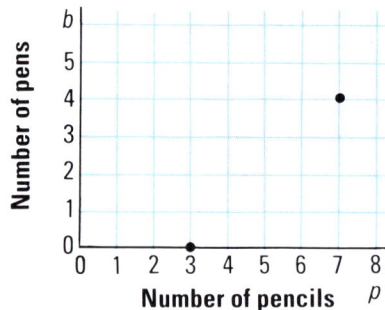

(i) $b = \ldots$ (ii) $\ldots = 3$

2 'Axis' also sells gift sets of notebooks and pencils. The graph represents the numbers in each set.

a) In a gift set with two notebooks, how many pencils are there?

b) In a gift set with nine pencils, how many notebooks are there?

c) In a gift set with five notebooks, how many pencils are there?

d) Copy and complete these two versions of the rule which connects p and n:

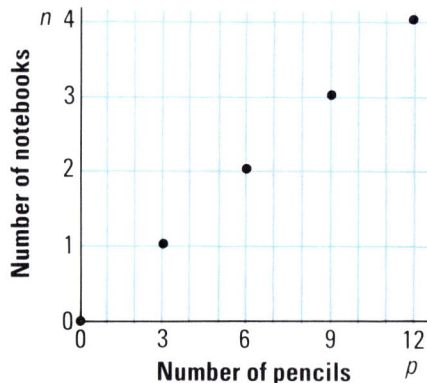

(i) $p = \ldots$ (ii) $n = \ldots$

CORE

Calculations from rules

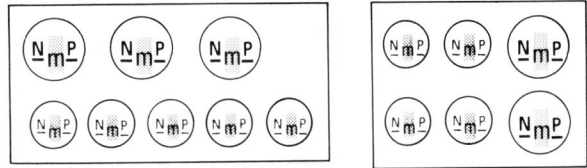

1   These are NMP stickers. This rule gives
    information about the numbers of large and
    small stickers on each card.

$$l \qquad = \qquad s - \ldots$$

$\uparrow$ $\uparrow$

number of large stickers   number of small stickers

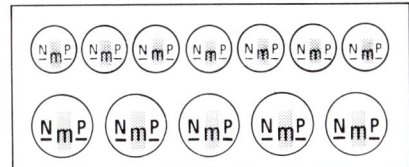

a) Copy and complete the rule.

b) Copy and complete these versions of the rule.
    (Use the drawings of the cards to help you.)

    (i) $s = l \ldots$ (ii) $s - l = \ldots$

c) The graph represents the rule for the
    stickers. Which axis (horizontal or
    vertical) should be labelled 'Number of
    large stickers (l)'?

d) What label should be used for the other axis?

e) For a large card of NMP stickers, l is 24.
    What is s?

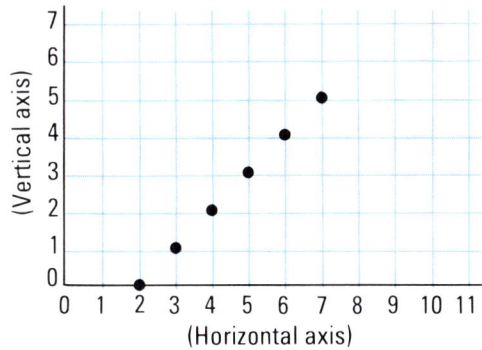

2   Asif packs picture frame hooks in different-sized boxes. Each hook has two holes for pins.
    He also includes some spare pins. This rule tells him how many pins he must include in each box:

$$p \qquad = \qquad 2 \times (h + 1)$$

$\uparrow$ $\uparrow$

number of   number of
pins   hooks

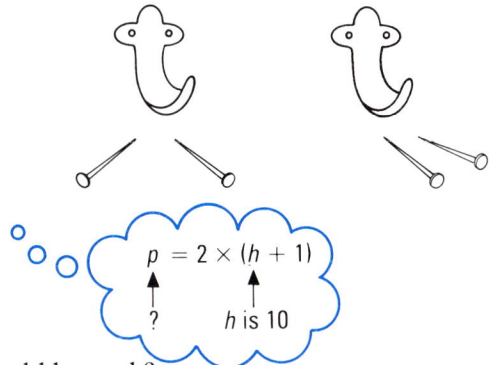

a) Check that a box of 10 hooks should have 22 pins.

b) How many pins should there be in a box of:

    (i) 12 hooks   (ii) 50 hooks?

c) Asif packs 42 pins into a box. How many hooks should he pack?

──────────────────── TAKE NOTE ────────────────────

Rules give us the connection between numbers of things (for example, the number of pins and hooks
in a box) or quantities (for example, the area of a circle and its radius). We can use the rules to make
calculations.

3   For a circle: $A \approx 3 \times r^2$ $C \approx 6 \times r$
                         $\uparrow$ $\uparrow$ $\nearrow$ $\nwarrow$
                      Area   Radius × Radius   Circumference   Radius

a) The radius of a circle is 5 cm (r is 5). Roughly, what is its area (in cm^2)?

b) Roughly, what is the circumference of a circle whose radius is 7 cm?

c) The circumference of a bicycle wheel is about 120 cm. What is its radius? Do you think the bicycle is for an 8-year-old, a 14-year-old, or a 17-year-old?

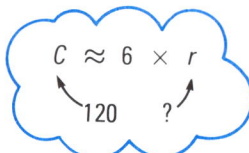

$C \approx 6 \times r$
120   ?

d) The area of a circular pond is 75 m^2. What is its radius?

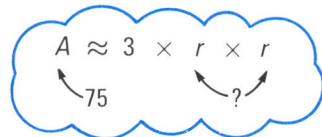

$A \approx 3 \times r \times r$
75   ?

4   All these triangles are equilateral. Use p cm to represent the perimeter of any equilateral triangle. Use k cm to represent the length of one of its sides.

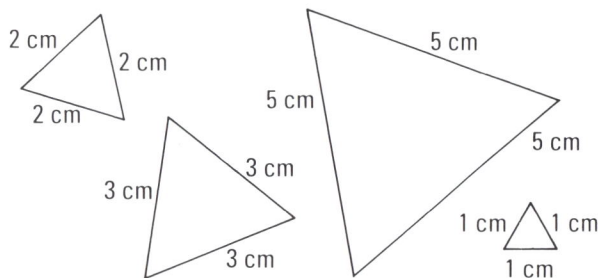

2 cm   2 cm   2 cm   5 cm   5 cm   5 cm

3 cm   3 cm   3 cm   1 cm   1 cm   1 cm

a) What is p when k is 2?

b) What is k when p is 15?

c) Two of these are correct rules. Which two?

   A $k = 3 \times p$ B $p = 3 \times k$ C $k = 3 \div p$ D $k = p \div 3$

d) Write your rules from part c) in words.

THINK IT THROUGH

5   Divers use this rule to work out how long they can stay under water:

Maximum time under water, in minutes $\longrightarrow T = \dfrac{12 \times V}{d}$ ← Volume of air needed, in cubic feet
← Depth, in feet

a) A diver has 40 cubic feet of air. She is 20 ft under water. How long can she stay there?

b) How many cubic feet of air does the diver need to stay for 10 minutes at a depth of 48 feet?

Rules at work

1   A car hire firm uses this rule to work out how much to charge per day for a van:

$$\text{Hire charge (£)} = 25 + \frac{\text{mileage}}{2}$$

a) Qumar wants to hire the van for a day. Why does he not know exactly how much it will cost him?

b) He hires the van and travels 120 miles. What does he pay for the day's hire?

c) About how much would it cost you to hire the van for an outing to your favourite place?

A18

CHALLENGE

d) Misbah hired the van for a day. It cost him £115. How many miles did he travel?

2   Here is how you can get a rough idea of the speed at which you can throw a ball.

- Throw a ball as high as you can straight up into the air.
- Ask a friend to time it from release to return.
- The approximate speed at which you released the ball is given by the rule:

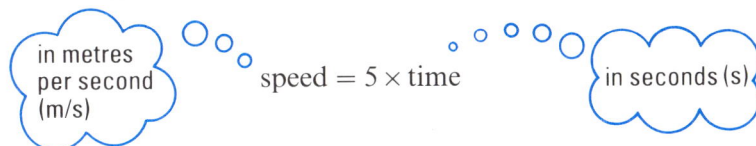

in metres per second (m/s) $\text{speed} = 5 \times \text{time}$ in seconds (s)

a) Jim's best attempt is timed at 3.5 s. Check that he threw the ball at about 17.5 m/s.

b) When next you have the chance, work out how fast you can throw a ball.

c) Rula can throw a ball at about 20 m/s. How long does the ball stay in the air when she throws it straight up?

3   Jack works in a store. He is paid £4 per hour. Last week he worked 32 hours.

a) What were his total wages for the week?

b) Check that your result in a) fits this rule:

weekly wage = pay rate × hours worked

c) Julie's weekly wage is £150 for a 30-hour week. What is her pay rate?

d) Copy and complete this rule for working out pay rates:
pay rate = weekly wage ...

CHALLENGES

4   Dave works for a tailoring firm. He cuts
    patterns out of large pieces of cloth.
    Before he cuts he has to work out the percentage of
    cloth he will waste.

   a) This is one of the patterns he cuts out.
      Roughly, how much cloth will he waste ...

      90%, 60%, 40%, or 20%?

   b) To calculate the percentage he uses this rule:

      $$\% \text{ of cloth wasted} = \left(1 - \frac{\text{area of pattern}}{\text{area of cloth}}\right) \times 100$$

      The area of the pattern in the example is $2\,\text{m}^2$. The cloth measures $2\,\text{m} \times 1\frac{1}{2}\,\text{m}$.
      What percentage is wasted?

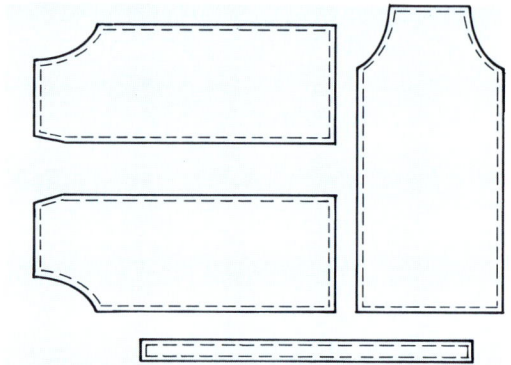

5   An electrical store buys audio tapes in packs of 100.
    To find the price to charge the customer, the store:

    ● works out how much it pays the supplier
      for the tape
    ● adds 50% 'mark up'
    ● adds 15% VAT (Value Added Tax) to the total.

   a) The store pays £107 for 100 tapes. Roughly,
      what price does the store charge the customer?

   b) This is how the store's selling price is worked out:

      $$\frac{\text{Cost to}}{\text{customer}} = \frac{\text{cost-price of 100-pack}}{100} \times 1.5 \times 1.15$$

      How much (to the nearest 1p) does a customer
      pay for a tape?

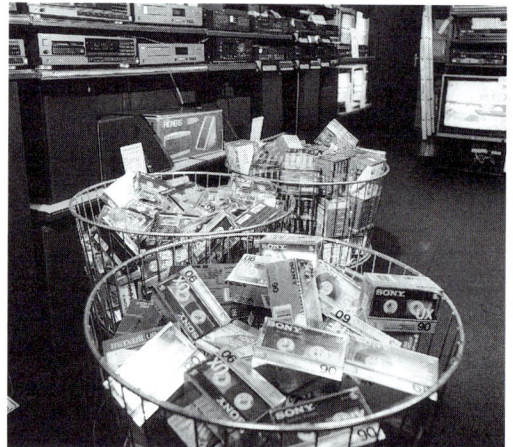

6   Overtime pay at Thames Tyre Factory is worked
    out like this:

    $$\frac{\text{Overtime}}{\text{wage}} = \frac{\text{basic wage}}{35} \times \frac{3}{2} \times \frac{\text{number of}}{\text{hours overtime}}$$

   a) In the rule, the basic wage is divided by 35.
      Why do you think this is?

   b) In the rule, $\dfrac{\text{basic wage}}{35}$ is multiplied by $\dfrac{3}{2}$.
      Why do you think *this* is?

   c) Moira's basic wage is £72. If she works 6 hours overtime, what is her total wage for the week?

REVIEW

The position of Carleston relative to Allerton is (5 km E, 2 km N). (We give East–West distances first, then North–South.)

■ What is the position of Allerton relative to Carleston?

Map indexes often name squares to help us to find places.

Square A 4

■ In which square is the Natural History Museum?

We can describe the position of points on a flat surface by drawing a grid. A is at the position $(3, 2\frac{1}{2})$.

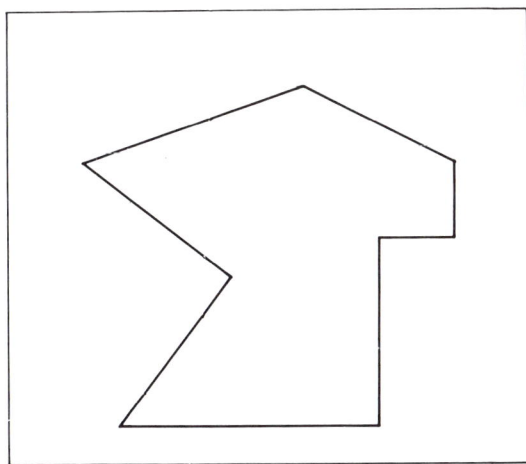

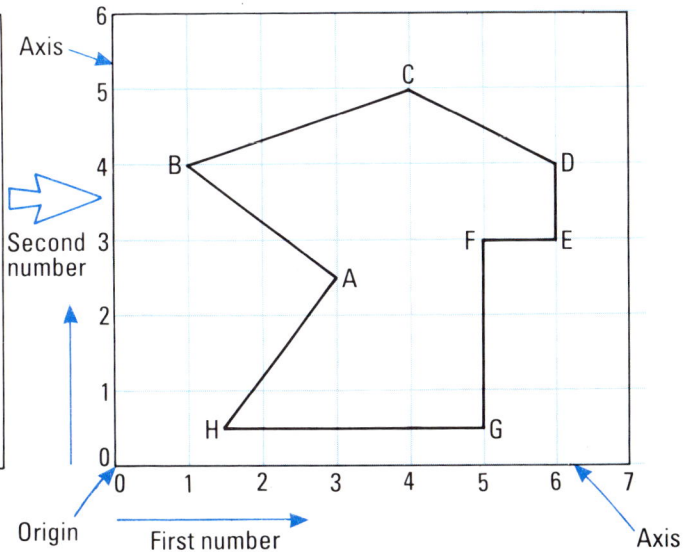

■ What is the position of a) C, b) H?

CONSOLIDATION

1   The map shows a part of Cardiff.

   a) Find the Keeper's cottage on Llanishen Reservoir, in square B2. Explain how to get to Cardiff High School, square B4, by the shortest route. Start like this:

      'Walk down the lane from the Keeper's cottage, away from the reservoir. At the T-junction, turn right ...'

   b) Each square on the map is $\frac{1}{4}$ mile $\times \frac{1}{4}$ mile. Roughly, how far is it from the Keeper's cottage to the High School?

   c) Find the railway track which runs from Llanishen (square A1) to Heath High Level. Roughly, how far is it from station to station?

   d) Name all the squares the railway track runs through.

A19

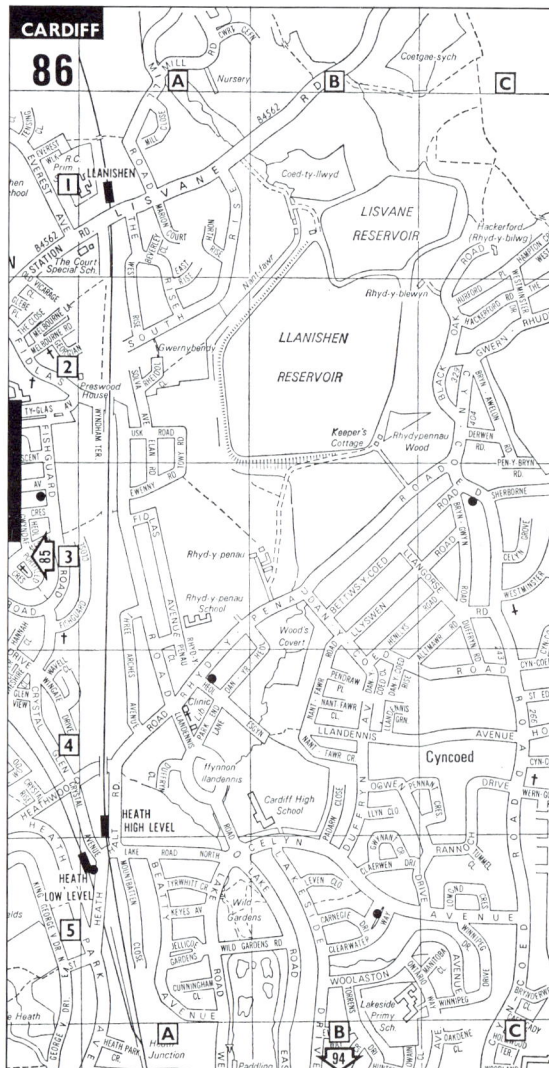

Llanishen Reservoir

Cardiff High School

WITH A FRIEND: ROAD MAPS

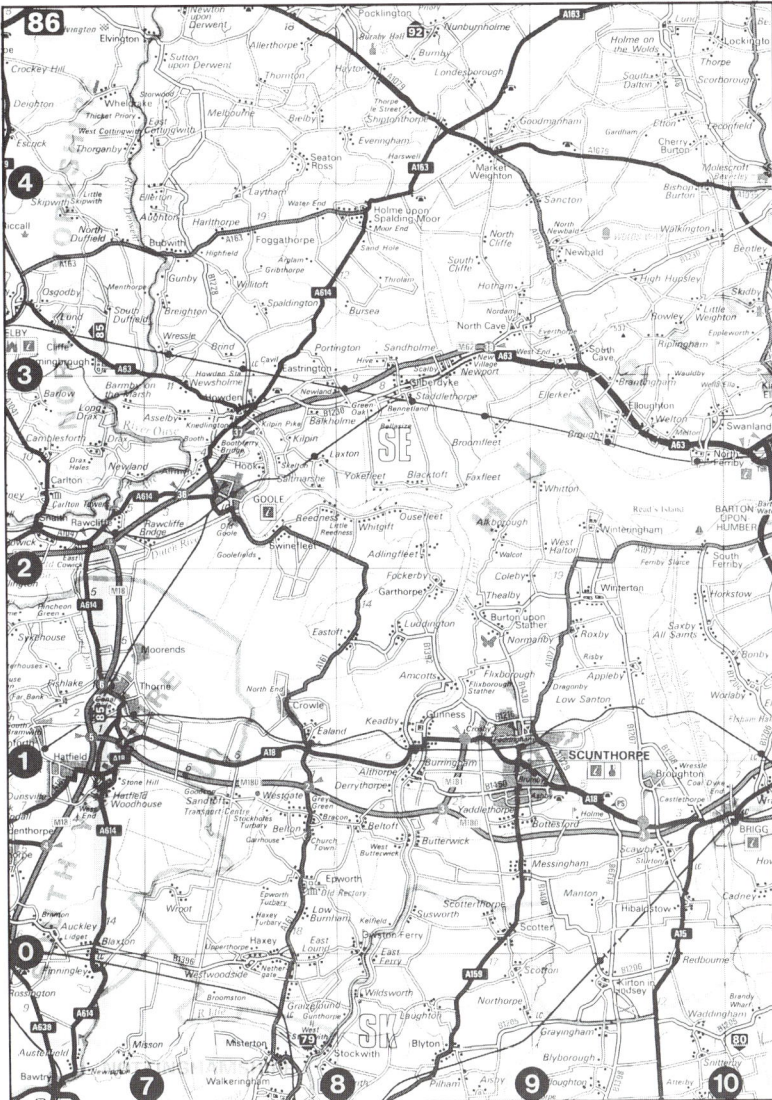

You are here

A19

2   This is page 86 of the *AA Great Britain Road Atlas*, and on the right are small sections taken from the index.

a) You can use the numbers in the index to find Foggathorpe, Butterwick, and Appleby. Decide between you how to do this.

b) Explain in your own words what **86** SE 7537 means. Both of you write down the name of the village at this place.

c) Find the Country Park (symbol 🦋) just North of Scunthorpe. Write down the position for the index.

Formerly St OW 1435
Foggathorpe **86** SE 7537
Foss 89 NE 5295

Butterstone 49 NE 0650
Butterwick **86** SE 8305
Butt Green 44 NW 5031

Appleby 100 NE 2400
Appleby **86** SE 9414
Applecross 54 NW 6005

Coordinates

1 This drawing is made on a computer screen.
The computer was given these instructions:

Write out instructions for the computer to draw this shape. Start and end at (4, 5). The shape should be drawn in a clockwise direction.

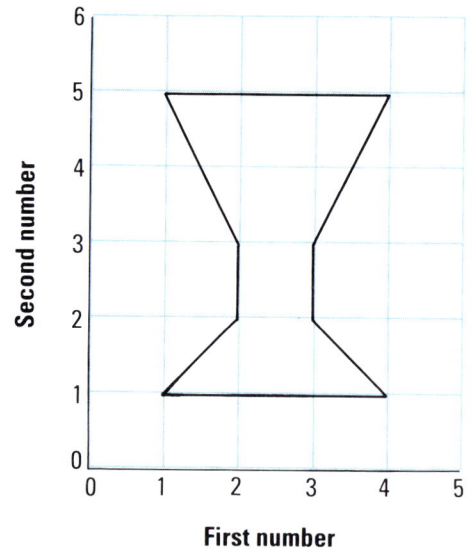

START : (5, 1)
THEN : (1, 1) (3, 4) (3, 5) (4, 5) (4, 4)
(5, 3) (5, 1) END

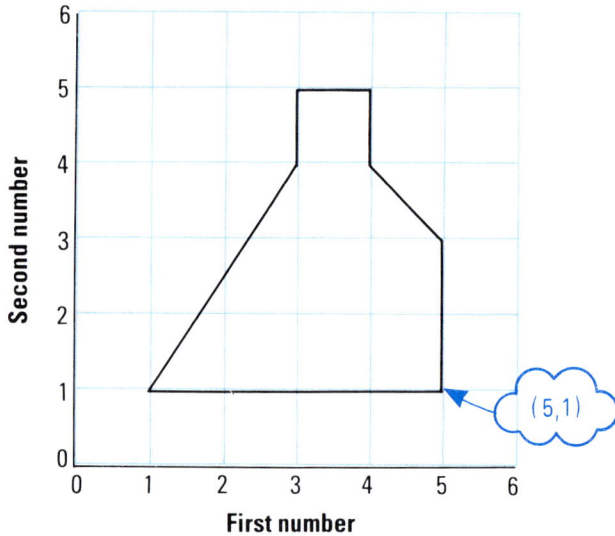

(5,1)

CHALLENGE

2 These patterns were made by an embroidery machine. The machine can be given instructions like those in question 1.

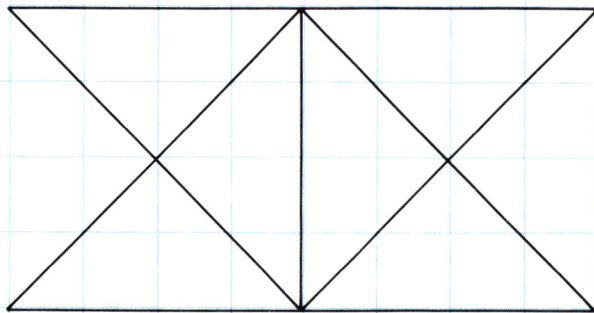

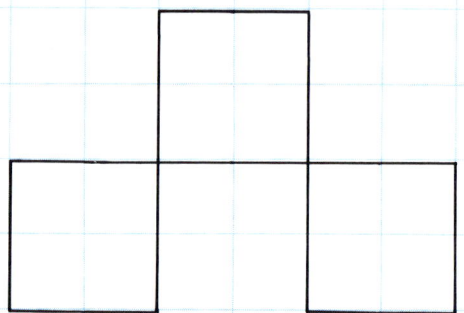

The instructions must be written so that the machine does not embroider the same line twice. Try to write down a set of instructions for each pattern. (Copy the shapes and draw in your own axes, then write down the list of coordinates.)

Pairs of numbers which give the position of a point on a grid are called *coordinates*.
The coordinates of A are $(3, 1)$. (We always give the horizontal reading first.)
The coordinates of B are $(2, 4)$.
The coordinates of D are $(0, 0)$.

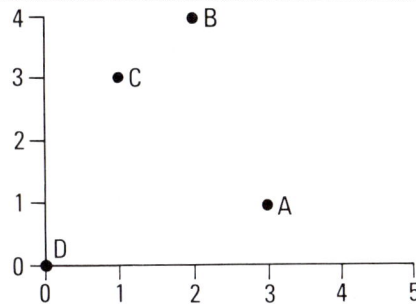

3 In the *Take note*, what are the coordinates of C?

4 You need 1 cm squared paper. Draw two axes. Number them.

a) Plot each of these points and join them up in order.

$A(1, 7)$ $B(1, 10)$ $C(3, 7)$ $D(3, 10)$

b) Start again. Plot:

$A(4, 3)$ $B(4, 6)$ $C(5, 5)$ $D(6, 6)$ $E(6, 3)$.

Join up the points in order.

c) Start again. Plot each of these points and join them up in order.

$A(1, 0)$ $B(1, 3)$ $C(3, 3)$ $D(3, 1)$ $E(1, 1)$.

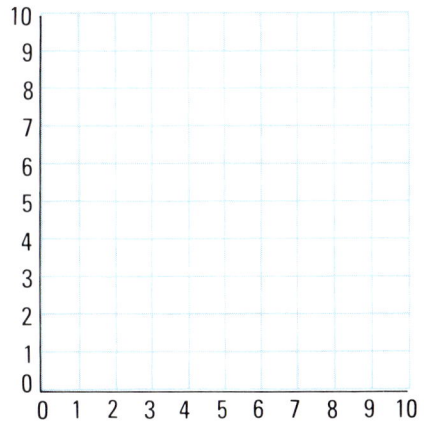

EXPLORATION

5 You need 1 cm squared paper.
There is a pattern in these coordinates:

$A(0, 1)$ $B(1, 2)$ $C(2, 3)$ $D(3, 4)$ $E(4, 5) \ldots$

The second coordinate is 1 more than the first coordinate.

a) Write down the next three coordinates.

b) Plot the points on a grid.

c) Write down the coordinates of the next three points in this sequence:

$(1, 2)$ $(2, 4)$ $(3, 6)$ $(4, 8) \ldots$ Plot the points on a grid.

d) Write down the coordinates of the next three points in this sequence:

$(1, 1)$ $(2, 4)$ $(3, 9)$ $(4, 16)$ $(5, 25) \ldots$ Plot the points on a grid.

e) Investigate some more sequences of coordinates. What is special about sequences of coordinates which give straight lines?

ENRICHMENT

EXPLORATION: FISHY COORDINATES

1 a) Think about the fish on the grid.

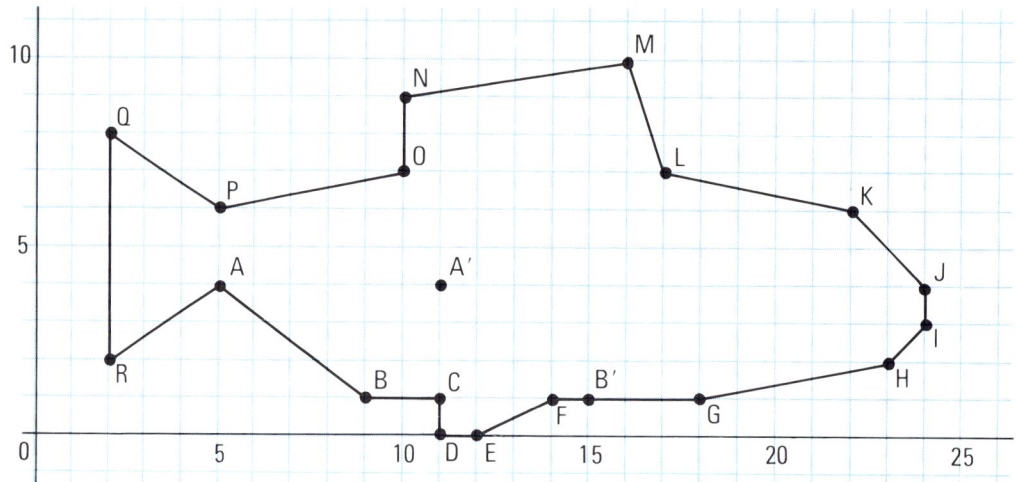

Increase the first coordinate of each point by 6 so that, for example:

A(5, 4) becomes A′(11, 4)
B(9, 1) becomes B′(15, 1)
and so on.

Redraw the fish using the points A′, B′, ..., R′.
Describe what happens to the fish.

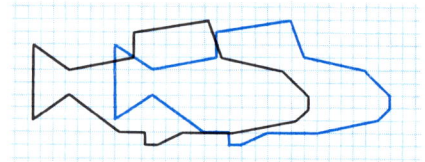

b) Increase *both* coordinates by 4. Describe what happens to the fish.

c) Explain how you can change the coordinates to make the fish:

   (i) move upwards
  (ii) move downwards
 (iii) move backwards
 (iv) grow to twice its size (twice its length and twice its height)
  (v) flip upside down.

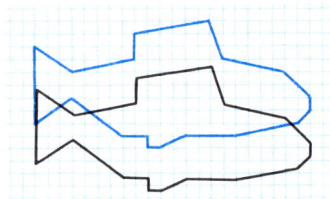

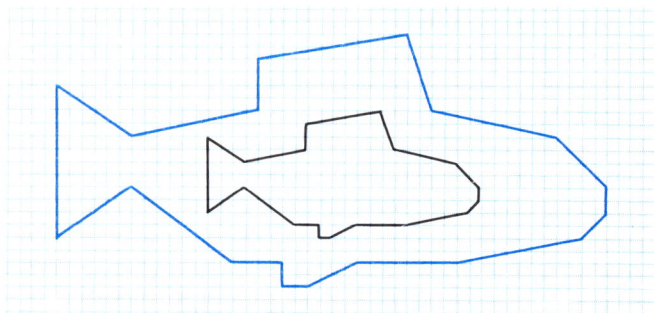

d) Make the fish change in some more ways. Explain what you do to the coordinates to make the fish change.

REVIEW

'75% of' means '$\frac{75}{100}$ of'.

We can calculate $\frac{75}{100}$ of £8.56 like this:

C 8 . 5 6 × 7 5 ÷ 1 0 0 =

or this:

C 7 5 ÷ 1 0 0 × 8 . 5 6 =

■ Check that each calculation gives the same result.

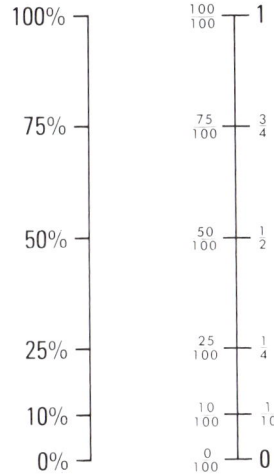

100%	$\frac{100}{100}$	1
75%	$\frac{75}{100}$	$\frac{3}{4}$
50%	$\frac{50}{100}$	$\frac{1}{2}$
25%	$\frac{25}{100}$	$\frac{1}{4}$
10%	$\frac{10}{100}$	$\frac{1}{10}$
0%	$\frac{0}{100}$	0

'20% OFF' means 'Save 20p in every £1'
or 'Save £20 in every £100' etc.

On this compact disc player you save 15p
in every £1.

■ How much do you save altogether on the
compact disc player?

£120

15% off

A20

■ What percentage of £1 is each British coin worth?

We can use percentages to make comparisons:

A mark of $\frac{13}{20}$ is $\frac{65}{100}$ which is 65%. A mark of $\frac{15}{25}$ is $\frac{60}{100}$ which is 60%. The English mark is worth more.

English Exam

$\frac{13}{20}$

Maths Exam

$\frac{15}{25}$

CONSOLIDATION

1   Alan works at the Keighley Sports Shop. During the sales it is his job to work out sale prices.

Was £30
Now £

Was £24
Now £

Keighley Sports

Sale

10p off in every **£**

Sale

a) What sale price should he give to:

(i) the squash racquet   (ii) the track suit?

b) By what percentage is everything reduced in the sale?

2   Here are some sale notices. What is the percentage saving in each sale?

a)
SUMMER SALE
45p off in the £

b)
Autumn Sale
For every **£1** you spend we return **20p**

c)
Christmas Sale
SHOES
SHOES
SHOES
30p in the £ reduction

d)
WINTER SALE *Save* £3 in every £10 you spend

e)
Spring **S**sale
Example Savings
Fridge was £100 now £85
Table was £50 now £42.50

f)
BOOKS
Clearance Sale
Example Savings
were £20 now £15

3   Copy and complete.

5% OFF means

5p off in every £1
or ? off in every £5
or 50p off in every £?
or £? off in every £100

20% OFF means

?p off in every £1
£? off in every £5
£? off in every £10
£? off in every £100

30% OFF means

? off in every £1
£? off in every £5
£3 off in every £?
£? off in every £100

4   These are some of the reductions in the
    Electricity Showrooms Sale. Which of the items
    have been reduced by 10% or more?

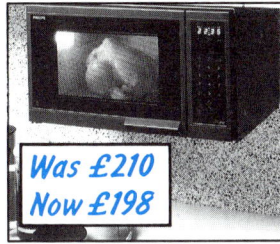

SALE SALE SALE
Massive reductions

Was £105
Now £90

Was £400
Now £370

Was £210
Now £198

Was £65
Now £60

5   Imagine you have to work out the sale prices in the Walling Shops Sale. Find the sale price of
    each item:

WALLING
SHOPS

SALE
20% OFF
ALL STATIONERY

A20

£1.00 for 50
Now

£1.50 each
Now

Dozen sheets 20p
Now

Was £2.30
Now £

THINK IT THROUGH

6   What is the percentage saving
    on each of these?

Was £200
Save £40

Was £300
Now £240

Percentages and decimals

1   Australia is about 25% of the size of Africa.

25% is $\frac{25}{100}$ which is 0.25.

Write information like this to compare the size of:

a) The British Isles and Australia

b) Wales and Scotland

c) The British Isles and the USA

d) The British Isles and continental Europe

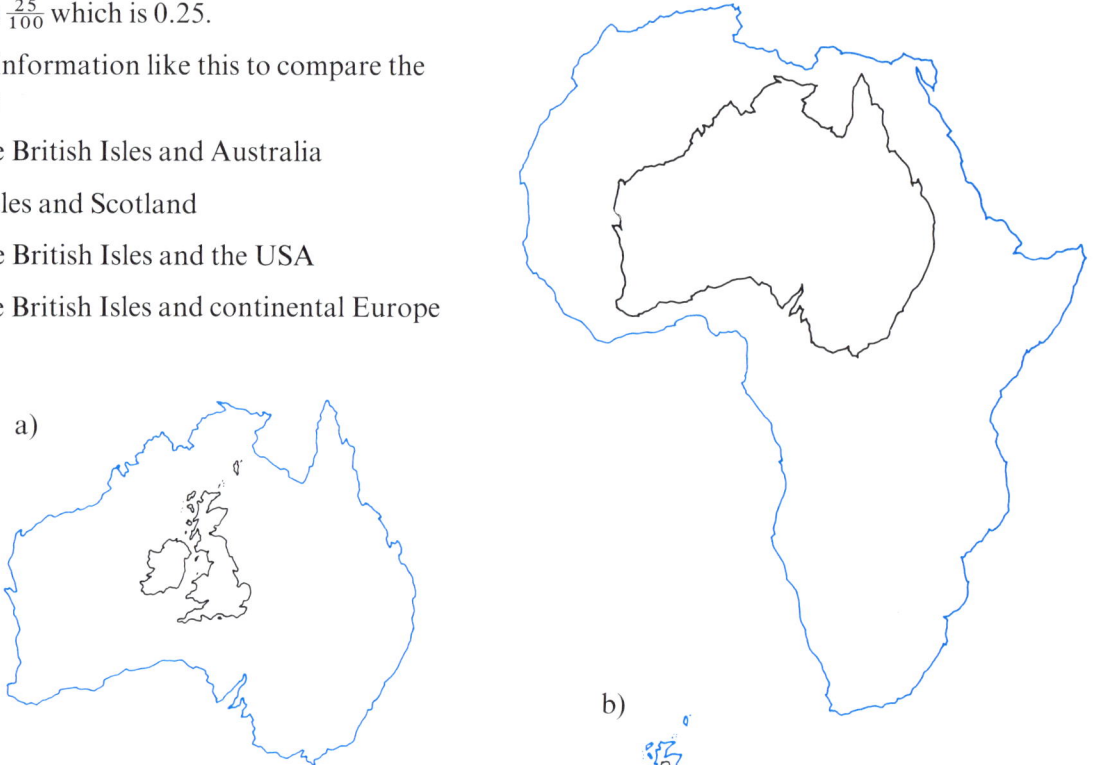

a)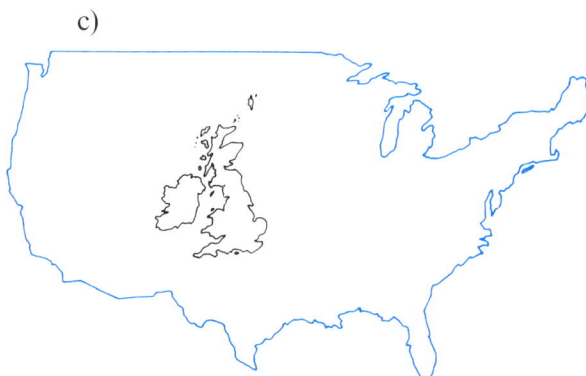

b)

c)

d)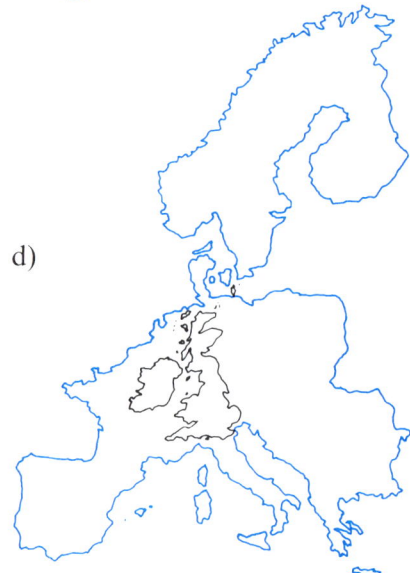

2   Copy and complete: a) $24\% = \frac{\square}{100} = 0.\square\square$ b) $\frac{72}{100} = \square\% = 0.\square\square$

c) $4\% = \frac{\square}{100} = 0.\square\square$ d) $0.09 = \square\% = \frac{\square}{100}$

3   These walls are being painted. Estimate what proportion of each is completed.

A

Write each proportion like this: $\square\% = \frac{\square}{100} = 0.\square\square$

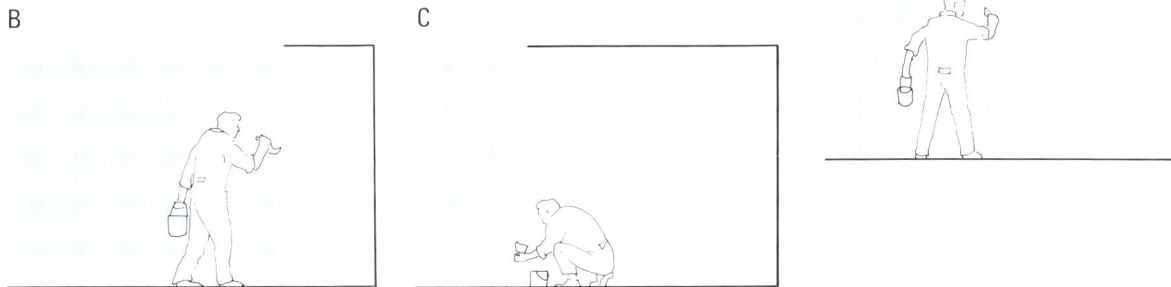

B   C

4   Copy and complete the chart for the bathroom tiles.

	Blue	White
Percentage		
Fraction $\frac{\square}{\square}$		
Fraction $0.\square\square$		

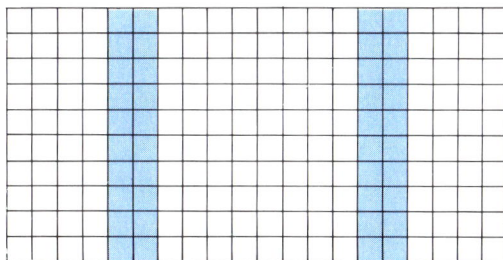

A20

5   a) Roughly, what percentage of people in your class have black hair?

b) Write your result in a):

(i) as a decimal fraction   (ii) as a vulgar fraction.

CHALLENGE

6   The pile of bricks on the right is being used to build the wall on the left. Estimate the proportion of bricks which have been used so far. Write the proportion like this:

$\square\% = \frac{\square}{\square} = 0.\square\square$

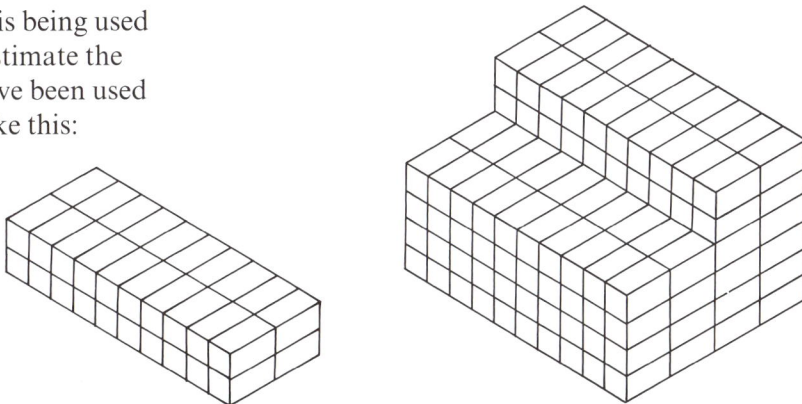

Calculating percentages

12% of £240 is 0.12 of £240, which is £28.80.   `C` `0` `.` `1` `2` `×` `2` `4` `0` `=` `28.8`

1   Calculate: a) 18% of £38, b) 76% of £148, c) $12\frac{1}{2}$% of £50, d) 32% of £75.

2   a) Estimate (i) the saving on each item, (ii) the sale price of each item.

A

Was £276
Save 30%

B

Was £960
18% off

b) Calculate the sale price of each item.

3   a) How many millilitres of
orange are in the new size can?

b) A junior nurse earned £7000
before the rise. What is the
salary after the rise?

October 13th 1988

**NURSES' PAY
INCREASE**

9% MORE

4   You need 1 mm squared paper.
This is a design for a fireplace
hood, seen from the side.

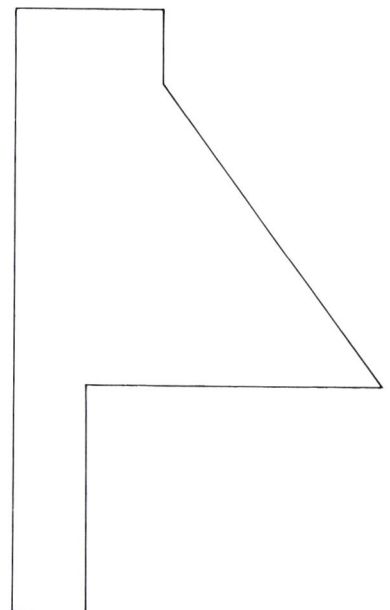

a) Draw a 20% enlargement of the hood. (Each length is 20%
longer than the lengths in the drawing.)

b) Make a smaller scale drawing, 80% of the size of the
original.

5 In the budget, the Chancellor of the Exchequer can increase the tax on goods. When the tax is increased the price goes up. Imagine you work in the corner shop. After the budget you have to work out the price changes. Here is your list:

	Present price	After budget
B ~ W Whisky 75cl	£7.95	
Old T Port 75cl	£4.45	
Siberian Vodka 75cl	£6.80	
Californian Red Table Wine 70cl	£4.64	
Liebfraumilch 70cl	£3.06	
Rancher bitter 4 cans	£2.15	
Bowman Cider 1l	.85p	
Heron Lager 4 cans	£1.95	
XL Cigarettes 20	£1.20	

a) The price of spirits goes up by 8%. Find the new price of whisky and vodka. Write the prices to the nearest penny.

b) Wine prices increase by 6%. Find the price of a bottle of red table wine after the budget.

c) Beer, cider, and lager prices go up by 12%. How much more does a customer pay for 2 litres of cider after the budget?

d) Cigarettes are increased by 18%. One of your customers buys 100 cigarettes each week. How much more per week does he spend on cigarettes after the budget?

e) Imagine you are Chancellor of the Exchequer. What percentage increase or reduction would you put on (i) cigarettes, (ii) spirits? What would be their prices after *your* budget?

REVIEW

- Area is measured in mm² (square millimetres)
  cm² (square centimetres)
  m² (square metres)
  hectares (1 ha = 1 hectare = 10 000 m²)

1 mm → ▫ 1 mm²
↑
1 mm

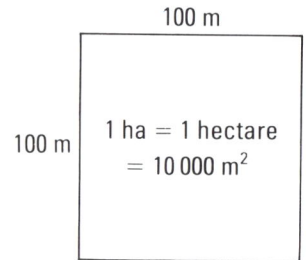

1 cm² | 1 cm

■ How many mm²?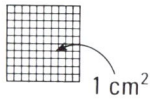

1 cm²

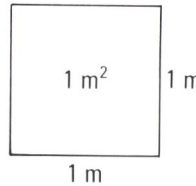

1 m² | 1 m

1 m

100 m

1 ha = 1 hectare = 10 000 m²

100 m

■ How many cm²?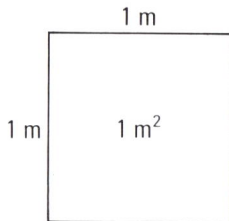

1 m

1 m | 1 m²

- Volume is measured in mm³ (cubic millimetres)
  cm³ (cubic centimetres)
  m³ (cubic metres)

1 mm³
⬡

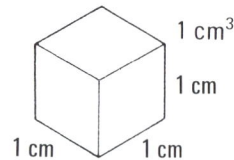

1 cm³

1 cm

1 cm   1 cm

■ 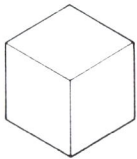 ⟹ 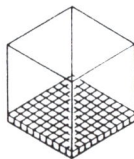 ⟹ 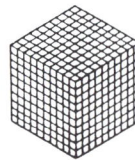 ⟹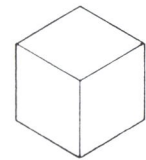

1 cm³

1 layer of 1 mm cubes.
How many 1 mm cubes?

10 layers of 1 mm cubes.
How many 1 mm cubes?

1 cm³

How many mm³?

■ How many cm³?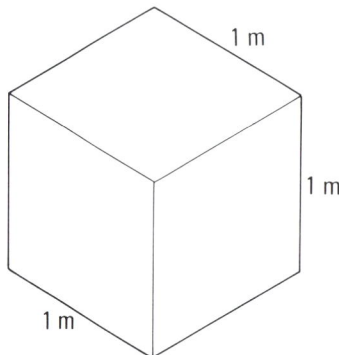

1 m

1 m

1 m

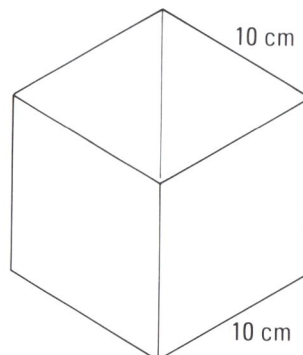

10 cm

10 cm

10 cm

A 1000 cm³ box holds 1 litre of water.

CONSOLIDATION

1   Which unit of area (mm^2, cm^2, hectare, ...) would you use for:

a) the area of a ceiling
b) the area of a potato field
c) the area of a fruit fly's wing
d) the area of an advertisement in a newspaper?

2   a) We might use mm^2 for the area of a microchip. Write down two more things for which we might use mm^2.

b) Write down two things for which we might use each of these units (i) cm^2, (ii) m^2, (iii) hectares.

3   a) Check that each of these shapes has an area of 2 cm^2. Draw a shape of your own which has an area of 2 cm^2.

b) Sketch three different shapes each having an area of 40 mm^2.

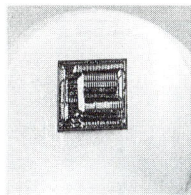

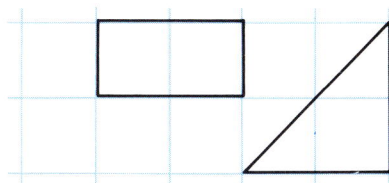

4   The area of glass in each of these windows is 2 m^2. $\frac{1}{2}$ m
Sketch two windows of your own each having an area of 6 m^2.

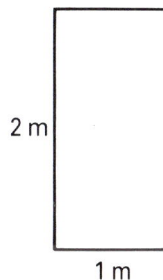

4 m

2 m

1 m

5   a) Guess the area of this stamp:

(i) in cm^2 (ii) in mm^2.

b) Check your guesses by measuring.

22ᴾ

A21

ASSIGNMENT

6   Roughly, how many cm^2 of toast can be buttered with 250 g of butter? In your report, explain how you made your estimate.

WITH A FRIEND

7   You need a large sheet of paper, or smaller sheets which you can stick together.

a) Draw a rectangle or square whose area is

1 MILLION SQUARE MILLIMETRES   (1 000 000 mm^2)

Decide between you what length and width to make your shape.

b) Inside your shape write down its area (i) in mm^2, (ii) in cm^2, (iii) in m^2.

ENRICHMENT

1   Choose the **Brilliant** tiles you like best from the advertisement. Choose a room at home (a bathroom or a kitchen). Sketch the walls of the room to show their sizes, and where you would put the tiles. Estimate how much it would cost to buy the necessary tiles.

A21

BRILLIANT
CERAMIC TILES

ALL PRICES INCLUDE VAT

KH6 Pattern   KH1 Plain   Waldorf   Geometric

Country Harmony   Ascot Pompadour   Radiant 19 Topaz   Orient 19 Topaz

Moortone   Cascade   Ascot KNT9 Pattern   Opal White

4¼" Pickwicke Tiles   Mayflower   Glade Fawn

For technical reasons connected with colour reproduction and tile manufacture, these patterns may not exactly match the tile colours.

Glade Pearl   Spot

BRILLIANT CERAMIC 4¼" × 4¼" TILES

Pickwicke Series in cartons of 50 Plain tile (fawn)   3.99
Patterned tiles – in Pearl, Avocado, Pampas and Sunking.   4.99

BRILLIANT CERAMIC 6" × 6" TILES

Moortone Series 6 colours available Carton of 25   6.99
Cascade Series Carton of 6   3.99
Kitchen Harmony KH1 Plain Carton of 75   18.99
KNT9 Pattern Carton of 6   1.99
KH6 Pattern Carton of 6   1.99
Opal White Pack of 75   13.99
In cartons of 25 except where stated Geometric series in Blue, Brown, & Avocado   4.99
Radiant/Orient Series in Topaz   7.99
Waldorf series pack of 75   15.75
Mayflower series carton of 12   11.99
Country Harmony Carton of 6   1.99
Ascot (Patterned tiles) 6 colours available   6.99
Glade series (Plain tiles)   18.99
Spot   7.99

REVIEW

- We say that the *chance* that a die will turn up five is 1 out of 6.

6 equally likely outcomes, one of which is ⚀

■ What is the chance that an even number will turn up?

In 60 throws we would expect five to turn up in about one-sixth of the throws, that is, 10 times.

This spinner has 5 equal sections, 2 of which are blue. The chance that blue will be scored is 2 out of 5.

We can also write this as $\frac{2}{5}$, or 0.4 $(\frac{4}{10})$ or 40% $(\frac{40}{100})$.

In 50 spins we would expect blue to be scored in about $\frac{2}{5}$ of the spins, that is, 20 times.

■ Match each chance of scoring blue with one of the spinners.

A   B   C   D   E   F   G

(i) 0.2   (ii) 50%   (iii) 1 in 4   (iv) $\frac{3}{4}$ (v) $33\frac{1}{3}$%   (vi) 1 in 2.

- When something is certain to happen, its chance of happening is 1. (100% certain, a 10 out of 10 chance.)
- When something is certain *not* to happen, its chance of happening is 0. (0%, $\frac{0}{10}$.)

■ With which of the spinners do you have a chance of a) 1, b) 0 of scoring blue?

- We can represent chances on a 'chance line'.

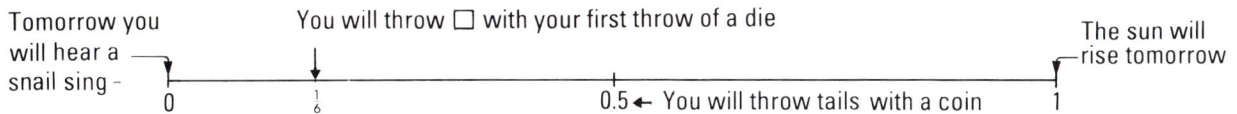

Tomorrow you will hear a snail sing –   You will throw □ with your first throw of a die   The sun will rise tomorrow

0    $\frac{1}{6}$ 0.5 ← You will throw tails with a coin   1

CONSOLIDATION

A22

============ CHALLENGE ============

1   Sketch a spinner for which the chance of scoring blue is more than that of spinner G but less than that of spinner C.

2   a) Which of these events do you think is (i) most likely to happen, (ii) least likely to happen?

A   It will rain tomorrow.
B   You will ride on a bus next week.
C   You will become a chiropodist.
D   You will go to the cinema during May next year.
E   You will visit Aldershot next year.
F   The next person to enter the room will be female.

b) Plot the events on a chance line.

Certain not to happen   Certain to happen

0   0.5   1

3   What is the chance that you will throw these with a die:

   a) a six b) an odd number c) a number greater than 7 d) a number between 1 and 6 inclusive?

4   a) You are going to the lake but the
       signpost has blown down. You cannot see
       the lake from the junction, so you guess
       which road to take.
       What is the chance you guess the right road?

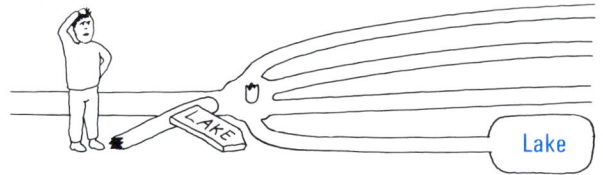

   b) Answer question a) for each of these road systems.

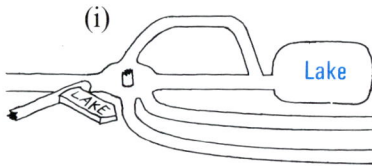

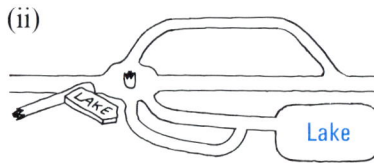

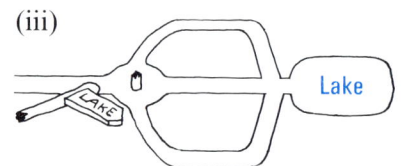

   (i)   (ii)   (iii)

5   You begin on 'START' and throw a die
     twice. The total score tells you how many
     squares to move.

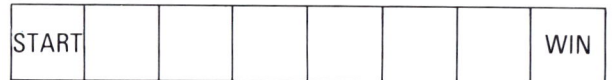

START						WIN

   a) How many different   b) You win only if you land   c) What is the chance that
      possible outcomes are   on WIN on the second   you will win?
      there altogether?   throw. In how many ways
      List them like this: can you do this? List them   d) What is the chance that
                                     like this: you will lose?

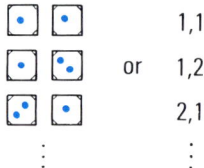

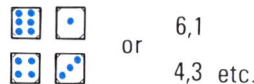

         1,1
      or 1,2   6,1
         2,1   or 4,3 etc.
          ⋮

6   a) You spin a coin 80 times. About how many *heads* would you expect? Why?

   b) You throw a die 120 times. About how many times would you expect a ⚅ ? Why?

   c) You spin this spinner 60 times. How many
      times would you expect to score (i) blue, (ii) white?

7   a) Design a spinner which, in 45 spins, should produce about 30 blue outcomes and 15 white
       outcomes.

   b) This is a dice board. You begin on START
      and throw the die just once. You win if
      your score takes you onto a WIN square.

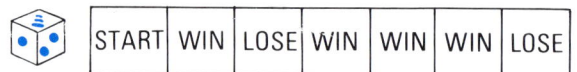

START	WIN	LOSE	WIN	WIN	WIN	LOSE

      (i) What is the chance you will win on this board?
      (ii) How many times would you expect to win in 300 goes?

   c) Design your own board on which you would expect to win about 30% of the time.

CORE

TAKE NOTE

When we spin a coin we can *calculate* the chance that it will turn up *heads*. There are other events for which we have to find a way of *estimating* chance. For example, we cannot calculate the chance that it will rain tomorrow ... but we might be able to make an estimate.

WITH A FRIEND

1   The table tells us about the weather over twenty years in Scarsdale, on 31 May each year.

Weather in Scarsdale on 31 May 1961–1980			
1961 warm/sunny	1966 cloudy	1971 warm/sunny	1976 cloudy
1962 rainy	1967 cold/sunny	1972 warm/sunny	1977 thunderstorms
1963 warm/sunny	1968 warm/sunny	1973 warm/sunny	1978 rainy
1964 warm/sunny	1969 cold/sunny	1974 warm/sunny	1979 rainy
1965 thunderstorms	1970 warm/sunny	1975 cloudy	1980 warm/sunny

a) Discuss between you how you might use the information to decide what the weather in Scarsdale might be like on 31 May next year.

b) Each of you estimate how likely it is that it will be warm and sunny in Scarsdale next year.

2   How would you estimate the chance that these events will occur? Write one or two sentences to explain.

a) It will snow on Christmas Day in your town or village next year.

b) The FA cup will be won by a Division 2 team next year.

TAKE NOTE

Sometimes we can estimate the chance that something will happen by collecting information about what has happened in the past.
For example, between 1951 and 1980 it snowed on Christmas Day in London twice. We might use this to estimate the chance that it will snow next year on Christmas Day in London as $\frac{2}{30}$ or $\frac{1}{15}$.

A22

ACTIVITY: DROPPING PINS

3   Emma drops a drawing pin onto the desk. 'I'll bet it lands point up' she says.

a) Do you think she is more likely to be correct or more likely to be wrong?

b) Do this experiment. Drop a drawing pin onto this page 100 times. Count how many times it lands point down. Use your results to estimate the chance that Emma wins her bet.

Sometimes we can estimate the chance that something will happen by carrying out an experiment.

4   Choose one of these assignments. Design an experiment for your assignment, and collect some information. When you have completed your assignment, write a report. In your report:

- describe your experiment and explain why you designed it in the way you did
- give the information you collected
- give your results.

A   What is the chance that the next person you meet in school lives less than $\frac{1}{2}$ km from you?

B   What is the chance that the number one record one week will still be number one the next week?

C   On a particular day, what is the chance that your daily newspaper will have a photograph of the Prime Minister on its front page?

D   On a particular day, what is the chance that the television evening news (BBC1 or BBC2 or ITV or Channel 4) will include an item about the President of the USA?

E   What is the chance that four coins turn up two *heads* and two *tails* when they are spun together?

A22

CORE

1 a) Helen works in a department store.
   At sale time she writes out sale cards like these.

   Sometimes she uses percentages. ○ ○ ○ ○ (33% off)
   Sometimes she uses fractions. ○ ○ ○ ($\frac{2}{3}$ off)

   Shoes 20% OFF

   $\frac{1}{2}$ price beddin

   Carpets why pay more when you can buy for $\frac{1}{3}$ of the normal price?

   The store produces these conversion charts to help. Write down the fractions and percentages missing from each chart.

   A

0%	10%	20%	30%	☐	50%	60%	☐	80%	90%	100%
0	$\frac{1}{10}$	☐ $\frac{1}{5}$	$\frac{3}{10}$	$\frac{2}{5}$	$\frac{1}{2}$	$\frac{3}{5}$	$\frac{7}{10}$	$\frac{4}{5}$	$\frac{9}{10}$	1

   B

0%	$12\frac{1}{2}$%	25%	$33\frac{1}{3}$%	50%	75%	100%
0	☐ $\frac{1}{8}$	☐ $\frac{1}{4}$	☐ $\frac{1}{3}$	☐	☐	1

   b) Draw your own 'conversion chart' for these fractions:

   $\frac{2}{3}$ $\frac{5}{8}$ $\frac{1}{3}$ $\frac{3}{8}$ $\frac{1}{2}$ $\frac{3}{4}$ $\frac{1}{8}$ $\frac{1}{4}$ $\frac{7}{8}$

WITH A FRIEND

2 a) You should both know these
   fraction–percentage pairs by heart. Test
   each other until you both *do* know them.
   For example, ask 10 questions like this:
   'What is 10% as a fraction'?
   'What is $\frac{1}{2}$ as a percentage?'

   b) Learn some more simple
   fraction–percentage pairs. Make a list of
   all the pairs that you know. Add new
   pairs to your list as you learn them.

   1% OFF $\frac{1}{100}$ OFF   About 33% OFF $\frac{1}{3}$ OFF

   10% OFF $\frac{1}{10}$ OFF   50% OFF $\frac{1}{2}$ PRICE

   20% OFF $\frac{1}{5}$ OFF   75% OFF $\frac{3}{4}$ OFF

   25% OFF $\frac{1}{4}$ OFF   100%   THE WHOLE AMOUNT: 1

A23

3   Estimate each of these. Write your estimate (i) as a percentage, (ii) as a fraction.

a)   The proportion of each day that you spend sitting down.

b)   The proportion of all the liquid you drink each day that is tea.

c)   The proportion of your money that you spend on clothes.

d)   The proportion of people in your class who are taller than you.

e)   The proportion of blue wool used for each sweater.

A   B

IN YOUR HEAD

4   Do these questions in your head. Write down only the answers.

a)   Some sausages contain 50% fat. How many grams of fat are there in 2 kg of these sausages?

b)   28% of all smokers die of lung cancer. How many is this in every 100?

c)   This is a sign in a sale.

> SALE Shoes 30% OFF

   (i)   What fraction of the original price do you save?
   (ii)   How much do you save on a £10 pair of shoes?

d)   In Britain during the 1980s one in every four marriages ended in divorce. What percentage is this?

e)   In some areas of Birmingham in 1987, 38 school leavers in every 100 were unemployed.

   (i)   What fraction were unemployed?
   (ii)   What percentage were *employed*?

f)   Which is larger: 80% or $\frac{3}{4}$?

g)   Which is larger: $\frac{1}{5}$ or 15%?

h)   Which is the best estimate for $\frac{1}{3}$: 30%, 35%, or 40%?

5   Try to do each of these in your head.

   a) 50% of 10 m
      (*Think*: $\frac{1}{2}$ of …)

   b) 25% of 12 m
      (*Think*: $\frac{1}{4}$ of …)

   c) 10% of 20 kg
      (*Think*: $\frac{1}{10}$ of …)

   d) 20% of 20 kg

   e) 10% of 200 kg

   f) 75% of 12 m

   g) 50% of £6

   h) 25% of £60

   i) 40% of 80%

6   Jack buys a car for £4000. He pays a 25% deposit. How much is this?

7   Jo Turner earns £100 per week. She gets a 10% pay rise. What is her new weekly wage?

8   75% of a 200 g beefburger is real beef. How many grams is this?

9   Students can get a 10% reduction on
     SAVER rail tickets.
     Senior Citizens get a 20% reduction.
     The normal SAVER return fare from
     Rugby to London is £11.

   a) How much do students pay?
   b) How much do Senior Citizens pay?

LONDON
£11 SAVER
RETURN

LONDON SAVER

FROM RUGBY

FOR RAIL INFORMATION RING:
RUGBY (0788) 60116

INTERCITY

━━━━━━━━━━ CHALLENGE ━━━━━━━━━━

10    $\frac{1}{10}$ is 10%. But $\frac{1}{5}$ is not 5%.

      Use your calculator to find out if there
      are any more fractions for which the first of these $\frac{1}{\square} = \square$%
      numbers (small square) equals the second.

      Write down any that you find.

A23

A24 GRADIENTS

1 This is a side view of part of a downhill ski run. It has been drawn on a grid so that the steepness of each section can be estimated.

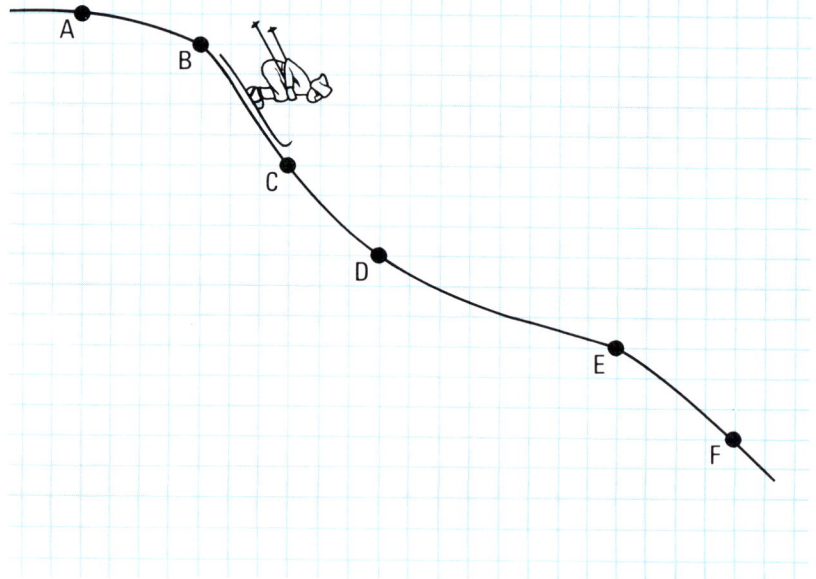

a) Decide between you which is the steepest section (AB, BC, ...?)

b) List the sections in order, steepest first.

2 This is a slate quarry. The road winds down the side of the quarry. The heights of points where the road turns are shown.

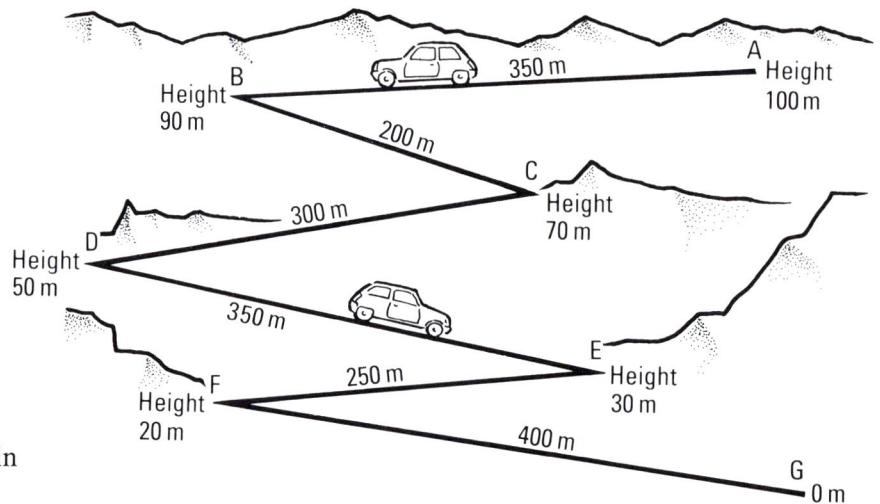

a) Decide between you which section of the road is steepest.

b) List the road sections in order, steepest first.

━━━━━━━━━━ ACTIVITY: WITH A FRIEND ━━━━━━━━━━

3   You need 1 cm squared paper. Mark your paper off in larger squares of about 4 cm × 4 cm or 3 cm × 3 cm.

Vertical grid

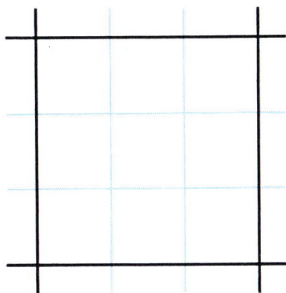

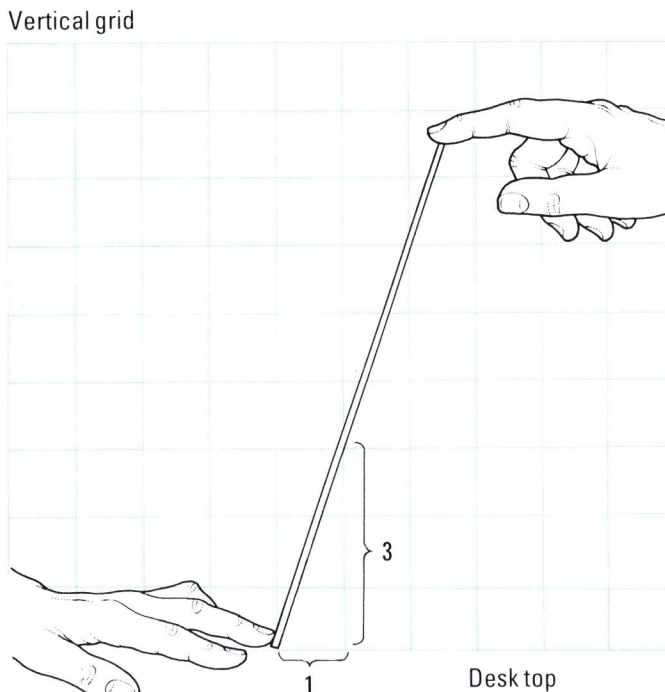

3

1   Desk top

The *gradient* of the ruler in the diagram is 3 in 1 (that is, 3 squares up for every square across). Decide which of you is person A and which is person B.

a) Person A holds the squared paper vertical. Person B holds the ruler so that its gradient is:

   (i) 2 in 1   (ii) 1 in 3   (iii) 1 in 4.

b) Another way of writing the gradient of the ruler in the diagram is 3:1. Switch tasks. This time person B holds the squared paper vertical. Person A holds the ruler so that its gradient is:

   (i) 1:2   (ii) 1:1   (iii) 5:1.

c) Discuss between you which of these gradients is the greatest (that is, the steepest). List the gradients in order, steepest first.

   1:1   1:2   3:1   2:1   1:4   4:1

━━

4   The RAC tested the performance of a car alone, and of a car and caravan together. Here are the results:
   a) Which can park on the steepest hills ... the car or the car and caravan?
   b) Which can start on the steepest hills ... the car or the car and caravan?

Hill restart

1:3   1:5

Handbrake hold test

1:3   1:4

A24

5 These road signs tell us the steepness of hills.
We read 1:8 as '1 in 8'.

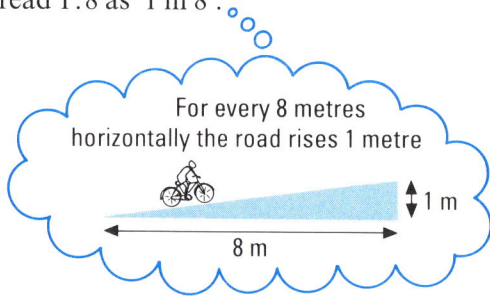

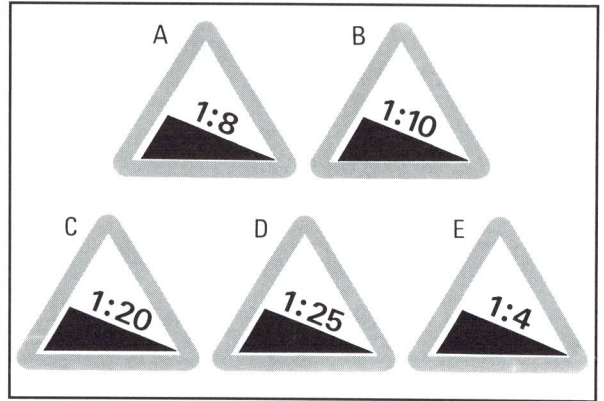

For every 8 metres
horizontally the road rises 1 metre

1 m

8 m

A 1:8

B 1:10

C 1:20

D 1:25

E 1:4

a) Which of the five signs warns of the steepest hill?

b) List the hills in order, steepest first.

c) You cycle up hill B. At the top you are 20 m higher.
   (i) How far away are you in a horizontal direction from your starting point?
   (ii) On 1 cm squared paper draw an accurate diagram of hill B. Use your diagram to estimate how far you have cycled up the hill.

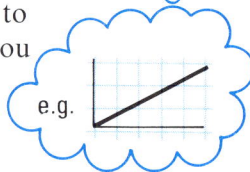

e.g.

THINK IT THROUGH

6 a) The council needs a road sign for this hill. This is the road sign.

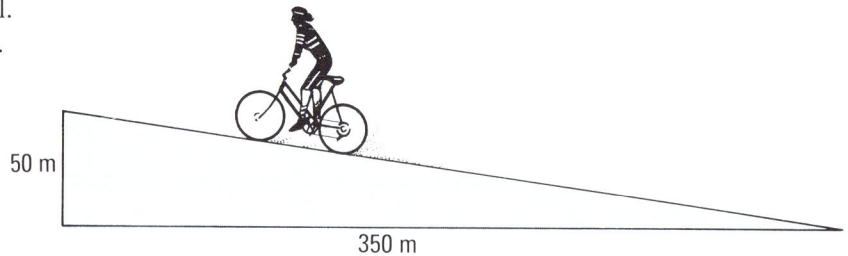

1:

50 m

350 m

What number is missing from it?

b) Draw your own road sign for this hill.

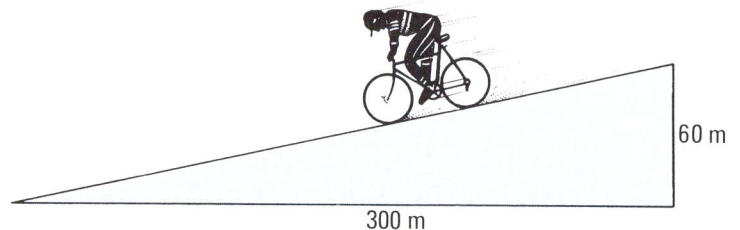

60 m

300 m

A24

7 More often today you will see road signs like these.
12% means '12 in 100'.

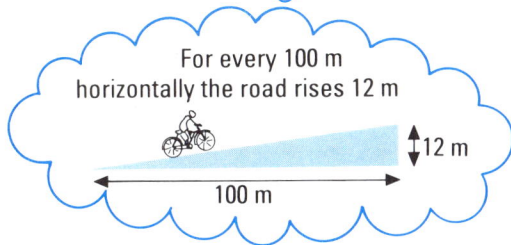

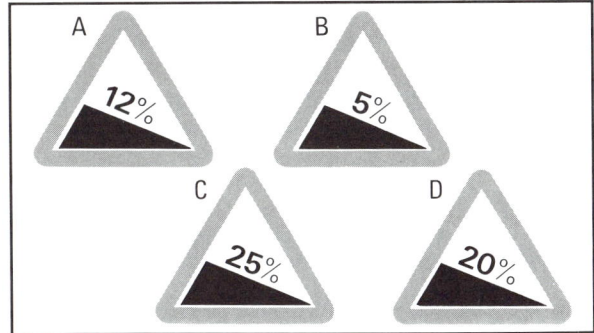

For every 100 m horizontally the road rises 12 m

12 m

100 m

A 12%
B 5%
C 25%
D 20%

a) Which of the four signs warns of the steepest hill?

b) List the signs in order, steepest first.

c) Sign B and one of the signs in question 5 show the same steepness. Which sign is it?

d) Sign C matches with one of the signs in question 5. Which sign is it?

e) You cycle down hill C. At the bottom you are 75 m lower than you were at the top.

(i) How far away are you in a horizontal direction from your starting point?

(ii) On 1 cm squared paper draw an accurate diagram of hill C. Use your diagram to estimate how far you have cycled down the hill.

CHALLENGE

8 This is taken from a walkers' map.

a) How many metres higher is the Hill Top Hotel than the Wishing Well Tavern?

b) Approximately, how far is it from the Wishing Well Tavern to the Hill Top Hotel, measured horizontally?

c) One of these hill signs is on the road from the Hill Top Hotel to the Wishing Well Tavern. Which do you think it is? Why?

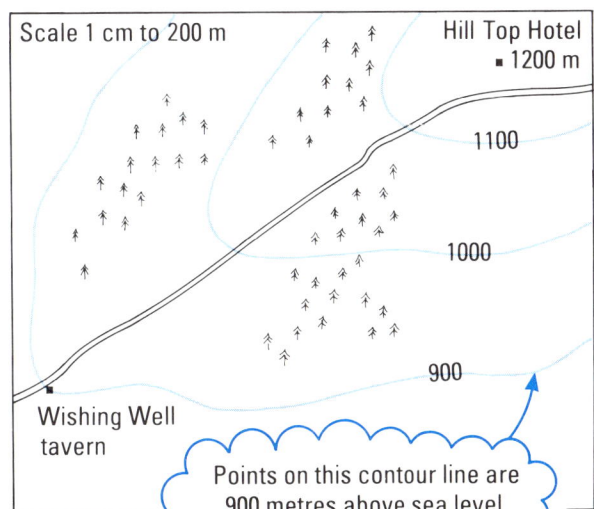

20% 10% 7%

Scale 1 cm to 200 m

Hill Top Hotel
■ 1200 m

1100

1000

900

Wishing Well tavern

Points on this contour line are 900 metres above sea level

A24

FIND OUT FOR YOURSELF

9 a) Normal railway systems cannot cope with steep slopes. The railway in the photograph takes tourists up to the top of Snowdon. Another railway in Switzerland takes tourists up to the ski slopes above Lake Geneva. The railways have to deal with gradients of 1:6 or 1:5. How do you think they do this? Find out, then write a short paragraph to explain.

b) You will find these signs on roads on an Ordnance Survey Map. What do they mean?

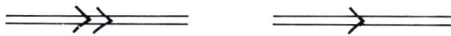

10 To move boats up or down on the waterways, systems of locks are used. The Bingley *Five* rise stairway on the Leeds–Liverpool canal has a 'lift' of 60 ft. Each lock is about 72 ft long.

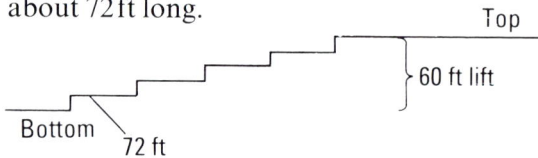

Top

60 ft lift

Bottom

72 ft

What is the 'gradient' from top to bottom?

Write the gradient like this 1: ☐
and like this ☐ %

Assume the water runs like this!

ACTIVITY

11 Find the 'gradient' of a set of stairs, at home or at school.

Write the gradient like this: 1: ☐ and this ☐ %

A24

REVIEW

- *Volume* is the amount of space taken up by something.

- The hot air balloon has a greater *volume* than the elephant (it takes up more space). But the elephant has a greater *mass* than the balloon (the elephant is heavier).

Volume is measured in cubic units. For example:
- cubic metres (m^3)
- cubic centimetres (cm^3)
- cubic millimetres (mm^3)

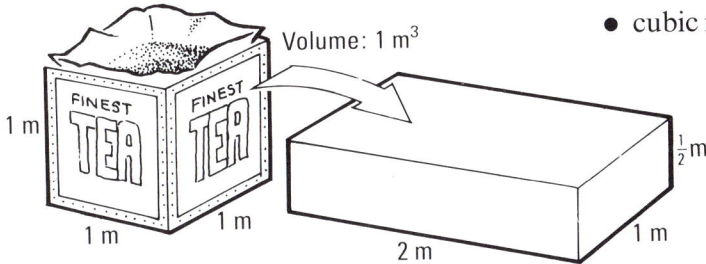

Volume: 1 m^3

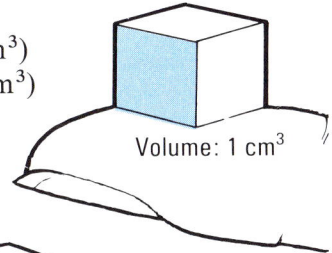

Volume: 1 cm^3

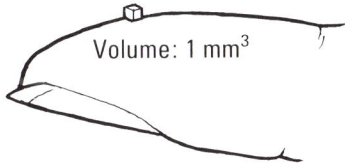

Volume: 1 mm^3

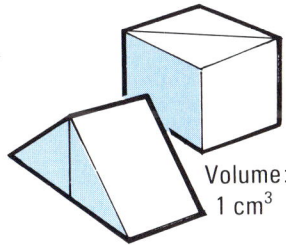

Volume: 1 cm^3

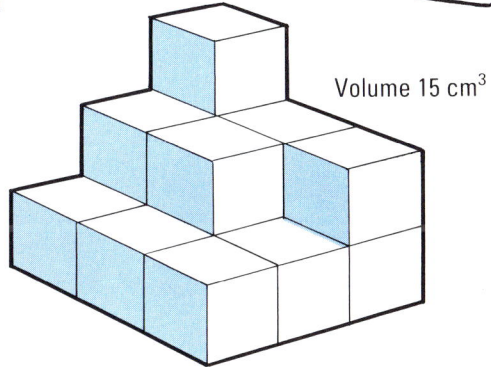

Volume 15 cm^3

- Volume of a prism = volume of one layer of cubes × number of layers

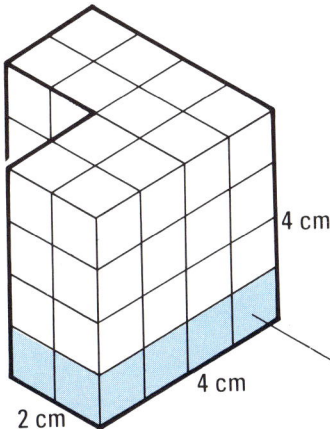

4 cm

4 cm

2 cm

Number of cubes in this layer is 12

Volume = 12 × 4 cm^3 = 48 cm^3

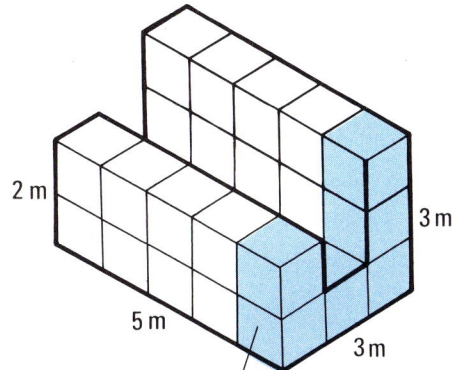

2 m

3 m

5 m

3 m

Number of cubes in this layer is 6

Volume = 6 × 5 m^3 = 30 m^3

A25

1   This is a normal-sized tennis ball. Between you, think of:

  a)   three objects which have a *larger volume* but a *smaller mass*.

  b)   three objects which have a *larger mass* but a *smaller volume*.

  Each of you write down what you decide.

2   a)   The volumes of these objects are increasing but their masses are decreasing. Suggest a possible fourth object.

  b)   Four models of a badger are all made from different materials. The volumes are the same but the masses are decreasing. Suggest what material could be used for each model.

3   Which unit of volume (mm³, cm³, m³) would you use to measure:

  a)   the amount of space in a warehouse
  b)   the amount of wood needed to make a matchstick
  c)   the volume of clay a potter uses to make a vase.

4   Estimate   a)   the total mass, in tonnes,
               b)   the total volume, in m³, of the total population of your school.

Density and volume

1   The replica cat is made from resin, the original from bronze.

   a) Which has the larger
       mass?
   b) Which has the larger
       volume?
   c) Which weighs more ...
       1 cm^3 of the original cat
       or 1 cm^3 of the replica cat?

Ancient Egyptian cat

Replica for display

TAKE NOTE

The original cat has a greater *density* than the replica, because ...

1 cm^3 of the original cat weighs more than 1 cm^3 of the replica.

2   Which object in each pair is more dense? Explain how you know.

*Tea in a
cup*

Telegraph pole

Apple

Steel nail

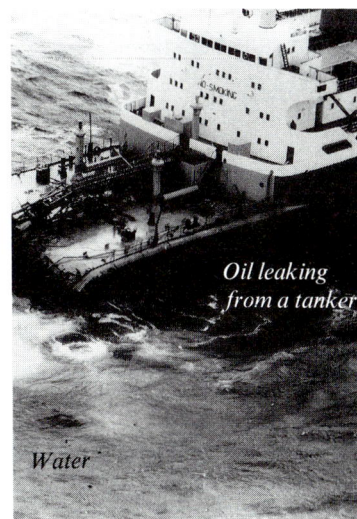

*Oil leaking
from a tanker*

Water

━━━━━━━━━━ CHALLENGE ━━━━━━━━━━

3   One log weighs 4000 kg. Its volume is about
    10 m³. Another log is from a different type of
    tree. It weighs 3600 kg and its volume is about
    8 m³.
    Rafts are built from each log. Which one
    sinks further into the water? Why?

Volumes of cuboids

1   These are the nets for three boxes.

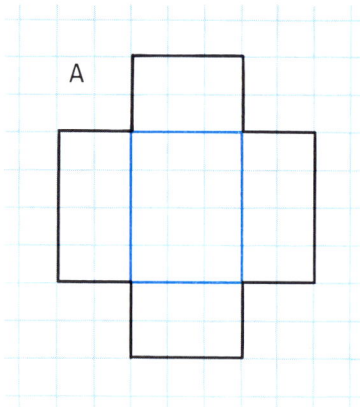

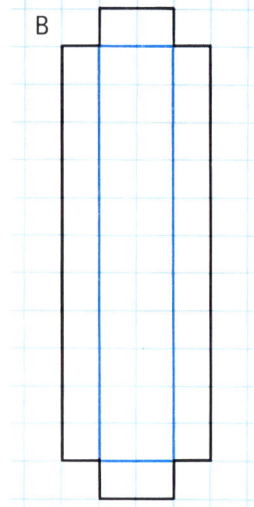

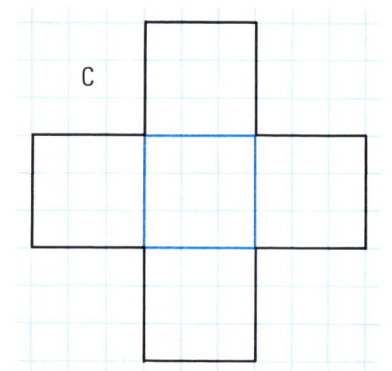

a) *Guess* which one holds:
   (i) most sand
   (ii) least sand.

b) Find out how many 1 cm cubes ⬡ fit into each box. (Each square represents a 1 cm square.)

c) Which box has:

   (i) the largest capacity
   (ii) the smallest capacity?

Were your guesses correct in a)?

A25

2 a) Cuboid A is made from 1 cm cubes.
       How many?

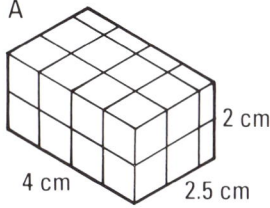

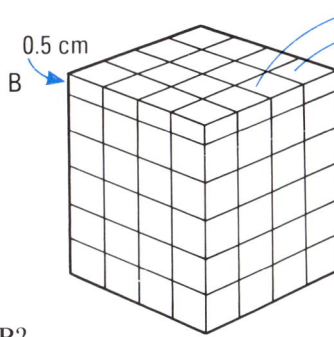

0.5 cm

A

B

2 cm

4 cm

2.5 cm

Notice: These two pieces make a whole 1 cm cube

b) How many cubes are used for cuboid B?

c) Cuboids C and D are also made from 1 cm cubes. Copy and complete the table. (Decide yourself which distance should be the length, width, and height for each cuboid.)

Cuboid	Number of 1 cm cubes needed	Length	Width	Height
A				
B				
C				
D				

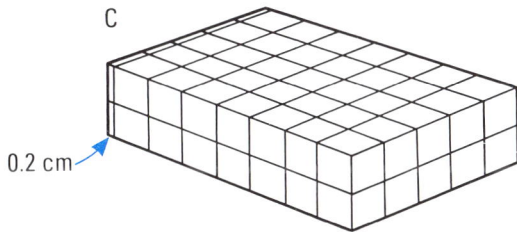

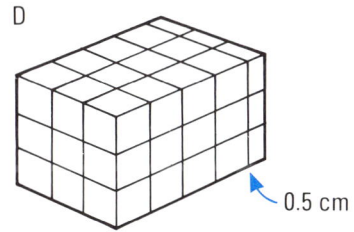

C

0.2 cm

D

0.5 cm

d) For each cuboid calculate:

length × width × height.

Write down what you notice about your results, and the results in your table.

e) Copy and complete the *Take note* below.

▨▨▨▨▨▨▨▨▨ TAKE NOTE ▨▨▨▨▨▨▨▨▨

Volume of a cuboid = length × ... × ...

$$V\,cm^3 = L \times W \times H\,cm^3$$

Length

....

Base

H cm

W cm

L cm

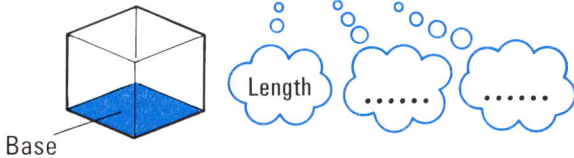

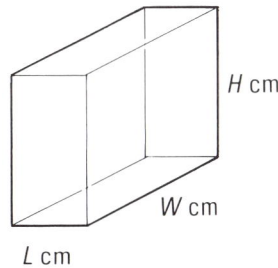

Volume of a cuboid = Area of its base × ...

ENRICHMENT

EXPLORATION: ECONOMY BOXES

1   You need 1 cm squared paper.

   a) Check that this net will make a box of capacity 24cm^3.

   b) Check that the area of card used for the net is 56cm^2.

   c) Draw a different net which will make a box (with a lid) of capacity 24cm^3.

   d) What area of card does your box need?

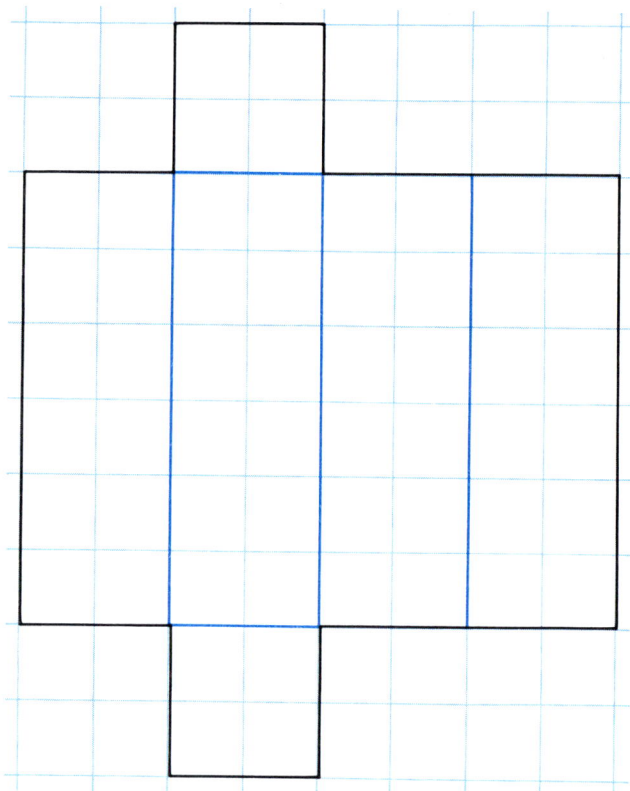

   e) Investigate more nets for boxes (with lids) which have a capacity of 24cm^3. Find which box uses the smallest amount of card, and make it.
   Write a short report to explain how you found which box used the smallest amount of card. Include a net of the box in your report.

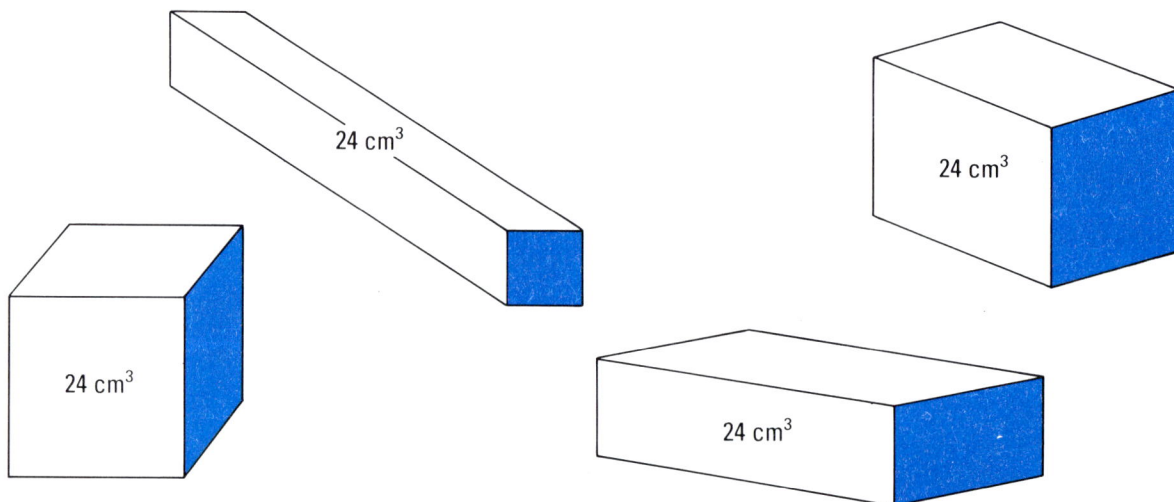

24 cm^3

24 cm^3

24 cm^3

24 cm^3

REVIEW

Data (information) can be represented in various ways:

In a table.

DATA	
French set	Mean test score
4	9
3	12
2	20
1	25

On a chart.

BAR CHART

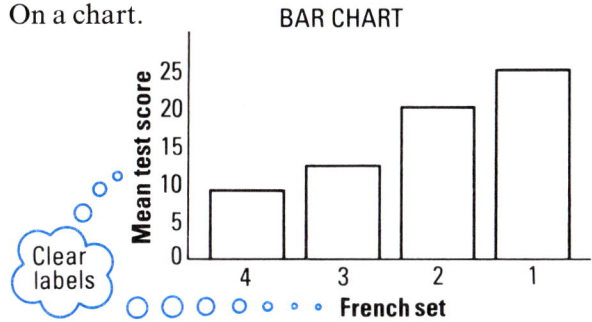

Clear labels

French set

■ Think of two advantages of representing data on a chart. Write them down.

In this example the weights of some babies are recorded. The exact weight of each one is not recorded. The weights are divided into *intervals*. We often do this when we are dealing with measurements (such as weight, height, amount of liquid, etc.) that can have lots of different values.

DATA	
Birth weight (kg)	Number of babies
2.01–2.25	15
2.26–2.50	22
2.51–2.75	17
2.76–3.00	19
3.01–3.25	4

The number in each interval is called the *frequency*

3.01–3.25 is an *interval*

HISTOGRAM

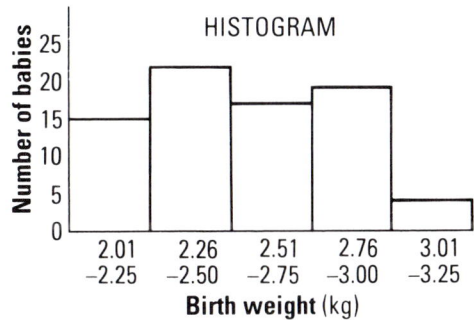

■ What might be a) the largest, b) the smallest number of babies weighing between 2.51 kg and 2.60 kg?

There are other ways of representing data, for example

Line chart

SALES CHART SCRIBE PENCILS

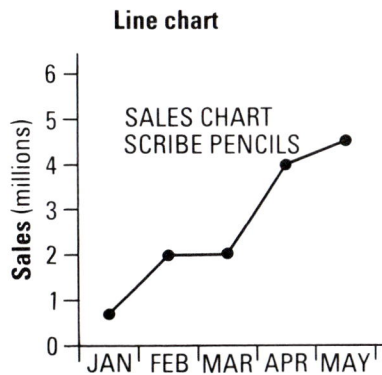

Pictogram

Class 4A's favourite fruits

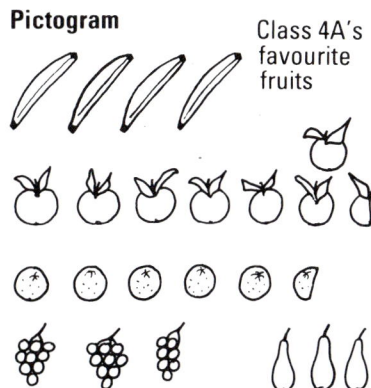

Pie chart

60% smokers

85% non-smokers

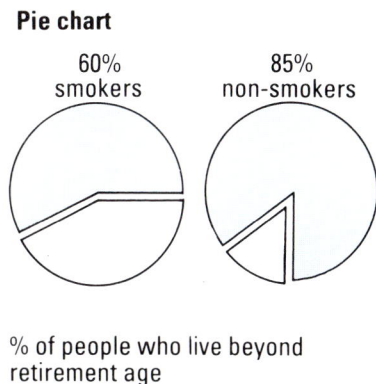

% of people who live beyond retirement age

■ Which method would you use for recording the ages of pupils in your class? Explain your choice.

CONSOLIDATION

1   A gardener wants to know which of two types of fertilizer is the better. He uses one fertilizer on one half of his potato patch and the other on the other half. When he harvests his potatoes he carefully weighs samples from each half of his potato patch. These are his results for fertilizer A. (He has already drawn a bar chart for fertilizer B.)

Weights for Fertilizer A (weights to the nearest gram)

42, 29, 30, 21, 28, 37, 56, 40, 31, 32
24, 26, 29, 35, 35, 34, 52, 40, 39, 37
28, 42, 55, 54, 36, 60, 55, 54, 64, 68
58, 72, 44, 21, 29, 37, 68, 75, 80, 44
20, 27, 36, 38, 44, 59, 63, 82, 64, 69
40, 42, 53, 63, 67, 72, 80, 38, 38, 39
38, 41, 27, 26, 25, 28, 34, 46, 55, 60
44, 48, 59, 67, 74, 82, 38, 76, 47, 50
86, 27, 38, 55, 59, 49, 56, 72, 67, 68
27, 34, 42, 51, 49, 51, 37, 32, 46, 55

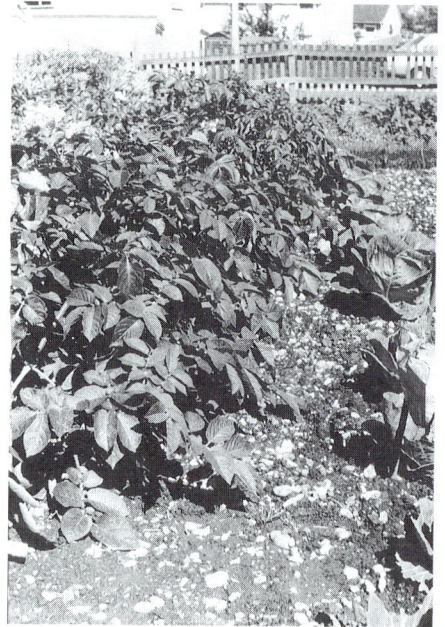

a) Copy and complete this tally chart for fertilizer A.

Weight (grams)	Tally	Frequency
15 – 19		
20 – 24	I	
25 – 29	II	
30 – 34	I	
35 – 39		
40 – 44	I	

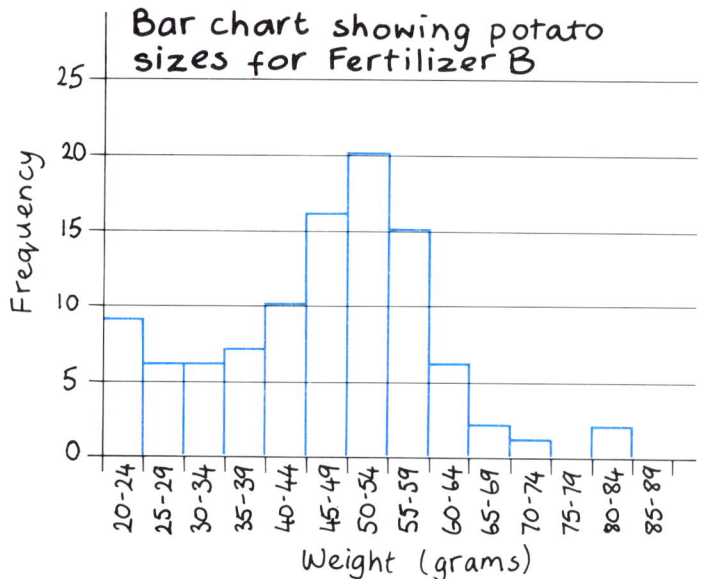

b) Draw a bar chart for fertilizer A, like the one for fertilizer B.

Bar chart showing potato sizes for Fertilizer B

c) Compare the two charts. Which do you think is the better fertilizer for potatoes? Why?

A26

2   This block graph shows information about apprenticeships in the UK.

a) Write in your own words what you think it is trying to explain about apprenticeships.

b) The graph shows that in 1964 there were 240 400 apprentices being trained. How many were being trained in 1986?

c) Do you think the graph gives a fair impression of how apprenticeship numbers have changed since 1964 (look at the year scale)? Write one or two sentences to explain what you think.

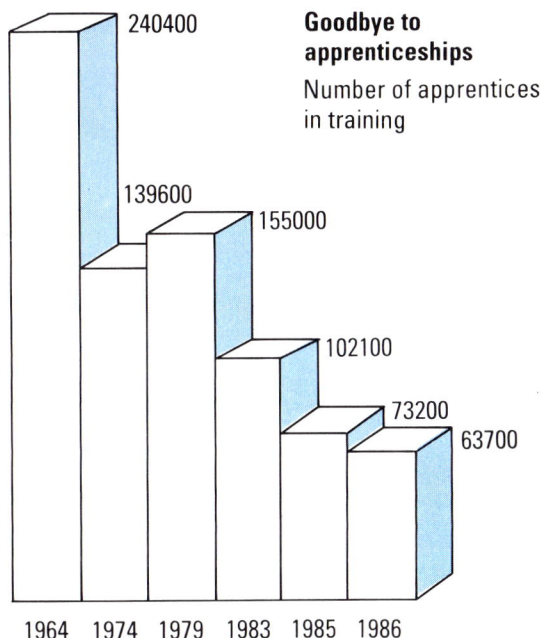

Goodbye to apprenticeships

Number of apprentices in training

240400
139600
155000
102100
73200
63700

1964 1974 1979 1983 1985 1986

3   The line chart shows how many staff were employed by British Rail between 1982 and 1987.

a) Write one or two sentences to explain how numbers changed between 1982 and 1987.

b) Check that the total number of staff decreased by about 20 thousand between 1983 and 1984/85. Roughly what was the decrease in railway staff in the same period?

c) Do you think this is true or false:

British rail employed a greater proportion of railway staff in 1986/87 than in 1982/83.

Write one or two sentences to explain your answer.

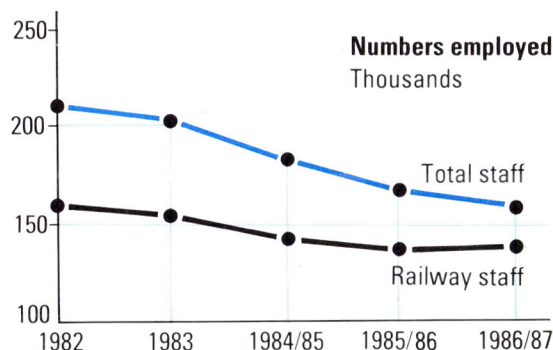

Numbers employed
Thousands

Total staff

Railway staff

250
200
150
100

1982 1983 1984/85 1985/86 1986/87

4   This table gives the number of divorces in the UK for 5 years between 1961 and 1980.

Number of divorces in Britain				
1961	1971	1972	1977	1980
27 200	79 200	124 500	137 800	157 300

a) Draw a bar chart of the information (or block graph like the one in question 2).

b) Write a short paragraph to explain what trends you can see from the chart or graph.

5   800 people were asked this question:

The man in the family should earn all the money. The woman should look after the house and the children. Do you agree?

The block graph shows the answers given in the survey.

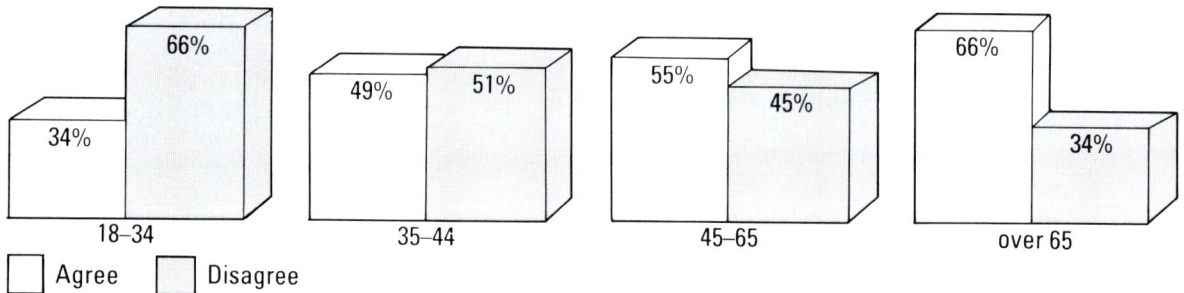

| 18–34 | 35–44 | 45–65 | over 65 |

☐ Agree   ☐ Disagree

Look carefully at the bars in the graph. The graph seems to show that older people think differently from younger people. Write a short paragraph to explain what you think the block graphs suggest.

6   This pictogram shows the number of people killed during the First World War.

Number of deaths in World War I

France	✝ ✝ ✝	1 500 000
Germany	✝ ✝ ✝ ✝	?
Britain	✝ ✝	?
Italy	✝	?
Austria–Hungary	✝ ✝	?
Russia		1 700 000

a) How many deaths does ✝ represent?

b) The pictogram shows that 1 500 000 French citizens were killed. Estimate the number of people killed from Germany, Britain, Italy, and Austria-Hungary.

c) Draw the symbols that should be added to the chart for Russia.

ACTIVITY: CLASS SURVEY

7   Draw a pictogram to show one of these pieces of information about your class.

A   Favourite colour   B   Favourite teacher   C   Favourite food

You will need to carry out a survey to collect the information.

ENRICHMENT

▦▦▦▦▦ ASSIGNMENT ▦▦▦▦▦

1 Ask your teacher if you should do your survey

- for the class, or
- for your School Year, or
- for the whole school.

Do *one* survey only. Draw whichever type of chart or graph you think is best for your survey.

Survey A   Find the number of pupils who come to school by these (or any other) means:

Draw a chart to represent your results. Write a report to explain how you collected the information. In your report explain what steps you took to make sure that the information you collected was not biased.

Survey B   Find the number of pupils who get up at these different times *at the weekend*:

Before 06:00   06:00–06:59   07:00–07:59
08:00–08:59   09:00–09:59   10:00 or later

Draw a chart to represent your results. Write a report to explain how you collected the information. In your report, explain what steps you took to make sure that the information you collected was not biased.

Survey C   Find out which brand of lipstick, cosmetics or deodorant is liked best by pupils in your class. Draw a chart to represent your results. Write a report to explain how you collected the information. In your report, explain what steps you took to make sure that the information you collected was not biased.

CHALLENGE

2   The information on this page was collected in the National Food Survey for Great Britain in 1982. Use the charts to answer these questions. (You will need to search around the charts – be patient!)

a) Each family in the survey was put into one of five groups named A, B, C, D, E. Families in Group A had a weekly income of less than £77. What was the weekly income for Group B families?

b) Families in Group D spent an average of £8.60 each week on food for each person in the family. On average, how much did each person in a Group A family spend on food each week?

A26

c) Which group (A, B, C, D, or E) spent the greatest amount per person on food?

d) On average, how much did each person in Group B spend on fruit per week?

e) Who tends to spend more on eggs each week – a person in Group E or Group A?

Who eats what
Average consumption per person per week in ounces, 1982–83

Cheese
Fresh fruit
Potatoes
Bread

☐ England
☐ Wales
☐ Scotland

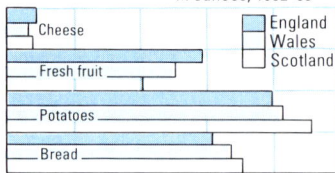

Figures: Great Britain   *Source*: National Food Survey

How our diet has changed
1982 as % of 1961

Liquid milk	80%
Eggs	75%
Mutton and lamb	53%
Poultry	283%

Butter	51%
Margarine	131%
Fresh potatoes	72%
Fresh fruit	108%

Bread	69%
Sugar	57%
Tea	71%

Figures: Great Britain   *Source*: National Food Survey

Who spends what on food

A	B	C	D	E	
Under £77	£77 and under £127	£127 and under £240	£240 and under £310	£310 or over	Income

Gross weekly income of head of household

A	B	C	D	E		A	B	C	D	E
84*	90	90	95	92		33	34	34	38	39

Milk and cream   Fish

22	20	18	19	18		20	17	15	16	16

Eggs   Sugar and preserves

227	254	256	279	257		34	41	51	71	73

Meat and meat products   Fruit (fresh, and fruit products)

* pence per person per week

7.49	7.88	8.03	8.60	8.59
A	B	C	D	E

All household food (£p)

Figures: Great Britain
Source: National Food Survey

The money we spend on food

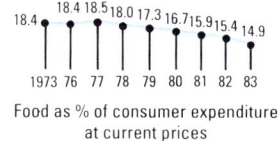

18.4   18.4 18.5 18.0 17.3 16.7 15.9 15.4 14.9

1973 76   77   78   79   80   81   82   83

Food as % of consumer expenditure at current prices

Figures: United Kingdom
Source: United Kingdom National Accounts

The rise of the superstore and hypermarket

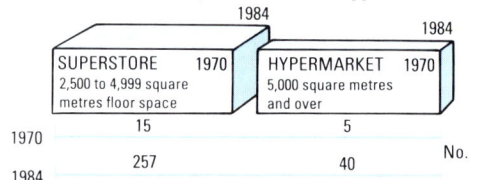

	1984		1984
SUPERSTORE 1970		HYPERMARKET 1970	
2,500 to 4,999 square metres floor space		5,000 square metres and over	

1970	15	5	
1984	257	40	No.

Figures: Food retailers in the UK
Source: Unit for Retail Planning Information

f) Look at the 'Who eats what' bar chart. Check that on average, each person living in England ate four ounces of potatoes each week. On average, how many ounces of fresh fruit did each person living in Scotland eat each week?

g) On average, who ate more cheese – a person living in Scotland, Wales, or England?

h) In 1982, people in Great Britain only ate 75% as many eggs as they ate in 1961. Did they eat more fruit or less fruit in 1982 than in 1961?

i) Write one or two sentences to explain how the use of butter and margarine changed between 1961 and 1982.

j) In 1970 there were 5 hypermarkets in Great Britain. How many were there in 1984?

A27 RATE, SPEED AND TIME

REVIEW

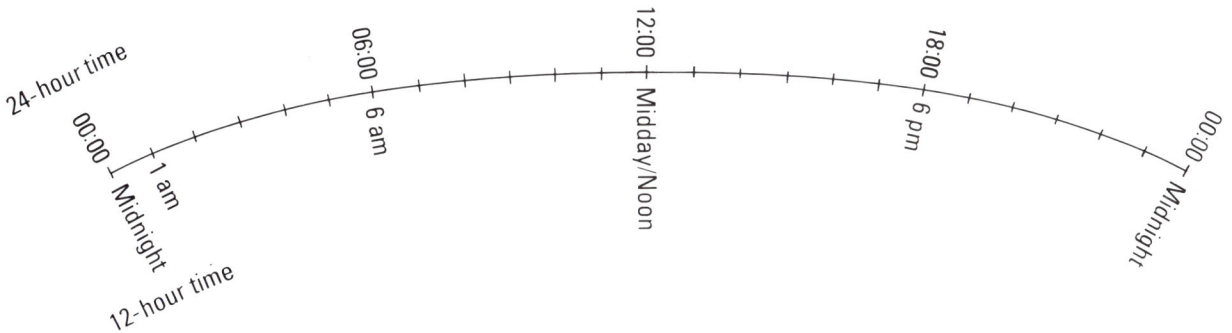

■ How do we write 1:25 am and 1:25 pm using the 24-hour clock?
How do we write 05:38 and 15:38 using the 12-hour clock?

These units measure *rates*.

beats per minute
kilometres per hour (km/h)
cars per hour
pounds per week
pence per litre
gallons per minute
revolutions per minute

■ Beats per minute might measure heart rate. What might each of the other units measure?

The steamroller is travelling 2 metres every second. Its speed is 2 m/s. It is travelling at a *steady speed*.

■ How far will it travel in 1 minute? How long will it take to travel 100 m?

■ Think of   a) two things which normally travel at a steady speed
             b) two things which do not normally travel at a steady speed.

List them.

CONSOLIDATION

1   At what time did you get up this morning? Write the time:

    a) using the 24-hour clock
    b) using the 12-hour clock.

2   A coach sets out from Newcastle at 14:35. It arrives in Edinburgh at 20:20. How long does the journey take?

3   An $8\frac{1}{2}$ hour working day starts at 06:15. At what time does it end? Write the time:

    a) using the 24-hour clock
    b) using the 12-hour clock.

4   A washing machine takes 48 minutes to complete its wash. At what time should it be started to finish at 18:05? Write the starting time:

    a) using the 24-hour clock
    b) using the 12-hour clock.

A27

5   This is the Birmingham to Bournemouth rail timetable.

    a) If you take the 14:10 from Birmingham, at what time do you reach Bournemouth?

    b) How many trains leave Birmingham for Bournemouth (without changing trains) in the afternoon, before 4pm?

    c) Which train completes the journey in the shortest time:

        the 12:10 or the 16:10?

        How many hours and minutes does it take?

Birmingham New St. — Birmingham International — Banbury — Oxford ⌙ — Reading ⌙ — Guildford — Portsmouth and Southsea ⌙ — Portsmouth Harbour — Winchester — Southampton ⌙ — Bournemouth ⌙

Birmingham New St.	1010	1210	1410	1610	1710	1810	1910	2121
Birmingham International	1021	1221	1421	1621	1721	1823	1922	2133
Banbury	1126	1326	1525	1714	1814	1921	2018	2226
Oxford	1205	1404	1604	1743	1841	1951	2049	2254
Reading	1239	1444	1639	1811	1909	2018	2117	2319
Guildford	1320	1608	1808	1915	1943	——	2308	2359
Portsmouth and Southsea	1411	1644f	1844f	——	2034	2218b	2318b	0053
Portsmouth Harbour	1415	1648f	1848f	——	2038	2222b	2322b	——
Winchester	1342f	1533	1742f	1909	2042f	2109	2209	——
Southampton	1409a	1553	1809a	1928	2110a	2128	2228	——
Bournemouth	1456a	1624	1856a	2002	2156a	2202	2300	——

NOTES
A   5 July to 30 August.
a   Change at Reading and Basingstoke.
b   Change at Southampton.
c   Change at Basingstoke and Eastleigh.
d   Change at Reading, Basingstoke and Eastleigh.
e   Change at Basingstoke.
f   Change at Reading.

6   'Voyager' circled the Earth non stop in December 1986 in 216 hours 3 minutes. Its journey began at 08:02 on Sunday December 14. On which day, and at what time did it arrive?
Write the arrival time:

    a) using the 12-hour clock
    b) using the 24-hour clock.

7 Match each of these petrol consumption rates (A), with a vehicle (B).

A	B
40 miles per gallon	Transporter
12 miles per gallon	Jet aircraft
0.2 miles per gallon	Small car
90 miles per gallon	Motorcycle

8 An 'extended play' record turned at 45 r.p.m. (revolutions per minute). At what rate do these turn (choose your own units):

a) the minute hand of a watch
b) the hour hand of a watch
c) the Earth, around the Sun
d) the Earth, on its own axis?

IN YOUR HEAD

9 Try to do these in your head.

a) A steamroller travels at a steady speed of 3 km/h.

(i) How far does it travel in half an hour?
(ii) How many minutes does it take to travel 1 km?

b) You have to cycle 8 miles. You need to get there in $\frac{1}{2}$ h. At what steady speed should you travel?

c) A conveyor belt moves baggage 24 metres at a steady 2 m/s.

(i) How long does the journey take?
(ii) At what speed should the baggage travel to arrive in 8 seconds?

d) An electric sewing machine can do 300 stitches every minute. What rate is this in stitches per second?

e) How many days would it take you to travel 300 km at 30 km per day?

f) How many kilometres per day should you walk to cover 96 km in 6 days?

10 Hilary earns £7.20 per hour.

a) How much does she earn in (i) 2 hours, (ii) $2\frac{1}{2}$ hours?

b) For how long must she work to earn £36?

c) She gets a new job. For each 8-hour day she earns £58. Is the rate of pay for this job better, or worse, than the rate of pay for her old job?

CHALLENGE

11 Estimate the rate (in revolutions per minute) at which the wheels of an exercise bike turn when the dial shows 25 km/h.

Average speed

1   The graph shows the
    distance a steamroller travels
    during a 9-second
    journey.

    a) Check that during the
       first second the
       steamroller travelled
       about $\frac{1}{4}$ m.

    b) Check that during the
       2nd second it travelled
       about $\frac{1}{2}$ m.

    c) About how far did it travel
       during:

       (i) the 4th second
       (ii) the 5th second?

    d) For approximately how
       many seconds did the
       steamroller travel at a
       steady 2 m/s?

    e) What was happening to
       the steamroller between
       the 7th and 9th
       seconds ... was it
       slowing down, or
       speeding up?

       Explain how you
       decided.

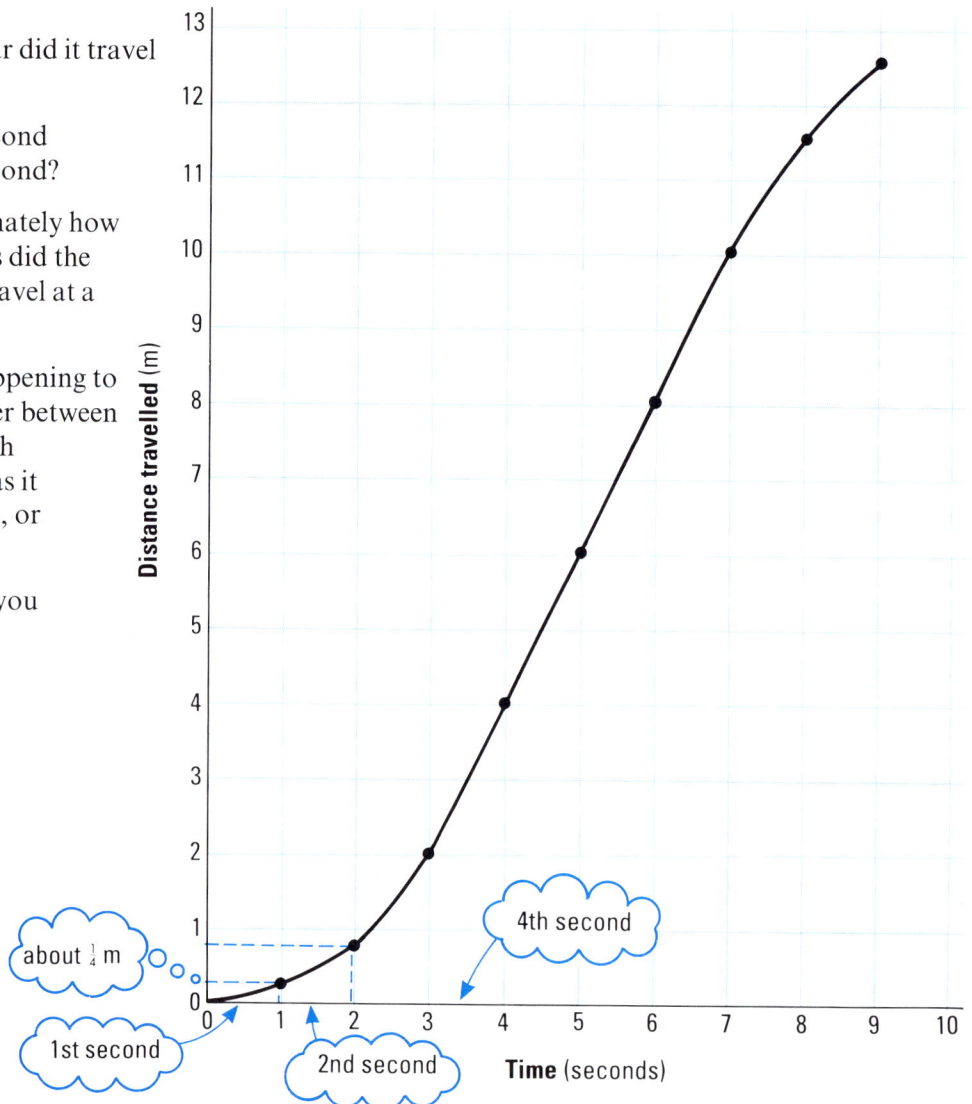

about $\frac{1}{4}$ m

1st second

2nd second

4th second

Distance travelled (m)

Time (seconds)

2

Rula stands on the moving pavement. It travels at a steady 3 m/s. This is a graph for her journey.

Alan starts alongside Rula. He is not on the moving pavement. Sometimes he walks. Sometimes he trots. They arrive together at the end of the moving pavement.

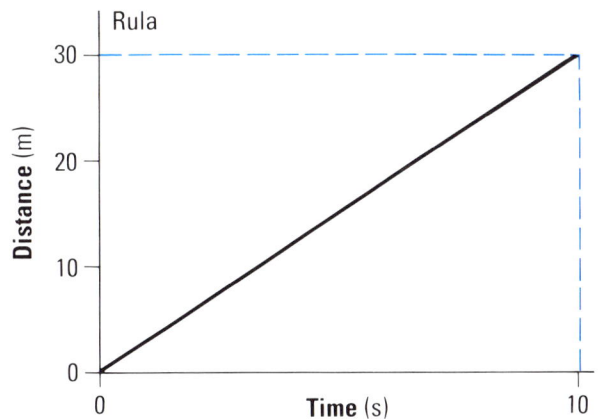

This is a graph for Alan's journey.

a) Was Alan walking or trotting at the end of the journey?

b) Was he walking or trotting at the beginning of the journey?

c) How many times did he break into a trot?

d) Do you think Alan trotted at more than 3 m/s or less than 3 m/s?

Explain your answer.

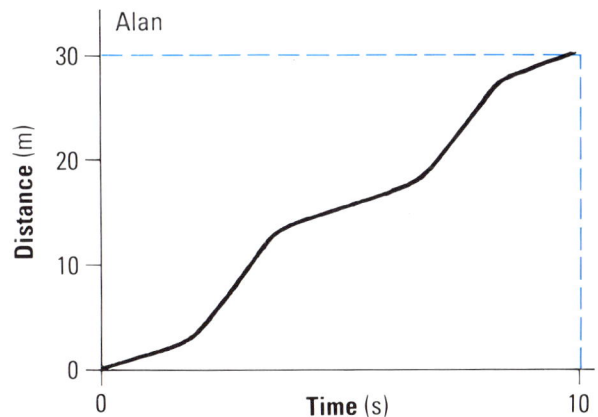

TAKE NOTE

Rula travelled 30 metres in 10 seconds, at a steady speed of 3 m/s. Alan travelled 30 metres in 10 seconds – but not at a steady speed. We say that Alan's *average speed* was 3 m/s.
An *average speed* of 3 m/s will take us the same distance in the same time as a *steady speed* of 3 m/s – but we might not travel at the same speed all the time.

3   Estimate the average speed, in m/min or miles/h, at which you travel to school.

4   The graph is for a 3-hour car journey. You can see that after one hour the car has travelled 20 km.

a) What is its average speed during the first hour?

b) How far has the car travelled after two hours?

c) What is its average speed during the first two hours?

d) What is its average speed during the second hour?

e) During the third hour the car travelled along a motorway. What was its average speed during this part of the journey?

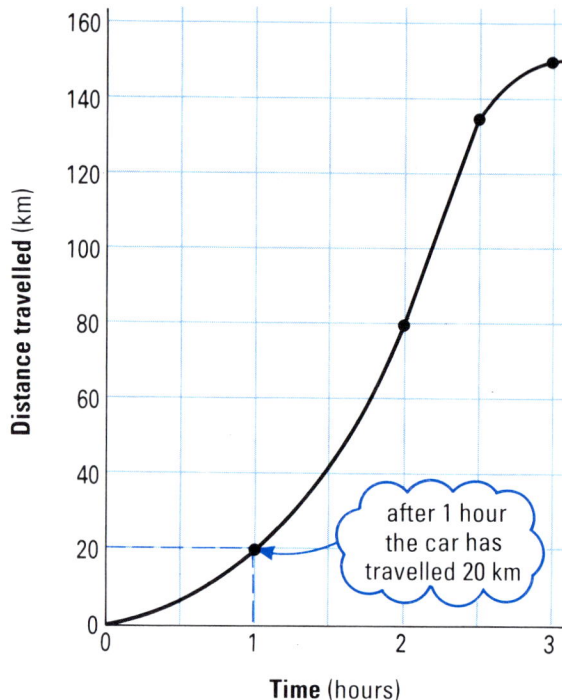

after 1 hour the car has travelled 20 km

A27

CHALLENGE

f) For a part of the motorway journey the car travelled at a steady speed. Roughly, what was the speed?

5   You travel 10 km in 2 hours. What is your average speed?

6   You travel 15 km at an average speed of 3 km/h. How long does the journey take you?

7   You travel at an average speed of 22 km/h for $2\frac{1}{2}$ hours. How far do you travel?

THINK IT THROUGH

8   a) These rules are about average speeds. Copy and complete them by filling in $\times \div -$ *or* $+$.

(i)   Average Speed ☐ Time Taken gives the Distance Travelled.

(ii) Distance Travelled ☐ Average Speed gives the Time Taken.

(iii) Distance Travelled ☐ Time Taken gives the Average Speed.

b) Check each of your rules using this information: Average Speed 12 km/h, Time Taken 3 h, Distance Travelled 36 km.

REVIEW

Capacity is the amount of space inside a container.

It is measured in
- millilitres (ml)
- centilitres (cl)
- litres (l)
- m³
- cm³
- mm³
- cc (cubic centimetres)

1000 millilitres = 1 litre
100 centilitres = 1 litre

The capacity of this bottle is 70 cl or 700 ml or 0.7 l.

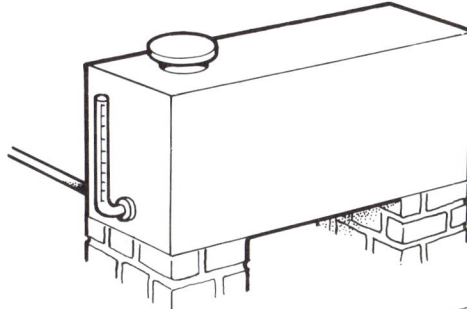

The capacity of this oil storage tank is 2000 l.

■ Name five objects with capacity less than 1 litre.

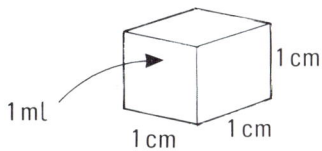

1 ml of liquid takes up 1 cm³ of space.

15 ml of liquid fills 15 cm³ of space.

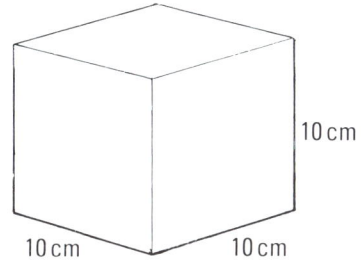

1 litre fills a
10 cm × 10 cm × 10 cm = 1000 cm³ container. It takes up 1000 cm³ of space.

■ Name five things with capacity more than 20 litres.

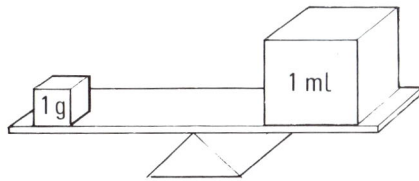

1 ml of water weighs 1 g.

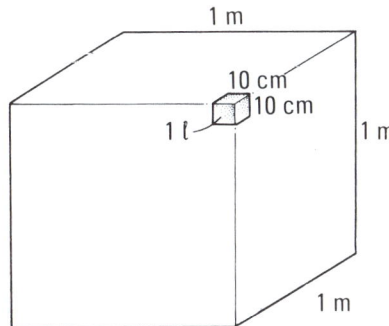

1 litre of water weighs 1 kg.

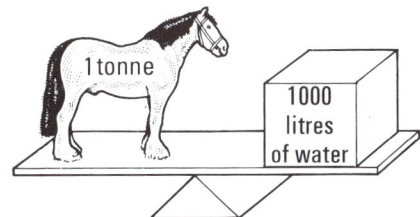

A 1 m cube holds 10 × 10 × 10 litres of water = 1000 litres.

This weighs 1000 kg or 1 tonne.

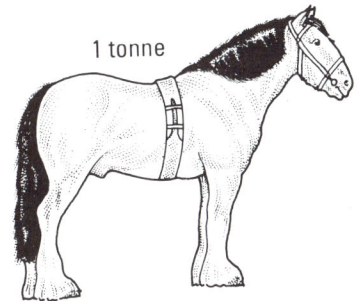

CONSOLIDATION

1 a) Which of the two freezers has the greater capacity?

b) Each freezer is to be packed with ice cream. How many litres of ice cream will each hold?

Capacity 5 cubic metres

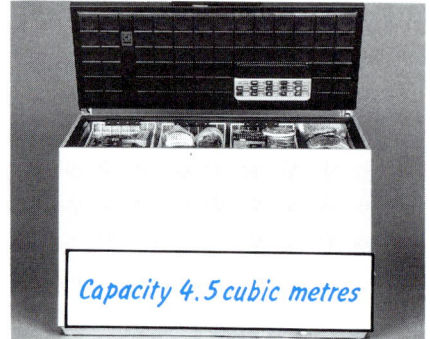

Capacity 4.5 cubic metres

2 Estimate the capacity of each of these (i) in m^3 or cm^3, (ii) in litres or centilitres.

a) a milk tanker
b) a tennis ball

A28

3 Explain in your own words what the newspaper headline means.

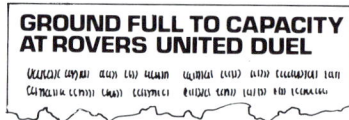

GROUND FULL TO CAPACITY AT ROVERS UNITED DUEL

4 Which units of capacity (for example, litres, cl, ml) would you use to measure:

a) the amount of wine in a bottle

b) the amount of water in a water tank

c) the amount of blood taken for a blood test.

5 You weigh approximately the same as ... how many litres of water?

6 a) What is the capacity of this large fish tank, in m^3?

b) How many litres of water are needed to fill it?

c) The tank weighs 20 kg when it is empty. Roughly, what does it weigh when full of water?

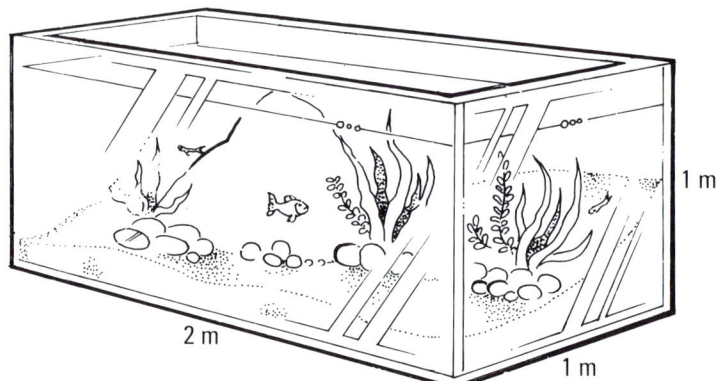

1 m

2 m

1 m

CORE

Litres and parts of litres

1 Write down how much orange the Tango can holds:

a) in centilitres   b) in litres.

2 a) Which of these bottles of hand cream contains the most?

b) How much hand cream is there altogether? Write your total:

(i)   in millilitres
(ii) in centilitres
(iii) in litres.

3 This box held a sample bottle of Claude Lamot eau de toilette.

a) What is the capacity of the box (i) in cm^3, (ii) in mm^3?

b) If the *box* could be filled with eau de toilette, how many millilitres would it hold?

c) Guess how many millilitres of eau de toilette the sample bottle held (your teacher knows the actual amount).

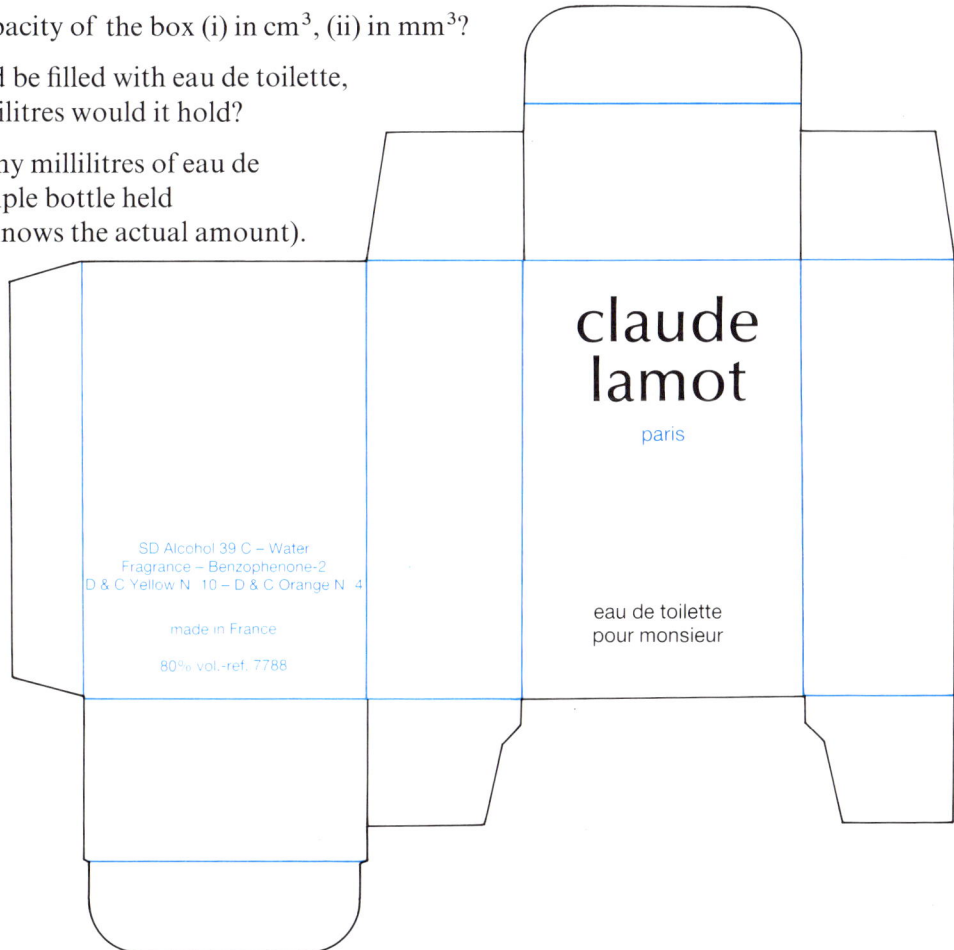

A28

claude
lamot

paris

SD Alcohol 39 C – Water
Fragrance – Benzophenone-2
D & C Yellow N 10 – D & C Orange N 4

made in France

80% vol.-ref. 7788

eau de toilette
pour monsieur

CHALLENGE

4   On average each person in a village uses
    30 litres of water each day. There are 4000
    people in the village. The village needs a
    water tank which holds 14 days' supply.

    a) How many litres of water should the
        tank hold?

    b) Design the tank yourself. Make it a
        cuboid shape.

Archimedes

ACTIVITY: HOW BIG IS YOUR FIST?

1   Work with a friend.
    You need: a measuring cylinder large
                enough to take your clenched fist,
                a tray to stand the cylinder in.

    a) Each of you estimate the volume of your
        fist in cm^3.

    b) Each of you, in turn:

        fill the measuring cylinder to the brim
        with water,
        gently plunge your clenched fist into the
        water,
        measure how many millilitres of water
        have overflowed into the tray.

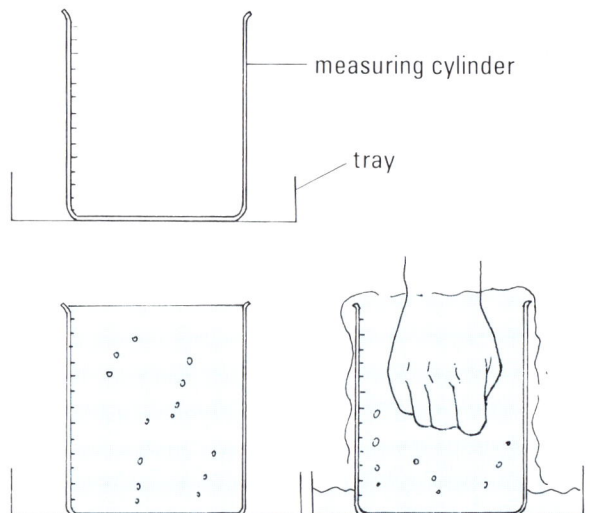

measuring cylinder

tray

        Who has the bigger fist? Discuss how you can use your results to find out. Each of you write
        down the volume of your own fist and your friend's fist, in cm^3.

2 Archimedes was born in about 287 BC and was killed by a
Roman soldier in 212 BC. He was a Greek scientist and
mathematician. He was set this problem by his king:

Find out whether my crown is made of pure gold ... without
damaging it.

Archimedes realised how he could do this one day, while he
was having a bath. What you have to find out:

Why is Archimedes running down the street shouting
'Eureka', and what was his solution to the problem?

(You might be able to solve the problem yourself ...
think about question 1.)

EUREKA!

READ:
"YEWREEKA"

ENRICHMENT

ASSIGNMENTS

A28

1 a) Collect eight stones, pebbles, pieces of wood, or other objects which have these volumes. (Find
two objects for each volume.)

Between 1 cm^3 and 3 cm^3 Between 9 cm^3 and 11 cm^3
Between 19 cm^3 and 21 cm^3 Between 49 cm^3 and 51 cm^3

b) Find the weight, in grams, of each of your objects.

c) Find the density in grams per cm^3
for each of your objects.
(For example, 1 cm^3 of this piece of
wood weighs $\frac{1}{2}$ g. Its density is
0.5 g per cm^3.)

Volume: 8 cm^3
Mass: 4 g

d) Write a report explaining how you measured the volumes of your objects. In your report,
list your objects in order from least dense to most dense.

2 This is a triangle-shaped container for ice-cream.
Design and make your own triangle-shaped
container. Your container should hold $\frac{1}{2}$ litre.
In your report:

• explain how you decided upon the
dimensions of your container.
• sketch a full-sized net of your container.

Ice cream

REVIEW

The white letter J is an *enlargement* of the blue letter J. Point O is the *centre of enlargement*. The *scale factor* is × 3.

■ How many times longer than BC is B′C′?

■ How many times further from O is A′ than A?

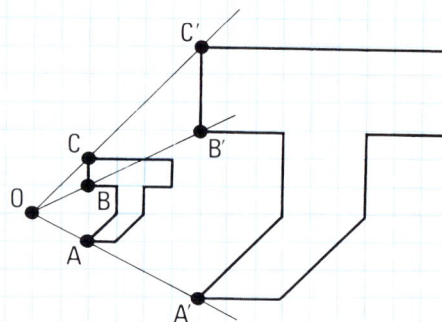

The white letter L is an enlargement of the blue letter L.

■ Which point is the centre of enlargement?

■ What is the scale factor?

The blue letter L is a *reduction* of the white letter L.

■ Write down the scale factor of the reduction
    like this ÷ ☐
    and like this × ☐

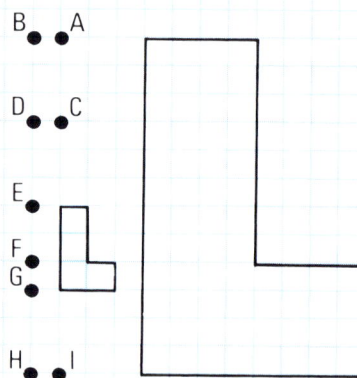

Shape B is an enlargement of shape A. We say shape B is *similar* to shape A.

■ Which other two shapes are similar to shape A?

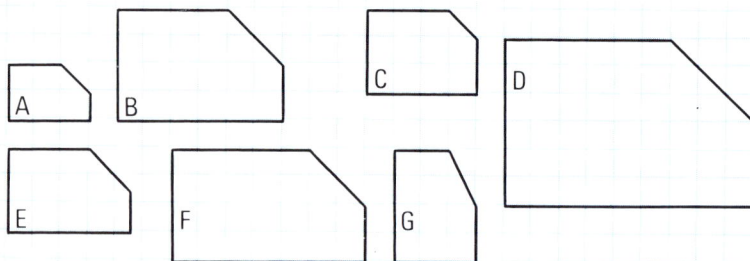

Triangles with corresponding angles equal are *always* similar.

■ Draw a pair of quadrilaterals with corresponding angles equal that are *not* similar.

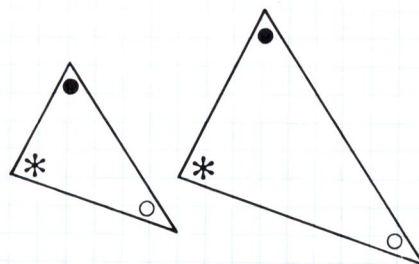

These triangles are similar.

■ What is the length of DE?
    (*Not 8 cm!*)

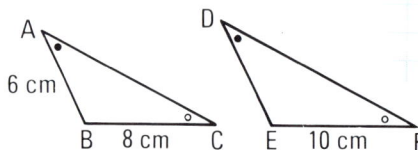

6 cm

B   8 cm   C   E   10 cm   F

CORE

B1

1   Esme has a $4'' \times 6''$ photograph which she wants to have slightly enlarged. At the photographic developer's she is told that the next size up is $5'' \times 7''$. Compared with the original photo, does this give a shape that is more elongated, similar, or more square?

2   a)   On squared paper, draw a $4'' \times 6''$ rectangle.

Choose a centre of enlargement inside the rectangle.

Draw these construction lines (in blue).

Produce an enlarged rectangle, with base $7''$ long.

centre of enlargement

$7''$

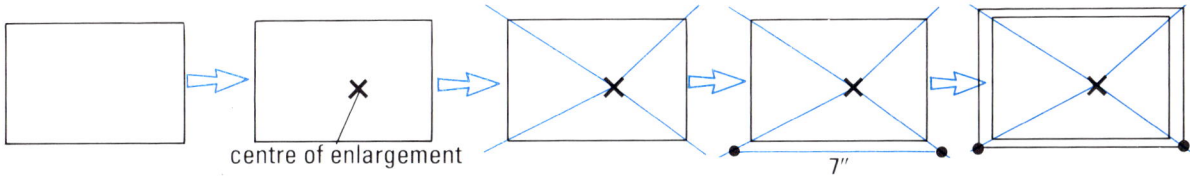

   b)   Measure the height of your new rectangle. Is it less than, equal to, or more than $5''$ long?

THINK IT THROUGH

3   a)   Dixons offer these photograph sizes for 35 mm negatives.

$3\frac{1}{2}'' \times 5''$ $4'' \times 6''$ $5'' \times 7''$ $8'' \times 10''$ $10'' \times 12''$ $11'' \times 15''$ $16'' \times 20''$ $20'' \times 30''$

Compare each size with a $4'' \times 6''$ photograph; decide whether you get a more elongated, similar, or more square shape.

   b)   Check your answers to part a) like this:

   (i)   Use a sheet of squared paper. Starting from the bottom left-hand corner, draw a $4'' \times 6''$ rectangle.

   (ii)   Draw this diagonal (the blue line) and continue it beyond the rectangle.

   (iii)   Starting from the bottom left-hand corner, draw rectangles for all other photograph sizes.

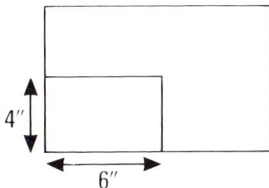

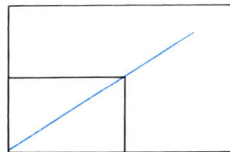

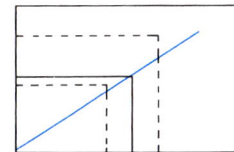

$4''$

$6''$

   (iv)   Do the top right-hand corners fall above, on, or below the diagonal line? What does this tell you about the shapes of the rectangles?

B1

4 a) (i) Copy this triangle on 1 cm squared paper.

(ii) Reduce the triangle. Use A_1 as the centre and use a scale factor of $\times \frac{1}{2}$.

(iii) Enlarge your new triangle. Use A_2 as the centre and use a scale factor of $\times 2$.

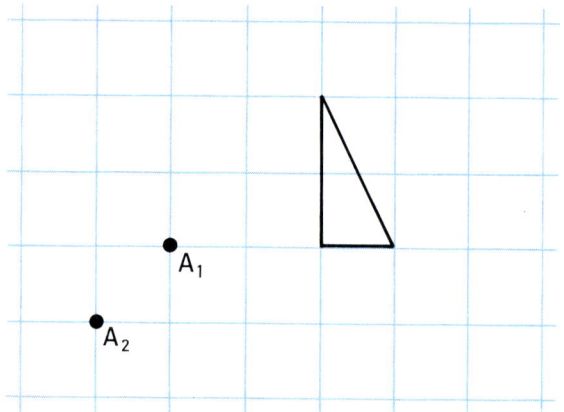

b) Repeat part a), but this time use B_1 and B_2 as the centres.

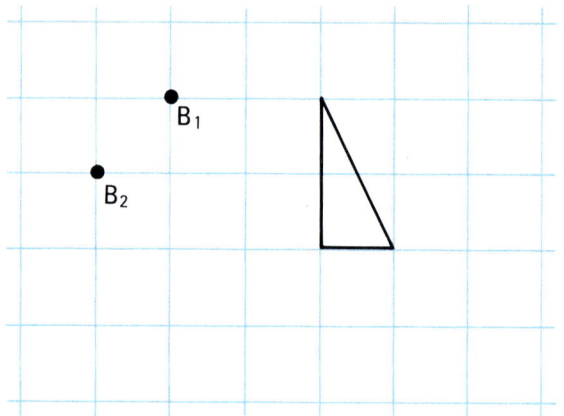

c) (i) Predict what would happen if you repeated part a) but with C_1 and C_2 as the centres.

(ii) Check your prediction.

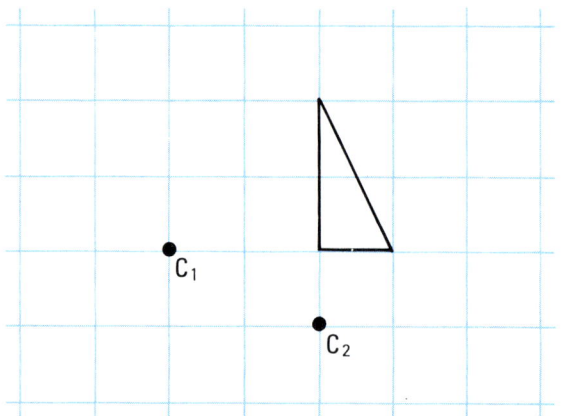

d) Try • some more pairs of centres.
• reversing the scale factors ($\times 2$ then $\times \frac{1}{2}$).
• other scale factors (for example, $\times 3$ then $\times \frac{1}{3}$).
• scale factors that do not multiply together to give $\times 1$ (for example, $\times 6$ then $\times \frac{1}{2}$).

Write about what you find out.

Blow ups

1 Get a small photograph (about the size of a postage stamp). Divide it into small squares, measuring about 2 mm × 2 mm, say. Transfer the photograph onto 1 cm squared paper.

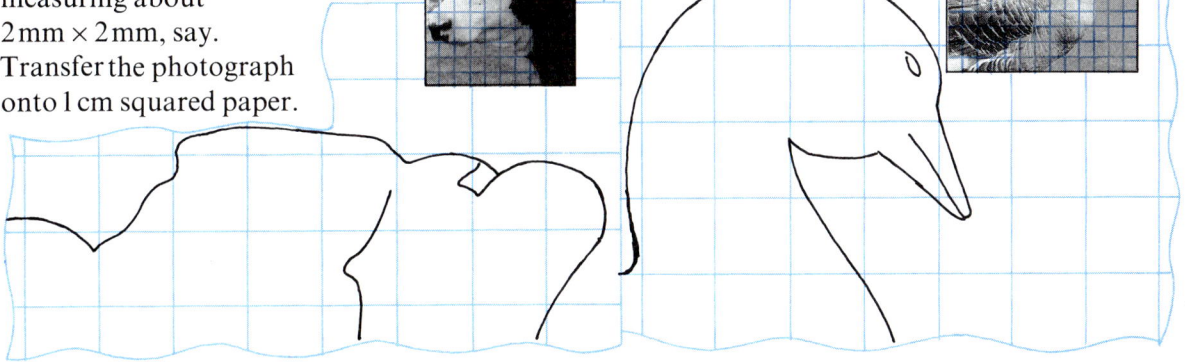

2 Get a normal size photograph. Divide it into 10 mm × 14 mm rectangles, and label them as shown on the right. Make a photocopy for each person in the class.

   Share out the rectangles among everyone in the class and transfer each one onto a plain sheet of A4 paper. Make a wall display.

ENRICHMENT

EXPLORATION: LIGHTS AND QUADRILATERALS

1   Here is a screen,
    a piece of card,
    and a lamp.

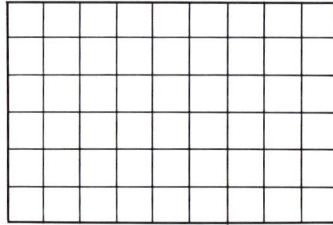 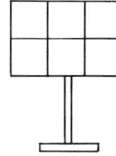

The card is placed midway between the screen and lamp.

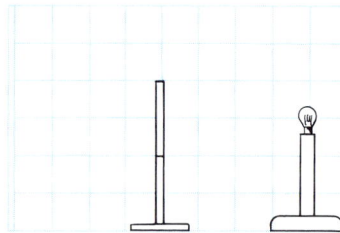 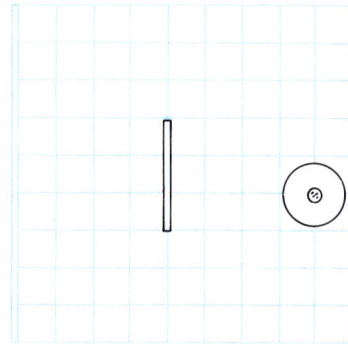

side view   top view

The shadow of the card on the screen looks like this.

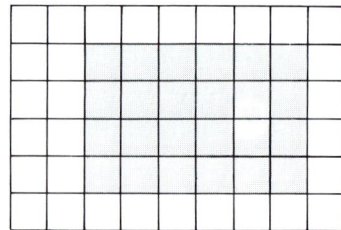

a) What happens to the size of the shadow when:

   (i)   the lamp is moved away from the card
   (ii) the card is moved nearer the lamp
   (iii) the card and the lamp stay the same distance
         apart but are moved away from the screen?

b) Choose one of the positions in the diagrams below as a starting point. Explore what happens
    to the shadow as the card, or the lamp, or both, are moved.

A 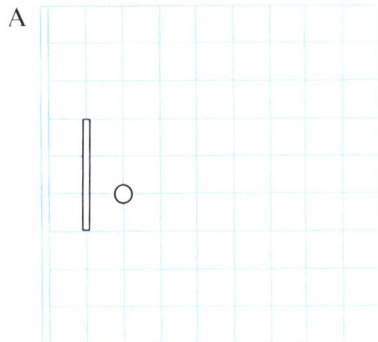   B 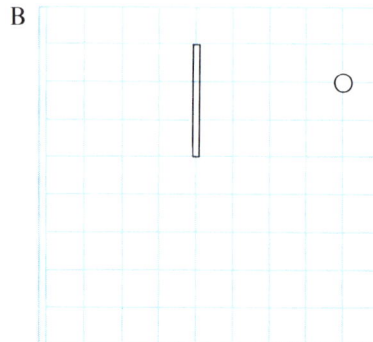   C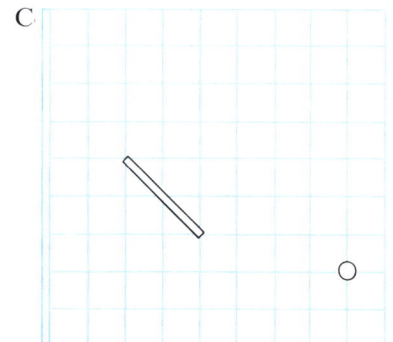

B1

2 Quadrilaterals A, B, C, D have all been made from rectangle X below, by choosing a starting point (*not* a corner of the rectangle) and increasing the distance of the vertices from the starting point by 2 cm.

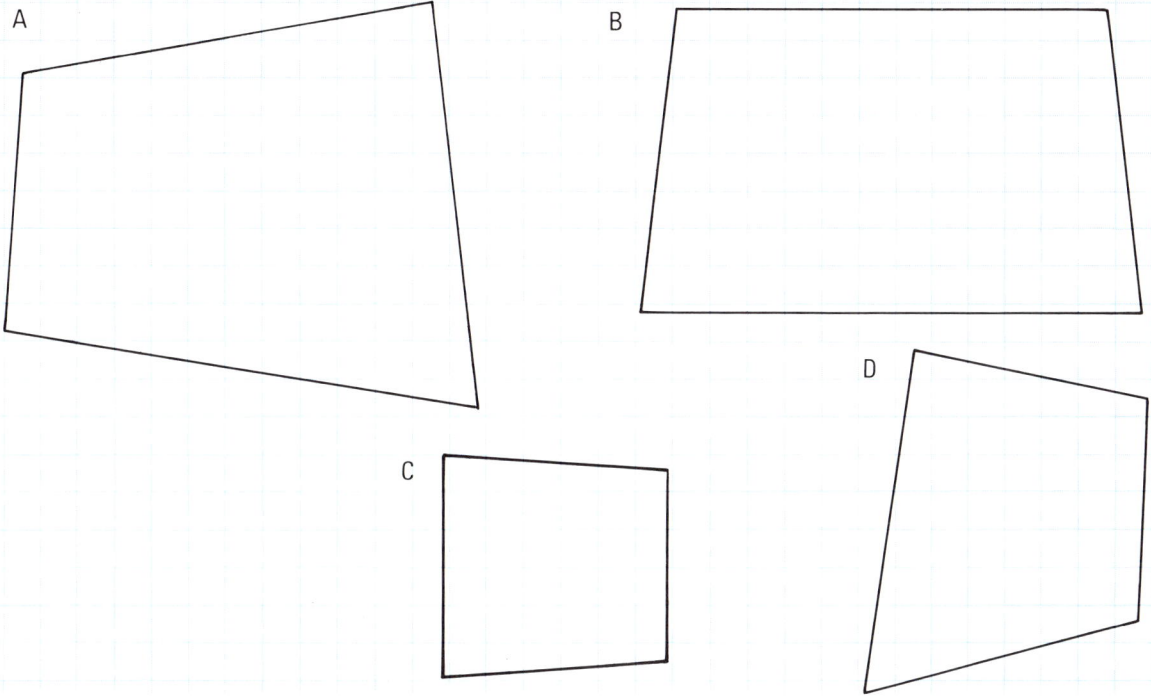

a) Draw the original rectangle on squared paper. Find the approximate position of the starting point for each quadrilateral.

b) Explore different starting points. Can you tell roughly where the starting point will be from the final shape of the quadrilateral?

c) Explore changes in the amount added on (for example, 5 cm instead of 2 cm).

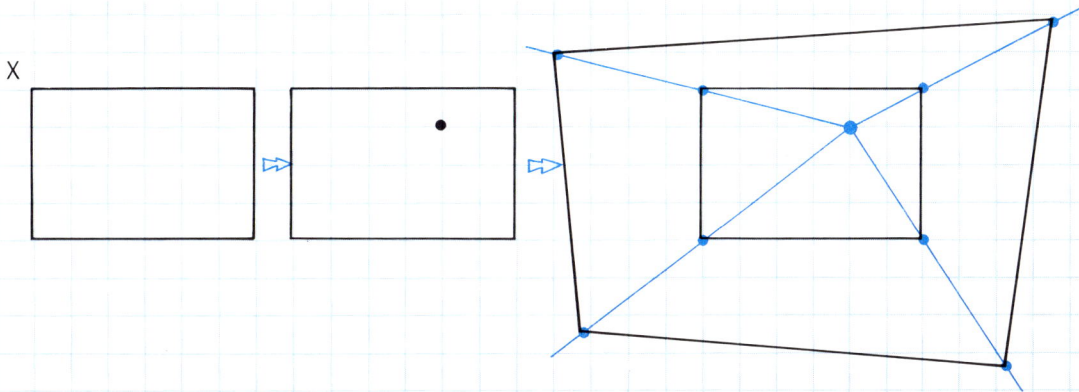

CORE

▨▨▨▨▨▨▨▨▨▨▨ EXPLORATIONS ▨▨▨▨▨▨▨▨▨▨▨▨▨▨▨▨▨▨▨▨▨

1 a) The perimeter of this rectangle is 12cm. What is its area?

   b) Draw some more rectangles with a perimeter of 12cm.
       Find the rectangle with the largest area.

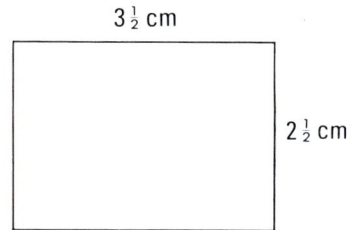

$3\frac{1}{2}$ cm

$2\frac{1}{2}$ cm

2 a) This rectangle has an area of 12cm². What is its perimeter?

   b) Draw some more rectangles with an area of 12cm².
       Find one:

       (i) whose perimeter is larger than 26cm
       (ii) whose perimeter is smaller than 14cm.

       (Use a calculator to help you.)

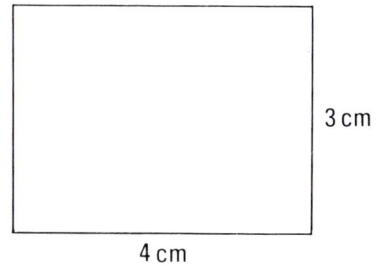

3 cm

4 cm

3 You need 5mm squared paper. Look at the sequence of shapes, A, B, C.

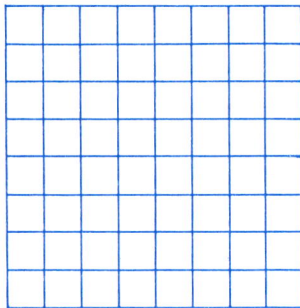

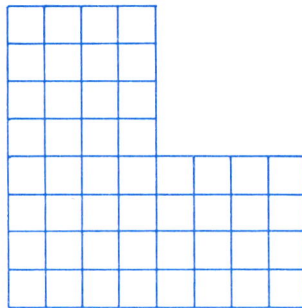

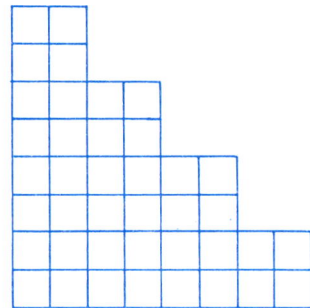

A   B   C

a) Draw the next shape. Call it D.

b) For each shape write down:

    (i) its perimeter   (ii) its area.

c) Imagine continuing the sequence of shapes.
    Describe what happens to:

    (i) the perimeter   (ii) the area.

d) Here is a conversation about the shapes.
    Who is right? Explain why.

YOU CAN MAKE THE AREA AS SMALL AS YOU LIKE.

Asif

THE AREA WILL NEVER GET SMALLER THAN 8 cm².

Jenny

EXPLORATION: THE LETTER H

4   You need 1 cm dotted squared paper.

   a)   Investigate the areas and perimeters of Hs which are 1 cm thick and as tall as they are wide.

   b)   Try to find formulas for their areas and perimeters.

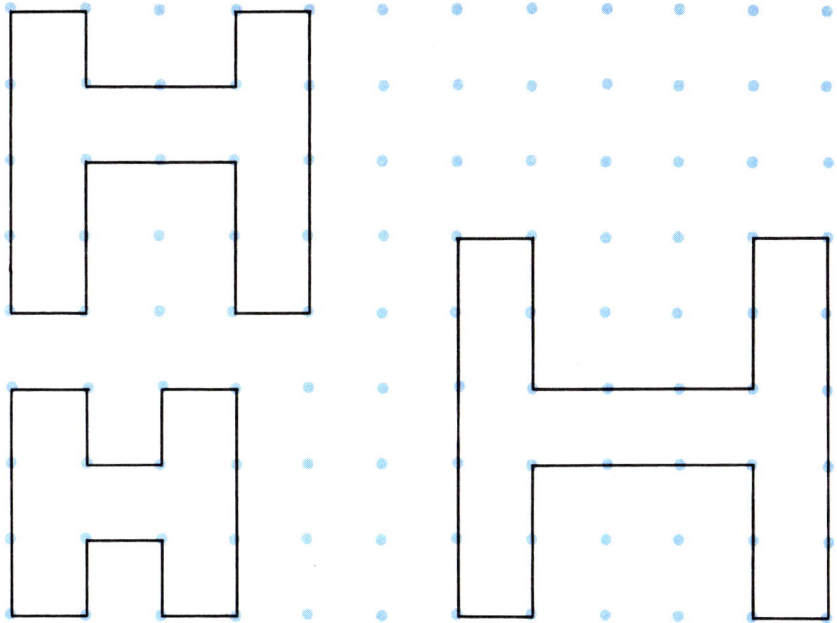

ASSIGNMENT: MORE LETTERS

5   Investigate the areas and perimeters of some more letters. Write a report about what you discover.

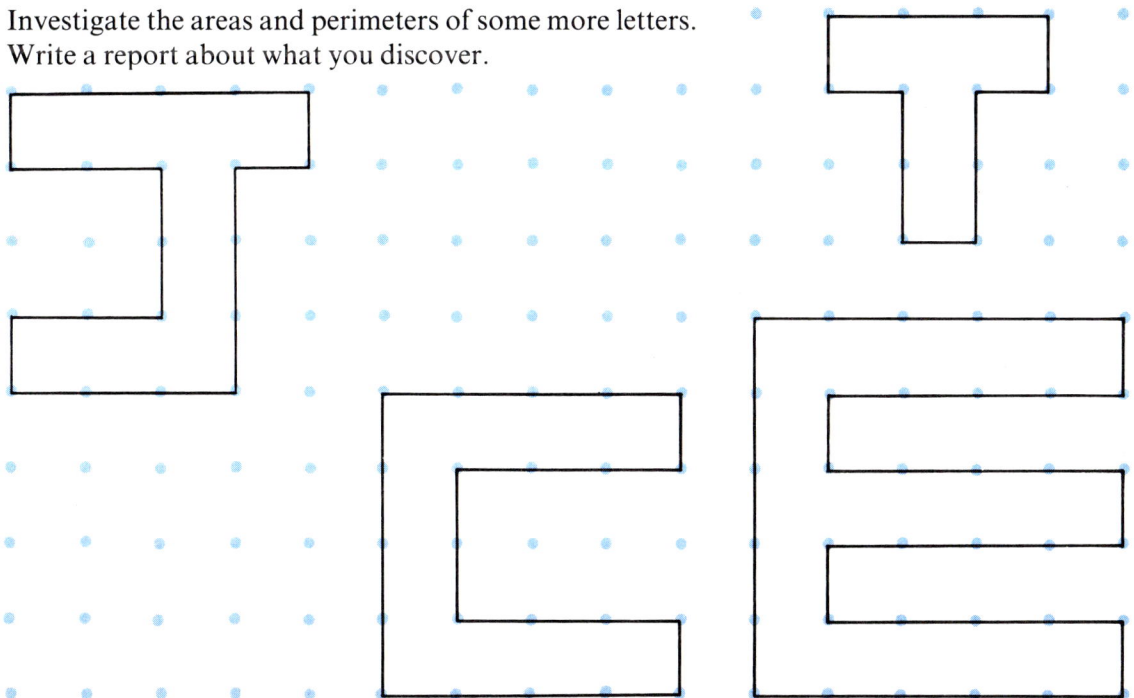

ENRICHMENT

1   a) Copy and complete this sequence of three shapes. Use 5mm squared paper.

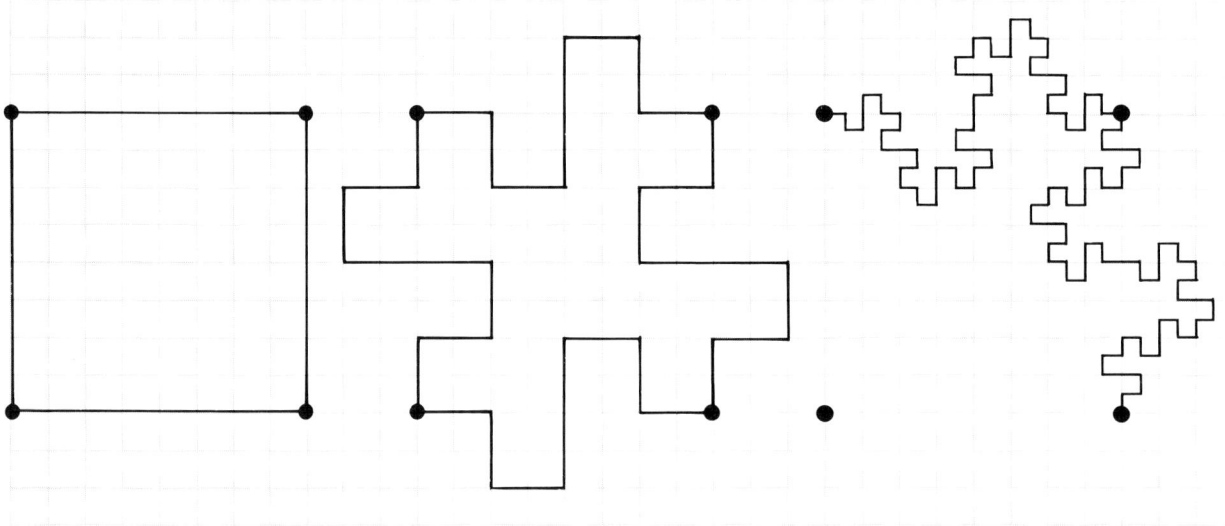

   (i)   Write down the area and perimeter of each shape.
   (ii) Think of the next shape in the sequence (that is, the fourth shape). What would its area
         and perimeter be?
   (iii) Which shape in the sequence would have a perimeter of about 40m?

   b) Draw (or complete) the next shape in these sequences. Explore what happens to their areas
       and perimeters. Make up your own sequences. Write about what you find.

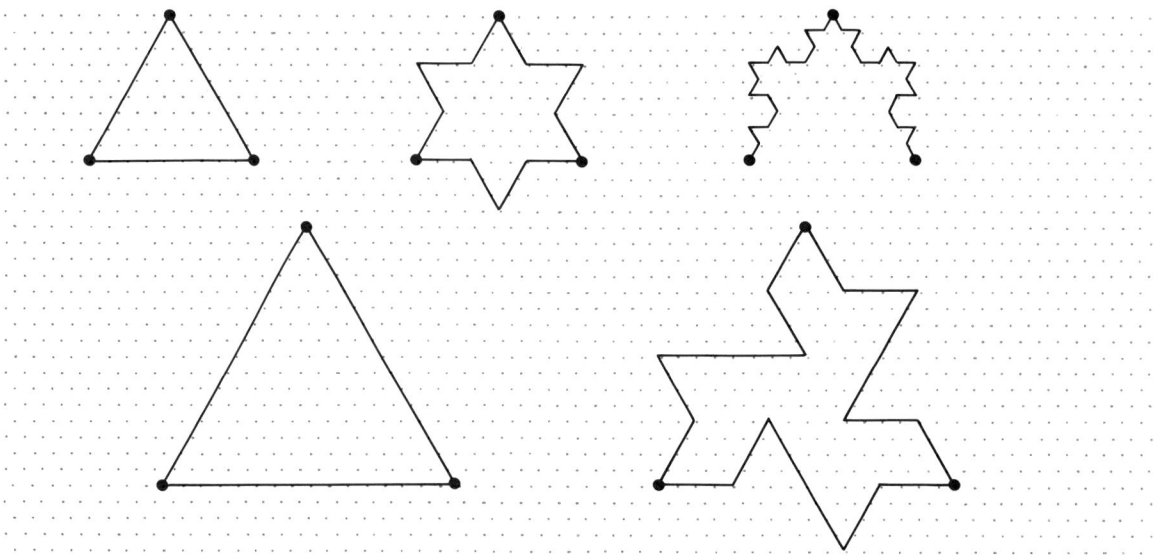

B3 WORKING WITH NUMBER

REVIEW

- Each decimal fraction goes with a whole family of divisions.

  $0.25 \longrightarrow \{1 \div 4, \quad 2 \div 8, \quad 3 \div 12, \quad 4 \div 16, \quad 0.1 \div 0.4, \quad \ldots\}$

  $0.7 \longrightarrow \{7 \div 10, \quad 0.7 \div 1.0, \quad 1.4 \div 2, \quad 14 \div 20, \quad \ldots\}$

■ Find five divisions which go with 1.4.

- Each decimal fraction goes with a whole family of fractions.

  $0.75 \longrightarrow \{\frac{3}{4}, \frac{6}{8}, \frac{9}{12}, \frac{12}{16}, \frac{15}{20}, \ldots\}$ $\qquad$ $0.4 \longrightarrow \{\frac{4}{10}, \frac{8}{20}, \frac{12}{30}, \frac{16}{40}, \ldots\}$

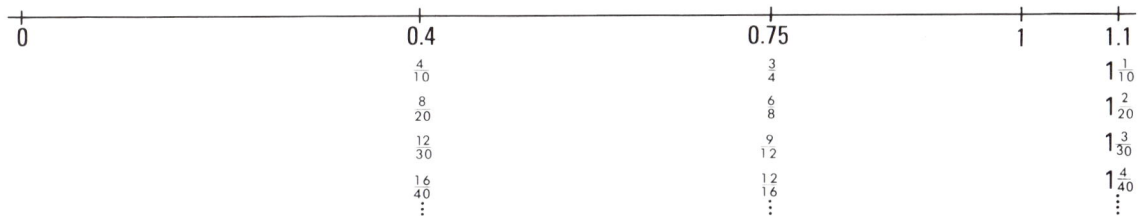

- We can find divisions in the same family like this:

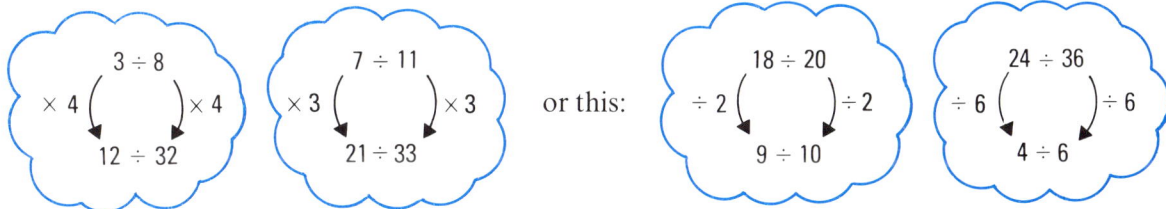

or this:

■ Copy and complete these three divisions from the same family:

  $15 \div 6, \quad \square \div 18, \quad \square \div 2$

- We can find fractions in the same family like this:

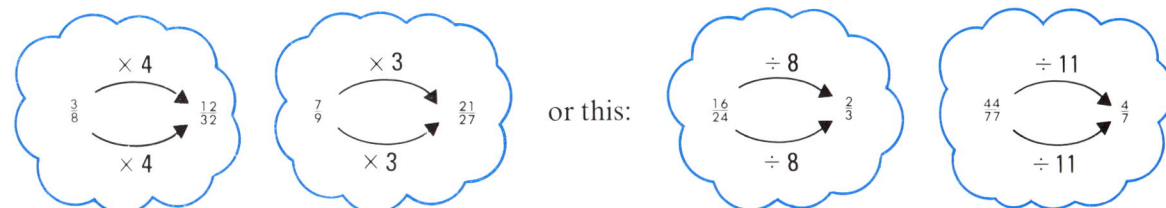

or this:

■ Copy and complete these three fractions from the same family:

  $\frac{32}{48}, \frac{4}{\square}, \frac{\square}{144}$

- $\frac{5}{8}$ as a decimal fraction is $\dfrac{0.\ 6\ 2\ 5}{8\overline{)5.\,{}^5 0\,{}^2 0\,{}^4 0}}$ $\qquad$ ■ Write $\frac{3}{8}$ as a decimal fraction.

CONSOLIDATION

CHALLENGE

1   Scale A uses fractions. It rounds weights down to the next $\frac{1}{8}$ ounce.

Scale B uses decimal fractions. It rounds weights down to the next 0.1 ounce.

A

B

a) Which scale gives the more accurate readings, A or B? Explain why.

b) Scale B shows this weight for a piece of cheese.
    What weight might scale A show for the same cheese?
    (There are two possibilities. Give both of them.)

    7.6 ounces

c) Scale A shows this weight for a tub of taramasalata.
    What weight might scale B show? (Again give both possibilities.)

    $7\frac{5}{8}$ ounces

IN YOUR HEAD

2   Write down only the answers.

a) Which weighs more, A or B?

A   B

b) You fill the glass
    from the bottle.
    Which holds more
    now, the glass
    or the bottle?

c) You need $1\frac{1}{2}$ kg of salt.
    Which two bags would you buy?

d) Which brush is widest?

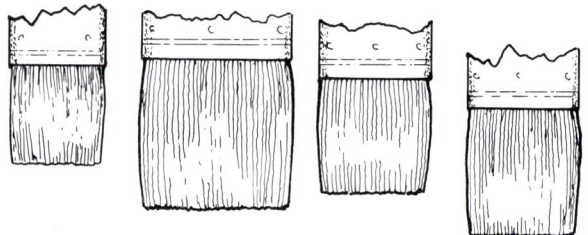

A $\frac{1}{2}''$ B 0.8''   C $\frac{3}{5}''$ D 0.75''   E $\frac{5}{8}''$

e) Your recipe says $1\frac{1}{2}$ litres of milk.
    Your milk bottles hold 0.8 litres.

    (i) How many bottles do you need?
    (ii) How much milk will you use
         from the last bottle?

f) Which is the largest: $\frac{2}{3}$, 0.5, or $\frac{5}{9}$?

3 Do not use a calculator.

a) Which is the higher rate of interest?

b) Which pack contains the longer panel pins?

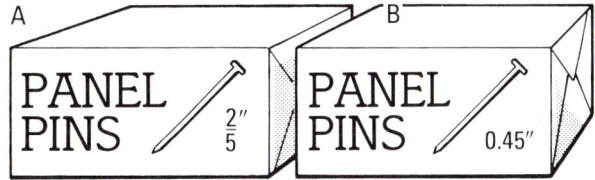

A **INTEREST** $9\frac{30}{8}\%$ YOUNG SAVER

B **INTEREST** **9.38%** 16+ SAVER

A PANEL PINS $\frac{2''}{5}$

B PANEL PINS 0.45''

c) Write the size of each paper clip as a decimal. Which paper clip is shortest?

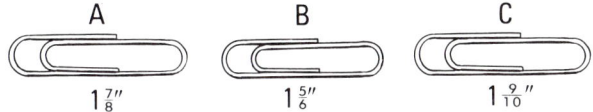

A $1\frac{7}{8}''$ B $1\frac{5}{6}''$ C $1\frac{9}{10}''$

4 Do not use a calculator.

a) Write $\frac{3}{5}$ as a fraction with denominator 25, like this: $\frac{\square}{25}$. Which is heavier, $\frac{14}{25}$ tonne or $\frac{3}{5}$ tonne?

b) Which is longer, $\frac{1}{8}$ km or $\frac{7}{40}$ km?

c) Which is faster, $62\frac{5}{16}$ km/h or $62\frac{11}{32}$ km/h?

5 Do not use a calculator.

A packet contains sixteen drill bits. List the drill bits in order of size, smallest diameter first.

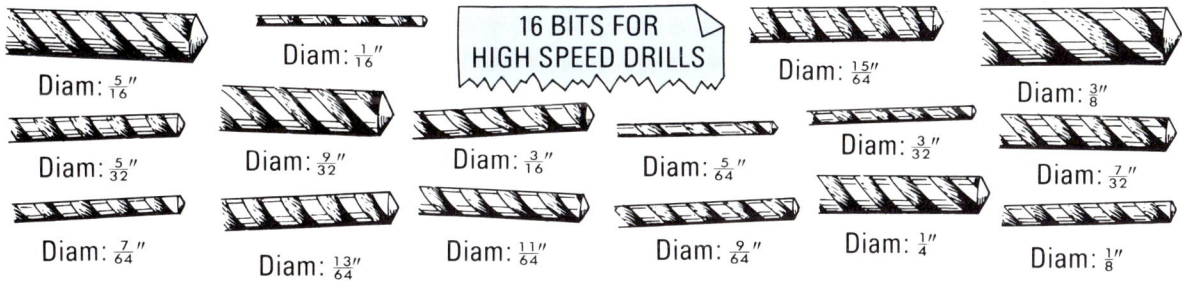

Diam: $\frac{5}{16}''$

Diam: $\frac{1}{16}''$

16 BITS FOR HIGH SPEED DRILLS

Diam: $\frac{15}{64}''$

Diam: $\frac{3}{8}''$

Diam: $\frac{5}{32}''$

Diam: $\frac{9}{32}''$

Diam: $\frac{3}{16}''$

Diam: $\frac{5}{64}''$

Diam: $\frac{3}{32}''$

Diam: $\frac{7}{32}''$

Diam: $\frac{7}{64}''$

Diam: $\frac{13}{64}''$

Diam: $\frac{11}{64}''$

Diam: $\frac{9}{64}''$

Diam: $\frac{1}{4}''$

Diam: $\frac{1}{8}''$

CHALLENGE

6 These are the diameters (in centimetres) of different copper pipes.

1 cm is about $\frac{2}{5}''$.

A plumber's price list has the approximate sizes in inches (as a fraction) as well as in centimetres.

Write an approximate size for each pipe like this.

0.5 cm
0.8 cm
1.0 cm
1.2 cm
2.5 cm
2.3 cm

Pipe sizes Metric and British		
Diameter cm	inches	
1.0	$\frac{2}{5}''$	

$1.0 \text{cm} \approx \frac{2}{5}''$
$0.5 \text{cm} \approx$
$0.8 \text{cm} \approx$

B3

7 a) Do not use a calculator. In each family there is a fraction which does not belong. Find it.

(i) $\left\{\frac{36}{90}, \frac{2}{5}, \frac{80}{200}, \frac{40}{110}, \frac{4}{10}, \ldots\right\}$ (ii) $\left\{\frac{6}{27}, \frac{10}{45}, \frac{14}{63}, \frac{16}{54}, \frac{20}{90}, \ldots\right\}$

b) Check your solutions with a calculator.

c) Write down seven fractions, one of which is an 'odd one out'. Ask your friend to find it.

8 0.23 is $\frac{23}{100}$ or $\frac{46}{200}$ or ...
Find three possible fractions for each of these. (You may use a calculator if you wish.)

a) 0.45   b) 0.842   c) 0.7777...

9 This family $\left\{\frac{4}{5}, \frac{8}{10}, \frac{12}{15}, \frac{16}{20}, \ldots\right\}$ is the 0.8 family. The member with the smallest (whole number) denominator is $\frac{4}{5}$. Find the member with the smallest denominator for these families:

a) 0.2   b) 0.72   c) 0.02

10 a) $\frac{40}{120}$ can be written as $\frac{1}{3}$. Explain why.

b) Write $\frac{260}{580}$ as a fraction with the smallest possible denominator.

c) Write each of these as fractions with the smallest possible denominator:

(i) $\frac{70}{135}$ (ii) $\frac{60}{400}$ (iii) $\frac{92}{180}$

CHALLENGE

11 Do not use a calculator.

a) Write $\frac{2}{3}$ and $\frac{11}{15}$ as fractions with the same denominator. Which is larger, $\frac{2}{3}$ or $\frac{11}{15}$?

b) Write $\frac{1}{3}$ and $\frac{1}{5}$ as fractions with the same denominator.
Find a fraction which is smaller than $\frac{1}{3}$ but larger than $\frac{1}{5}$.

c) Find two different fractions which lie between $\frac{1}{4}$ and $\frac{1}{3}$.

d) Find two different fractions which lie between 0.4 and 0.41.

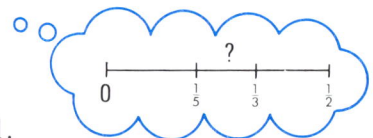

ENRICHMENT

CHALLENGE: DIGIT PATTERNS

1   Use your calculator.

a) Check that $\frac{1}{8}$, as a decimal fraction, is $\boxed{0.125}$

b) Check that for fractions with denominator 8 ($\frac{1}{8}$, $\frac{2}{8}$, $\frac{3}{8}$, …)
we get only these digit patterns after the '.'

$\boxed{.125}$

$\boxed{.25}$

$\boxed{.375}$

$\boxed{.5}$

$\boxed{.625}$

$\boxed{.75}$

$\boxed{.875}$

$\boxed{.}$

c) Find all the possible digit patterns after the '.' for fractions with denominator 4.

d) Try to explain why some of them are the same as the patterns for 'eighths'.

e) Find which digit patterns 5 and 15 share.

f) Try to explain why 5 and 15 share these digit patterns.

EXPLORATION: DIGIT PATTERNS

2   Explore the digit patterns that different denominators give.

For example, a denominator of 2 only gives $\boxed{.}$

and $\boxed{.5}$

a denominator of 4 only gives $\boxed{.}$

$\boxed{.25}$

$\boxed{.5}$

$\boxed{.75}$

a) Which denominator between 1 and 20 gives the largest number of digit patterns?

b) Which two denominators between 1 and 20 share the largest number of digit patterns?

B3

CORE

EXPLORATION

1   You need 1 cm squared paper.

a) Squares have been drawn on the
sides of this right-angled triangle.

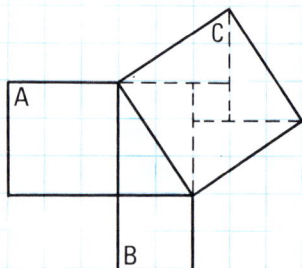

What is the area of:

(i) square A
(ii) square B
(iii) square C? (The square has been
broken up to help you.)

b) Investigate some more right-angled
triangles. Make a table of your results
like this. (Square C is always the largest.)

Area (cm²) of		
Square A	Square B	Square C
9	4	

c) Try the experiment for some triangles
without right angles. Is there a connection
between the areas for these triangles?

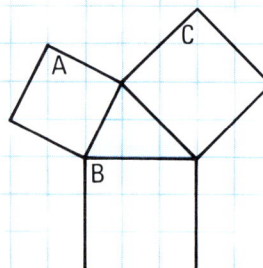

THINK IT THROUGH

2   Use what you discovered in the *Exploration* to find the area of
the square drawn on the longest side of this triangle.

How long are the sides of the square? (Use the √ key
on your calculator. Write your result to the nearest
millimetre.)

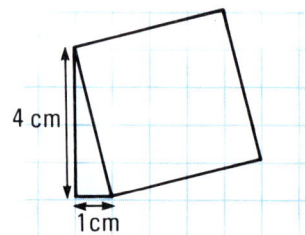

4 cm

1 cm

▓▓▓ TAKE NOTE ▓▓▓

For *right-angled triangles*, the
area of the largest square = the
total area of the other two
squares.

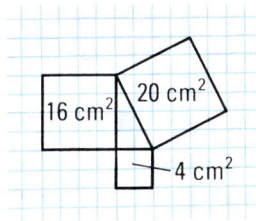

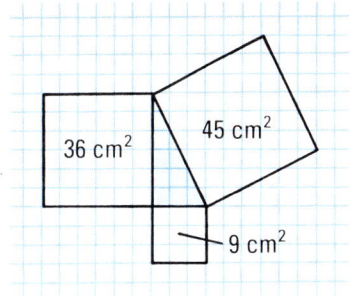

The Greek mathematician Pythagoras (about 580–500 BC) learned about this result during his
travels in Babylonia. It is called *Pythagoras's rule* or *Pythagoras's theorem*.

▓▓▓▓▓▓▓▓▓▓▓▓▓▓▓▓▓▓▓▓▓▓▓▓▓▓▓

3   Sketch the diagram.

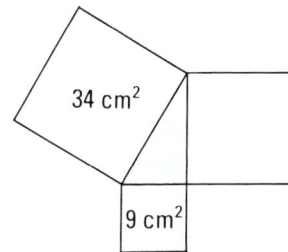

a) What is the area of the third square?

b) How long is each side of each square? (Write the length of
the longest side using '$\sqrt{}$', for example $\sqrt{30}$, $\sqrt{26}$.)

4   Make sketches like these, using your own lengths and areas.
Make at least two sketches.

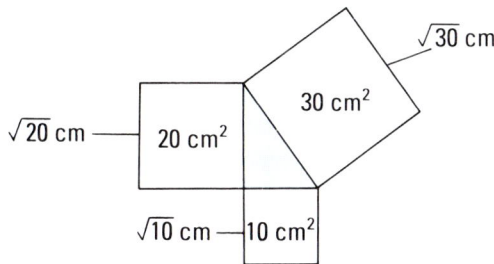

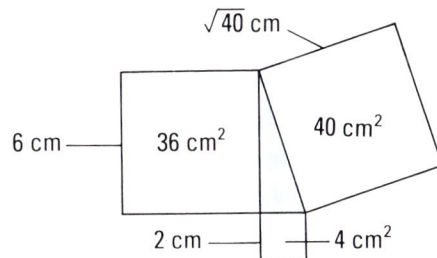

5   Copy and complete each sketch. If distances are not whole numbers, write them using '$\sqrt{}$'.

a)

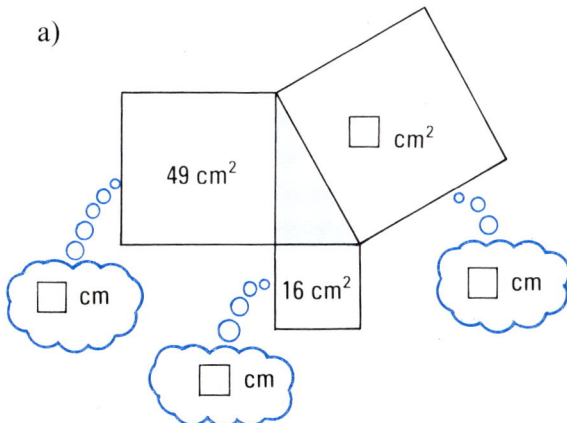

b)

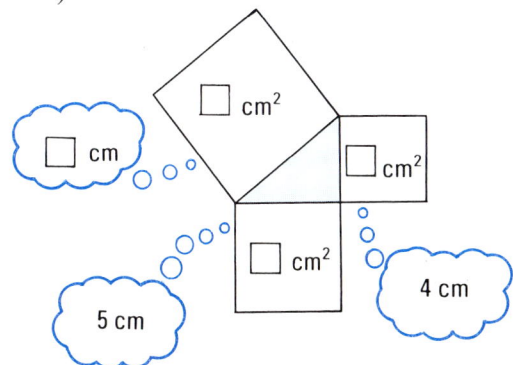

B4

6 Copy and complete each sketch. If distances are not whole numbers, write them using '$\sqrt{}$'.

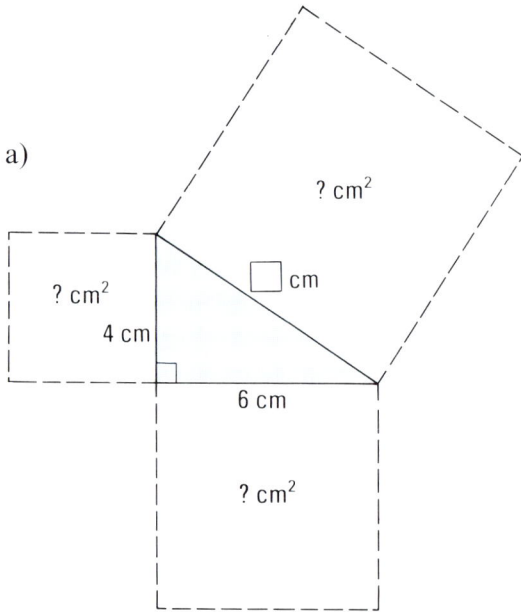

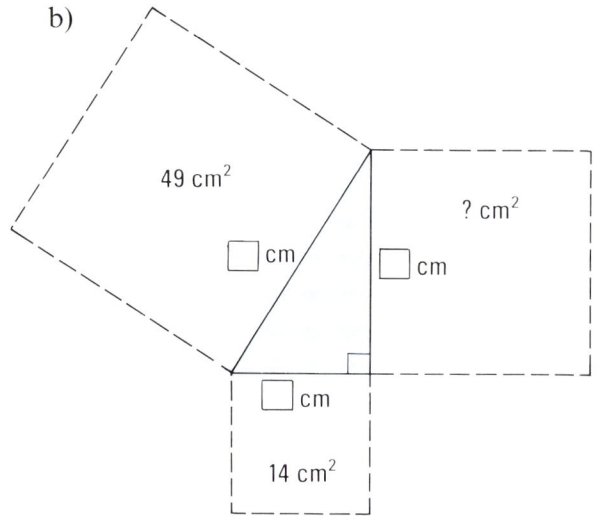

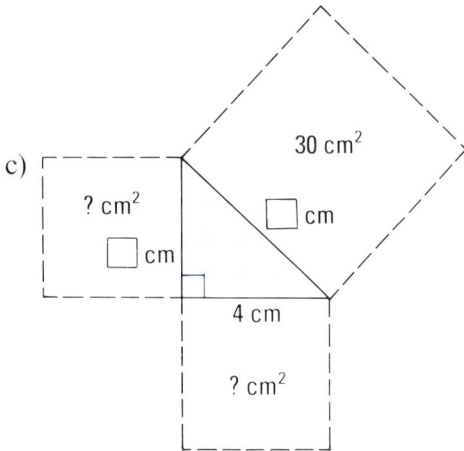

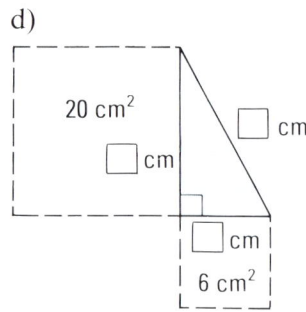

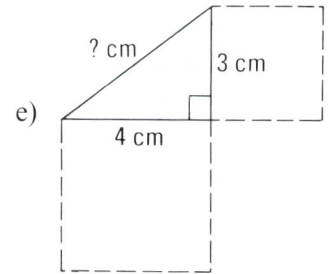

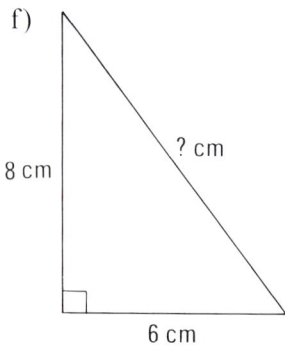

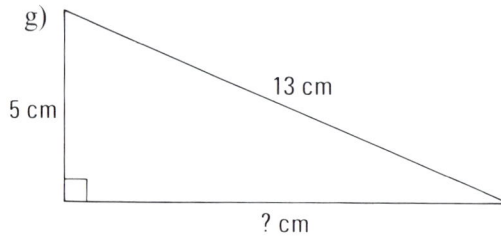

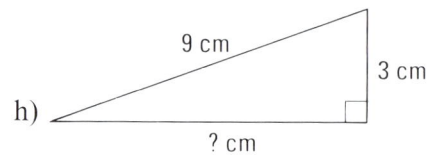

a)

? cm^2

$\square$ cm

? cm^2

4 cm

6 cm

? cm^2

b)

49 cm^2

$\square$ cm

$\square$ cm

? cm^2

$\square$ cm

14 cm^2

c)

30 cm^2

? cm^2

$\square$ cm

$\square$ cm

4 cm

? cm^2

d)

20 cm^2

$\square$ cm

$\square$ cm

$\square$ cm

6 cm^2

e)

? cm

3 cm

4 cm

f)

8 cm

? cm

6 cm

g)

5 cm

13 cm

? cm

h)

9 cm

3 cm

? cm

Right-angled triangles

EXPLORATION

1   We know that the blue triangle
(sides 3 cm, 4 cm, 5 cm) has a right angle because
the area of X + the area of Y = the area of Z.

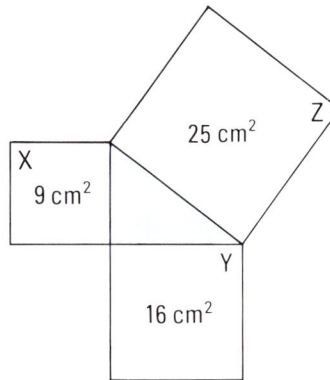

25 cm²   Z

X

9 cm²

Y

16 cm²

a) Which of these triangles have a right angle?

    A   sides: 1 cm, 2 cm, 3 cm.   B   sides: 6 cm, 8 cm, 10 cm.
    C   sides: 14 cm, 48 cm, 50 cm.   D   sides: 12 cm, 14 cm, 20 cm.

b) Check that these four triangles each have a right angle.

Lengths of sides (cm)		
3	4	5
5	12	13
7	24	25
9	40	41

    Notice that:

    ● the length of the shortest side is an odd number of
       centimetres;
    ● the other two sides differ in length by 1 cm.

    Find four more examples which follow the same rules.

c) Find four examples of right-angled triangles which do not fit the pattern in b).

d) Check that this is a right-angled triangle:

    sides: 6 cm, 5 cm, $\sqrt{11}$ cm   (*Remember:* $\sqrt{11} \times \sqrt{11} = 11$.)

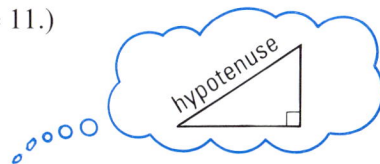

hypotenuse

    Find four more right-angled triangles whose hypotenuse is 6 cm long.

CHALLENGES

2 a) Midge is making shelf brackets.
She is using metal strips like this.

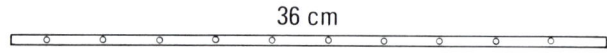

36 cm

She bends each strip into a right-angled triangle and
then welds the ends.

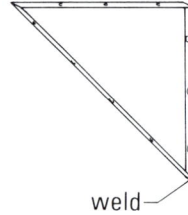

Sketch a bracket which
Midge might make.
Mark the lengths of its
sides on your sketch.

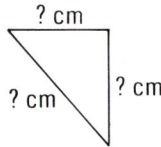

? cm

? cm   ? cm

weld

b) This is a bracket Midge made from a
60 cm strip of metal. Find the length of
the part which fits against the wall.

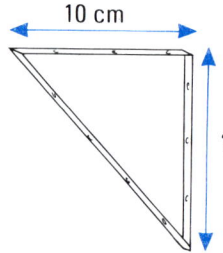

10 cm

?

c) Midge makes another bracket from a
60 cm strip of metal. All three sides of the
bracket are more than 10 cm long. Sketch
what Midge's bracket might look like.
Mark the lengths of its sides on your sketch.

d) Here are some more of
Midge's brackets.
Find the length marked '?'
for each one, to the
nearest centimetre.

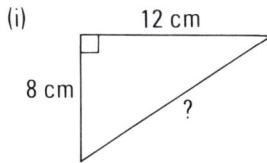

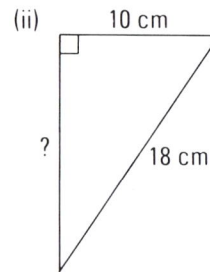

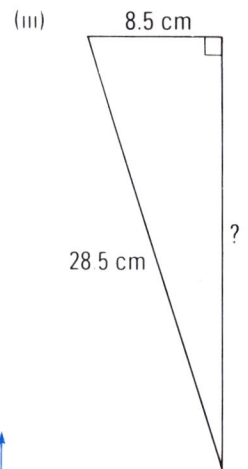

(i)   12 cm

8 cm

?

(ii)   10 cm

?   18 cm

(iii)   8.5 cm

?

28 5 cm

3 Here is the net for the
'triangular' box.

How wide is the edge
marked '?' on the net?
Write the length to the
nearest centimetre.

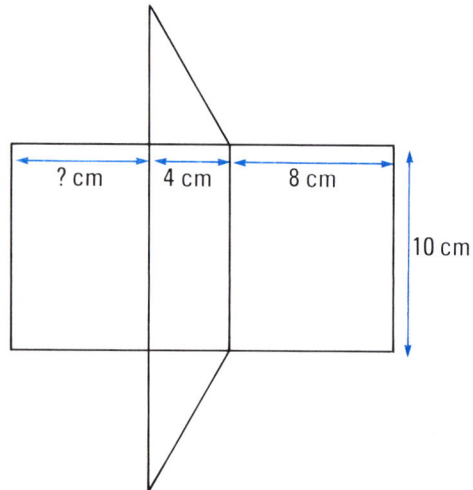

? cm   4 cm   8 cm

10 cm

OBELTONE

Using Pythagoras

1   Use Pythagoras's rule to calculate
the length of the diagonal of:

a) the rectangle
b) the square.

Write your results to the nearest centimetre.

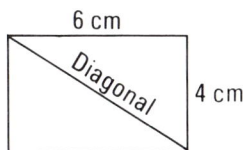

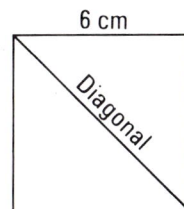

2   ABCD is a kite.

a) What is the length of AB?

b) What is the length of BC,
to the nearest $\frac{1}{2}$ cm?

c) What is the perimeter of the kite,
to the nearest centimetre?

d) Sketch a kite whose perimeter is 30 cm.

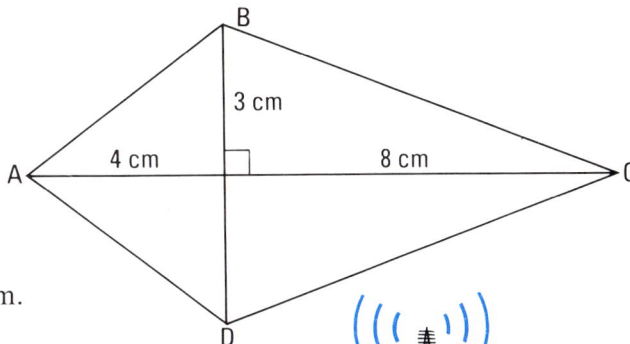

3   This radio mast has to be held vertical by four cables. Each
cable will be fixed to the mast, at a height of 36 m above the
ground. Each cable will be anchored to the ground at a
distance of 15 m from the foot of the mast.
Use Pythagoras's rule to calculate how long each cable must
be.

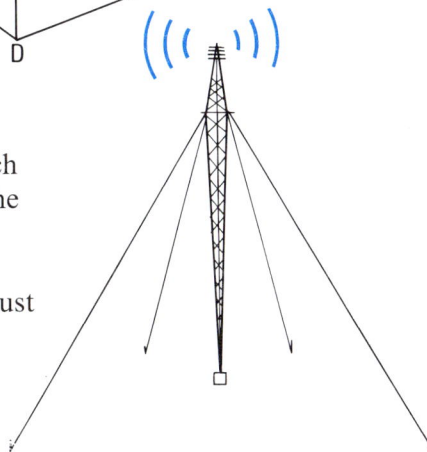

EXPLORATION

4   These are three semicircles drawn on a right-angled
triangle. Does Pythagoras's rule work
for semicircles? (Is Area A = Area B + Area C?)

Investigate other shapes on the three sides
(for example, equilateral triangles).

Write a report about what you discover.

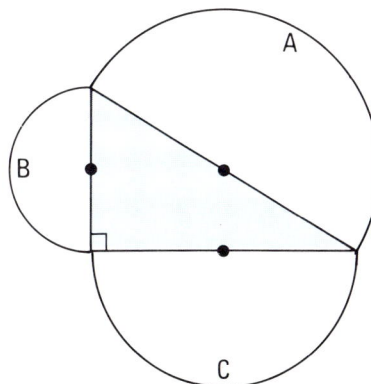

CORE

1 A gardener uses Pythagoras's rule to calculate the length of fence he needs to fit diagonally across his garden. This is what the calculator shows (part of the display is covered up):

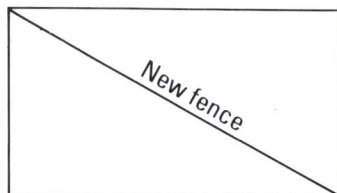

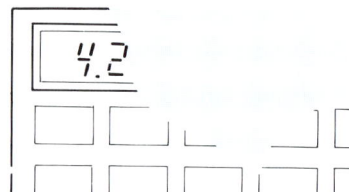

a) Is the length of the new fence nearer to 4 m or to 5 m?

b) How much more of the display would you need to see to decide if the length is nearer 4.2 m or nearer 4.3 m? Explain why.

c) Here are three different possibilities for the next digit in the display:

A `4.24` B `4.25` C `4.25`

   (i) Suppose A shows the correct display.
       Is the length of the new fence nearer to 4.2 m or nearer to 4.3 m?
   (ii) Suppose B is the correct display; now answer part (i).
   (iii) Suppose C is the correct display; now answer part (i).

d) The correct display in part c) is A. Here are three different possibilities for the third digit in the display:

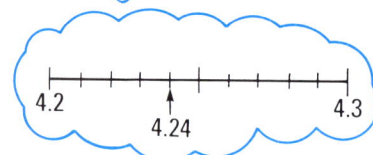

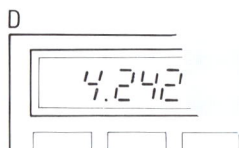

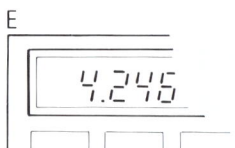

(number line: 4.2 ———— 4.24 ———— 4.3)

D `4.242` E `4.246` F `4.245`

   (i) Suppose D shows the correct display. Is the length of the new fence nearer to 4.24 m or nearer to 4.25 m?
   (ii) Answer (i) if E is the correct display.
   (iii) Answer (i) if F is the correct display.

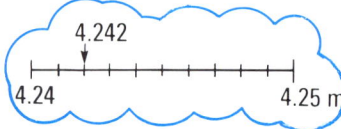

(number line: 4.24 —— 4.242 ———— 4.25 m)

e) The gardener calculates the circumference of the circular pond as 12.566371 m. Is this:

   (i) nearer to 12.5 m or 12.6 m
   (ii) nearer to 12.56 m or 12.57 m?

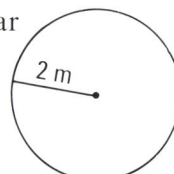

(circle with radius labelled 2 m)

━━━━━━━━ TAKE NOTE ━━━━━━━━

This is the calculator reading for the length of the new fence in question 1.   `4.2426407`

We say that 4.2426407 is:

4.2   rounded to 1 decimal place (1 DP)
       (4.24 is nearer to 4.2 than to 4.3)

4.24   rounded to 2 decimal places (2 DP)
       (4.242 is nearer to 4.24 than to 4.25)

4.243 rounded to 3 decimal places (3 DP)
       (4.2426 is nearer to 4.243 than to 4.242)

For 'halfway' readings we round up, for example,
4.25 is 4.3 (correct to 1 DP).

---

2   The weight of a bar of gold is measured accurately to 4 decimal places as 0.8456 kg. Write the weight correct to (i) 1 decimal place, (ii) 2 decimal places, (iii) 3 decimal places.

3   The average bee humming bird weighs about 0.0016 kg.

   a) Scientists studying bee humming birds do not record their results in kilograms rounded to 2 DP. Why not?

   b) With what accuracy do you think they record their results?

4   Use Pythagoras's rule to find the length of the diagonal of the square room. Write your result correct:

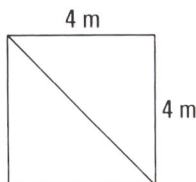

   a) to the nearest metre   b) to the nearest 10 cm
   c) to the nearest centimetre   d) in metres to 1 DP
   e) in metres to 2 DP.

5   Barnard's Star is about 56 573 960 000 000 km away from Earth. Astronomers do not record distances in kilometres rounded to 2 DP. Why not?

6   Round these numbers correct to a) 1 DP, b) 2 DP.

   (i) 7.7894   (ii) 0.60034   (iii) 1.1784   (iv) 5.5555
   (v) 1.00055   (vi) 3.9999

7   a) The heights of bridges are normally written with one figure after the decimal point. Do you think the height has been rounded to 1 DP? Why?

   b) Sign posts on country walks are often written with 1 figure after the decimal point. Do you think the distance has been rounded to 1 DP? Why?

8   The length of a section of railway track is given as 2.1 m (1 DP). What do you think is:

   a) the greatest length the section might actually be   b) the shortest length the section might be?

Significant figures

1 There are thousands of people at the carnival. These four cards hide the exact number.

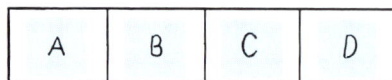

a) You are allowed to remove one card. Which one should you pick up to get the best idea of how many people there are?

b) Now you want an even better idea of how many people there are. Which card should you pick up next?

c) ... and next?

d) ... and next?

▓▓▓▓▓▓▓▓▓▓▓▓ TAKE NOTE ▓▓▓▓▓▓▓▓▓▓▓▓

There were | 6 | 4 | 7 | 3 | **people at the Carnival.**

The 6 is called the *most significant figure*. Of all the four numbers this one gives us the best idea of how many people there are at the Carnival.

Fourth most significant figure.
Third most significant figure.
Second most significant figure.

2 This is the length in centimetres of the virus in the photograph. The three cards hide part of the number. Card A hides the first figure that is *not* zero.

a) You are allowed to remove one card. Which one should you pick up to get the best idea of the length of the virus?

b) Now you want an even better idea. Which one should you pick up next?

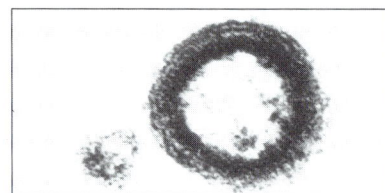

c) ... and next?

▓▓▓▓▓▓▓▓▓▓▓▓ TAKE NOTE ▓▓▓▓▓▓▓▓▓▓▓▓

The length of a virus is | 0.000702 | **centimetres.**

The 7 is called the *most significant figure*. Of the figures which follow the beginning zeros this one gives us the best idea about the length of the virus.

The 2 is the third most significant figure.
This 0 is the second most significant figure.

3 a) Which is the most significant figure in each of these numbers?

(i) 17064   (ii) 0.4176   (iii) 10.47   (iv) 0.00340

b) Which is the third most significant figure in each number?

4   We can use the idea of significant figures for rounding. To help us we give each figure a name (SF = Significant Figure):

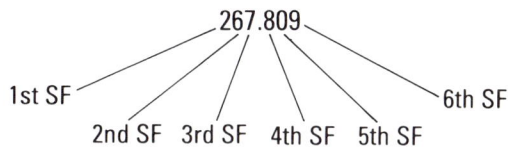

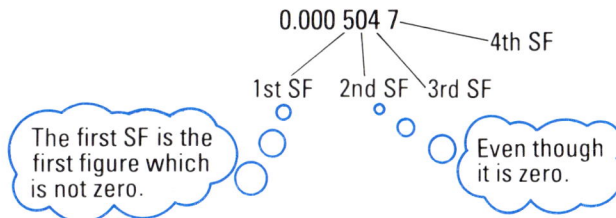

267.809

1st SF   2nd SF   3rd SF   4th SF   5th SF   6th SF

0.000 504 7

1st SF   2nd SF   3rd SF   4th SF

The first SF is the first figure which is not zero.

Even though it is zero.

We say that 267.809 rounded to 2 significant figures (we write 2 SF) is 270.

a) Copy and complete each of these:

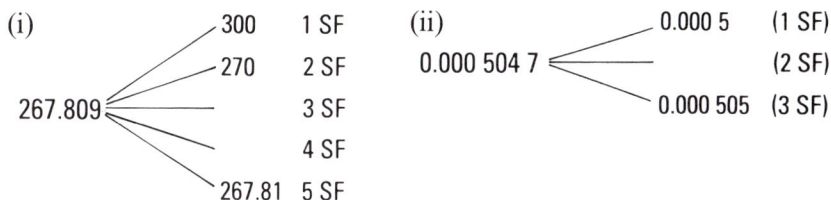

(i) 267.809

300   1 SF
270   2 SF
—   3 SF
—   4 SF
267.81   5 SF

(ii) 0.000 504 7

0.000 5   (1 SF)
—   (2 SF)
0.000 505   (3 SF)

b) Write down the first significant figure in each of these (i) 0.047 3   (ii) 6001   (iii) 50.50.

5   Do each of these calculations. Write each result correct to (i) 1 SF (ii) 2 SF.

a) 3.76×0.04 b) $0.83 \div 17$ c) $6.71 + 14.06793$

6   a) The heights of five basketball players are 2.15 m, 1.73 m, 1.97 m, 2.08 m, and 1.93 m. Find their mean height correct to 1 SF.

b) Find $\sqrt{3}$ correct to 2 SF.

IN YOUR HEAD

7   Estimate the result of each of these by first rounding each number to 1 SF.

a) 3.7×0.51 b) $8.3 \div 1.7$ c) $\dfrac{516 \times 2.3}{98}$

Check your estimates with your calculator.

8   Use rounding to 1 SF to estimate each of these:

a) The cost of 18 pens at £2.16 each.

b) The average price of each egg in a box of twelve. Total price £1.73.

c) £7.64 + £113.72 + £57.60.

d) 398.17 kg − 73.8 kg.

9   Which of these calculations must be incorrect?

A $72.16 \times 13.7 = 9885.92$
B $0.02 \times 14.76 = 29.52$
C $890.18 \div 0.47 = 1894$

B5

ENRICHMENT

Keeping track of error

1   The measurements given for the bedroom are correct to 1 DP.

a) Copy and complete this sentence to show the greatest and least possible length of the bedroom.

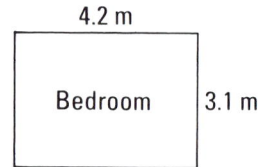

$$\square \text{ m} \leqslant \text{Length} < 4.25 \text{ m}$$

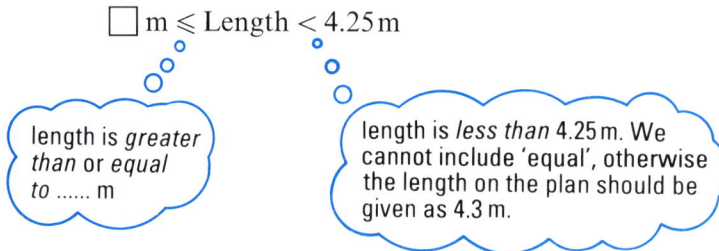

> length is *greater than* or *equal to* ...... m

> length is *less than* 4.25 m. We cannot include 'equal', otherwise the length on the plan should be given as 4.3 m.

b) Write a sentence similar to the one in a) for the width of the bedroom.

c) The greatest possible perimeter of the bedroom must be less than
4.25 m + 3.15 m + 4.25 m + 3.15 m.
So, perimeter < 14.8 m.
Copy and complete: Perimeter $\geqslant \square$

d) Copy and complete: $\square \leqslant$ Area of the bedroom $< \square$

2   For each of these, copy and complete the sentences to give the whole range of possible values.

a) Time for 1500 m race = 218.75 s (2 DP)
$\square$ s $\leqslant$ Race time $< \square$ s

b) Population of the UK = 60 000 000 (1 SF)
$\square \leqslant$ UK population $\leqslant \square$ (*Careful!* Both signs are $\leqslant$.)

c) Number of telephones in USSR in 1982 = 23 707 000 (to 5 SF)
$\square \leqslant$ Number of phones in USSR in 1982 $\leqslant \square$

d) Mass of a bee humming bird = 0.001 58 kg (5 DP)
$\square$ kg $\leqslant$ Mass of bee humming bird $< \square$ kg

3   a) The length of side of a cube of marble is measured as 2 m 40 cm to the nearest 10 cm. Calculate (i) the greatest, (ii) the least volume the cube might have.

b) What is the difference between the greatest and least volume in a)?

c) The cost of the marble is £300 per m³. What is the most you might be overcharged if the price is calculated using the 2 m 40 cm approximation?

4   You calculate the circumference of your bicycle wheel as 2.7 m (to 1 DP). You want to know how many times the wheel turns in a 100 km journey. What is the greatest error you could make by basing your calculation on the 2.7 m approximation for the circumference of the wheel?

B5

B6 AREAS OF SHAPES

REVIEW

The area of the rectangle = length × width

■ The area of the parallelogram = ... × ...

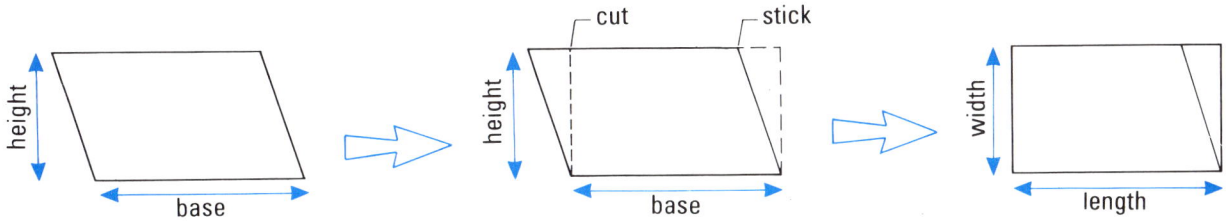

CORE

1   Copy and complete the final sentence of the *Review*.

2   Find the area of each parallelogram. Measure any distances you need.

a)

b)

3   The area (A cm²) of the rectangle is given by
$A = p \times t$

Write an expression for the area of the parallelogram.

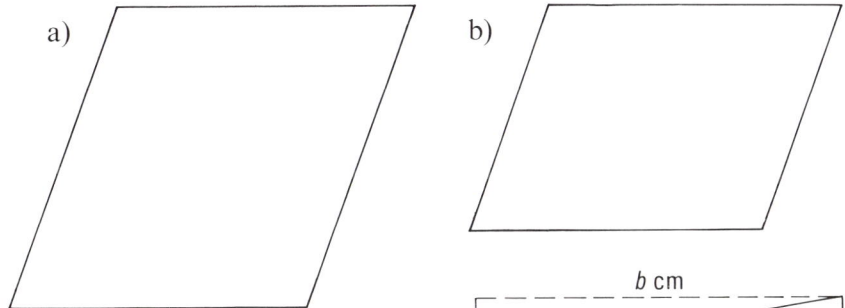

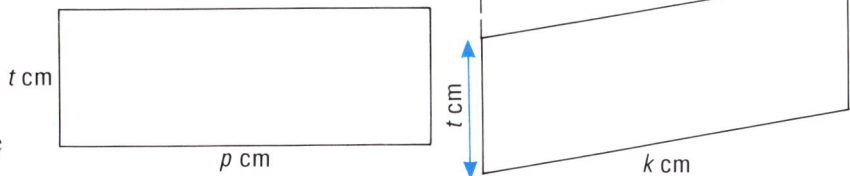

CHALLENGE

4   These bathroom tiles are used to cover an area 1.2 m wide, 0.8 m deep.

How many tiles and how many half-tiles are needed, if they are used:
(i) this way up   (ii) this way up?

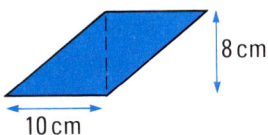

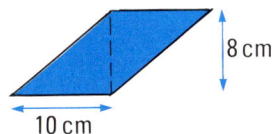

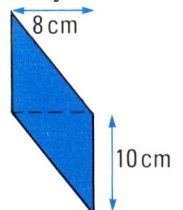

5 a) Copy the parallelogram onto squared paper.

   b) These are trapeziums (each has at least one pair of parallel sides).

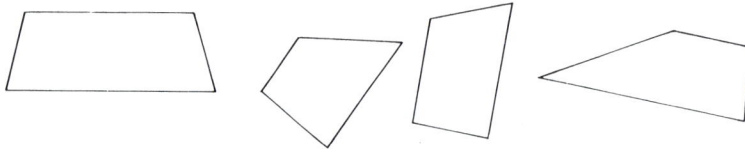

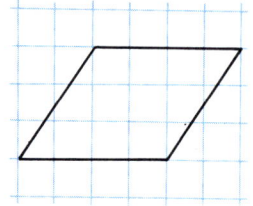

   Draw a line across your parallelogram to divide it into two
   identical trapeziums. What is the area of:

   (i) the parallelogram   (ii) each trapezium?

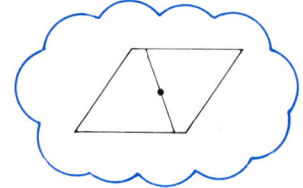

6 The parallelogram has been divided
   into two identical trapeziums.

   What is the area of:

   a) the parallelogram
   b) each trapezium?

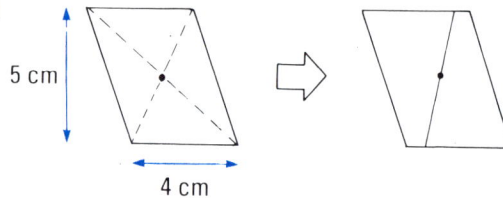

7 Use squared paper. Draw a parallelogram and halve it to make a trapezium whose area is 12cm^2.

8 a) Copy each of the drawings A and B. (A freehand sketch will do.)
       Mark in the missing lengths on each drawing.

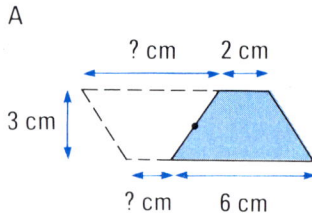

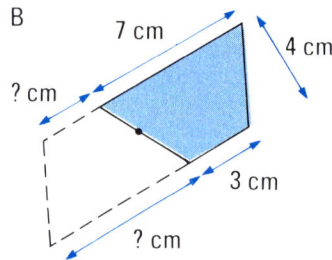

   A

   B

   b) Write down the area of (i) each full parallelogram, (ii) each shaded trapezium.

9 Find the area of each trapezium.

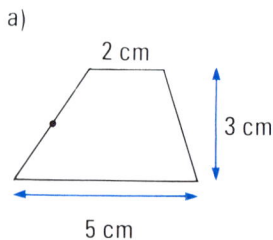

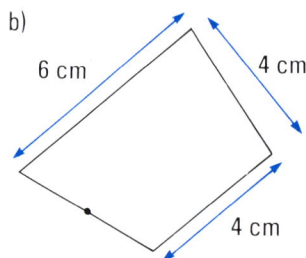

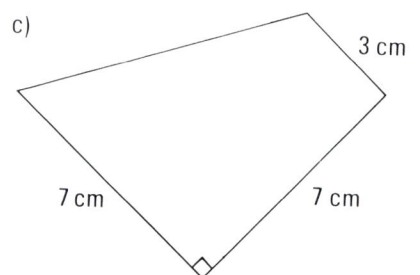

   a)   b)   c)

10 Make a sketch of each trapezium. Write in dimensions of your own so that:

a) the area of A is 26 cm².
b) the area of B is 36 cm².

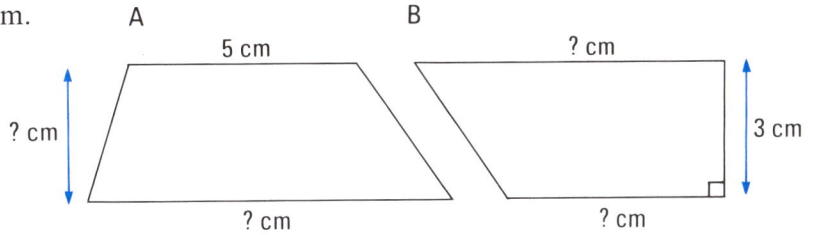

A
5 cm
? cm

B
? cm
3 cm
? cm
? cm

11 a) One of these is correct for the area of parallelogram A. Which is it?

- Area = $(b-a) \times h$ cm²
- Area = $(b+a) \times h$ cm²

b) Check that the rule you chose is correct for parallelogram B.

c) One of these is correct for the area of the shaded trapezium in A. Which is it?

- Area = $\frac{1}{2} \times h \times (a+b)$ cm²
- Area = $2 \times h \times (a+b)$ cm²

d) Check that the rule you chose in c) is correct for each of these trapeziums.

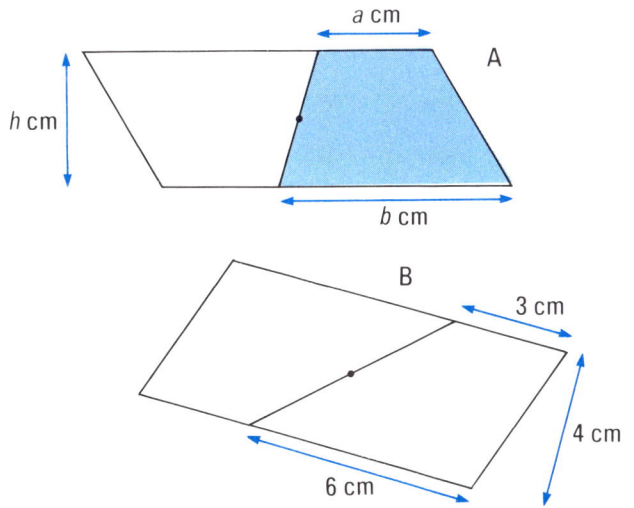

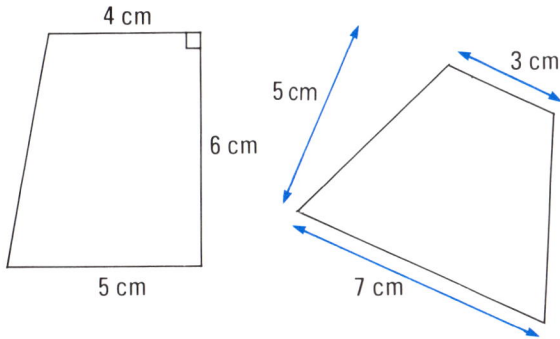

a cm
A
h cm
b cm

B
3 cm
4 cm
6 cm

4 cm
5 cm
6 cm
5 cm

3 cm
7 cm

12 This is a folded parallelogram. What is the area of:

a) trapezium BCDE
b) trapezium ACDE?

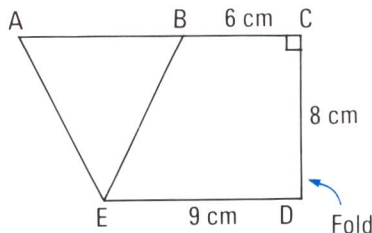

A   B   6 cm   C
8 cm
E   9 cm   D   Fold

TAKE NOTE

The area of the trapezium is $\frac{1}{2}(a+b)h\,\mathrm{cm}^2$ (one half the sum of the two parallel sides multiplied by the height).

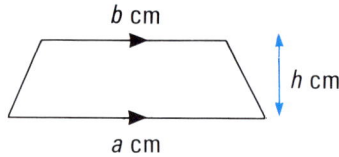

13 Calculate the area of each trapezium.

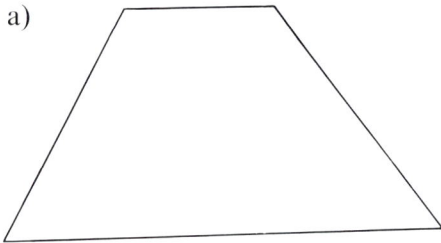

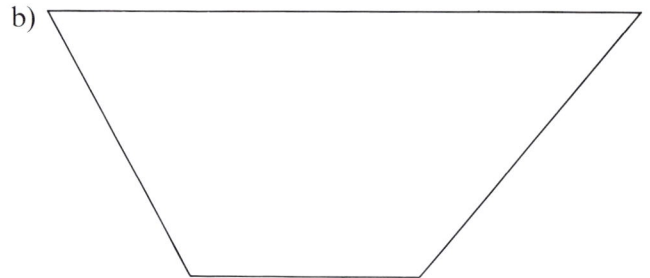

a)

b)

14 a) Calculate the area of the side of the shed.

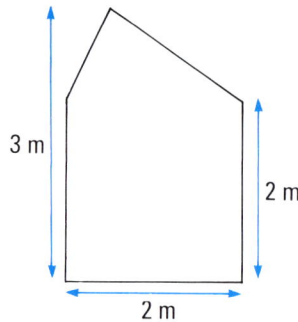

3 m

2 m

2 m

b) Use squared paper. Sketch a shed with a side view shaped like this, of area $12\,\mathrm{m}^2$. Mark in the height of the front and back of the shed, and its width.

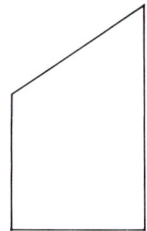

15 A, B, and C are three vertices of a trapezium of area $12\,\mathrm{cm}^2$. What are the coordinates of the fourth vertex, D:

a) if AC and BD are parallel
b) if AB and CD are parallel?

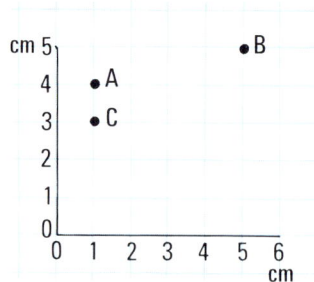

16 The drawing on the right shows the size of each two-tone floor tile.

a) Find the area of each tile.

b) Each tile is cut from a $0.4\,\mathrm{m} \times 0.4\,\mathrm{m}$ square. What area is wasted?

c) Approximately how many tiles are needed for a rectangular kitchen measuring 4 m by 3.5 m?

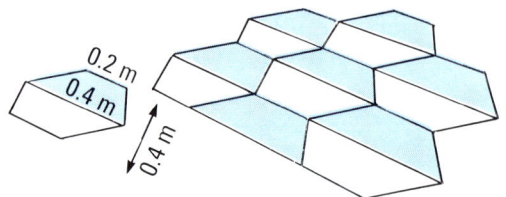

ENRICHMENT

CHALLENGE

1   This is a cross-sectional view of a river bed.

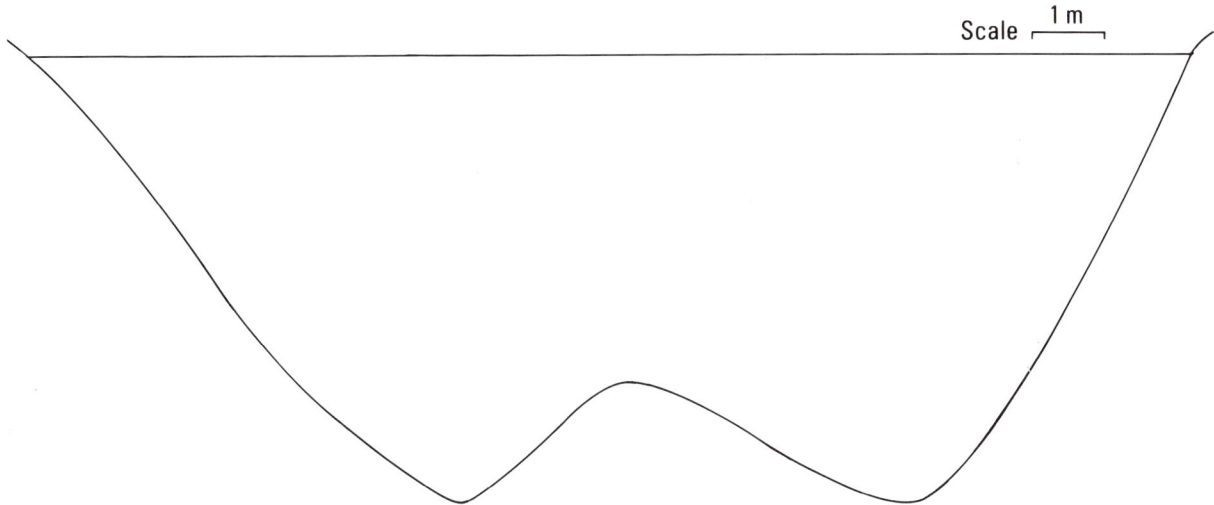

Scale ⊢—1 m—⊣

a) Measure with a ruler to find the width of the river to the nearest metre.

b) What is the greatest depth of the river?

c) Here is the cross-section divided into trapeziums and triangles:

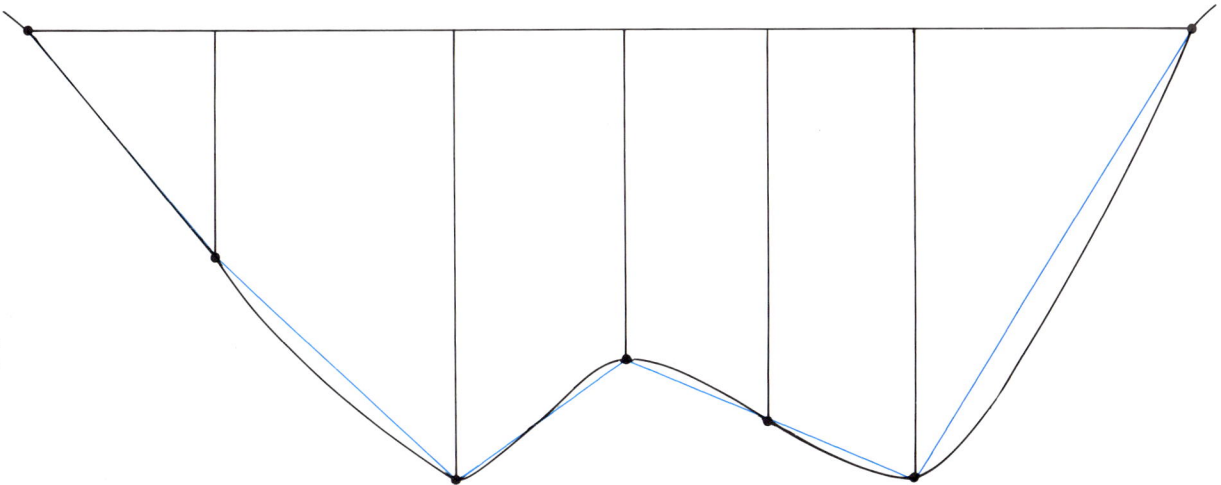

Use the drawing to estimate the area of the cross-section.

Remember that: Area of a triangle $= \frac{1}{2} \times$ base $\times$ height
Area of a trapezium $= \frac{1}{2} \times h \times (a + b)$

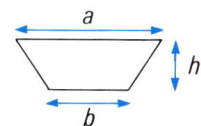

d) How many cubic metres of water flow past this part of the river every hour when the rate of flow is 2 m/s?

B6

CORE

(You will find all the information you need for questions 1–16 on this page.)

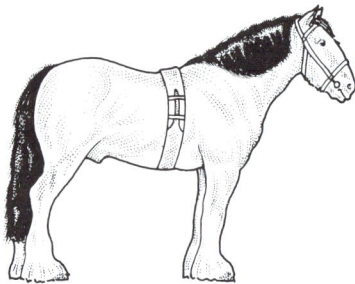

About 1 tonne
About 1 ton

N m p

SUGAR

1 kg   2.2 lb

ounce

2.2 pounds

1 oz ≈ 28 g

3 cm

1 inch
(1″)

1″ ≈ 2.5 cm

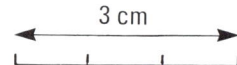

Lemonade

1 l
1.76 pts

1 pt ≈ $\frac{3}{5}$ l

1 gall

4.5 l

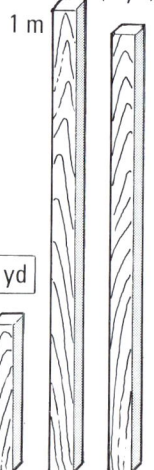

1 yard
(1 yd)

1 m

1 m ≈ 1 yd

1 foot
(1 ft)
(1′)

1 ft ≈ $\frac{1}{3}$ m

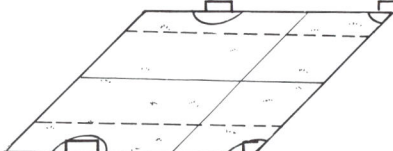

1 hectare ......
about 1$\frac{1}{2}$ hockey pitches

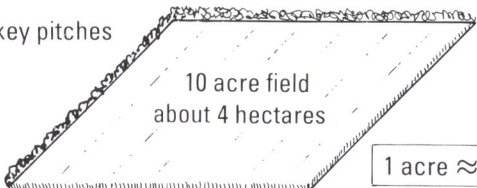

10 acre field
about 4 hectares

1 acre ≈ 0.4 hectare

1   Which bag holds more charcoal?

CHARCOAL

4 kg

CHARCOAL

8 lb

2   Who is
heavier –
Imran or
Alan?

1 STONE = 14 lb

3   Which of the 'metric' pipes in the table is nearest in size to the pipe in the drawing?

Copper pipes Metric sizes						
Internal diam. (cm)	1.0	1.0	1.2	1.2	1.5	1.5
External diam. (cm)	1.4	1.5	1.6	1.8	1.8	2.0

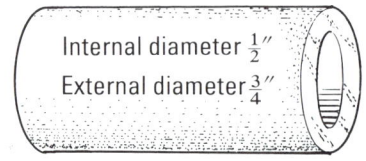

Internal diameter $\frac{1}{2}$″
External diameter $\frac{3}{4}$″

4   Approximately, how many fully grown Shire horses will balance the truck?

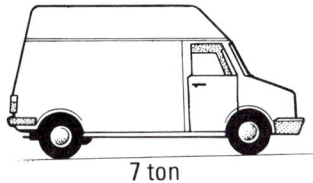

7 ton

5   Which balls of wool are heavier?

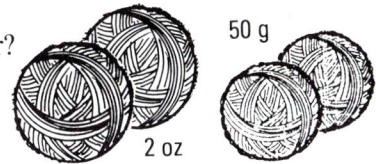

50 g

2 oz

6   Jane is travelling at about 45 mph. How many km/h is this?

7   Which car has the larger petrol tank?

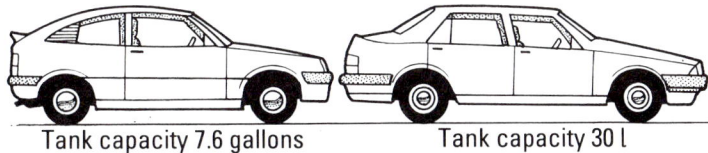

Tank capacity 7.6 gallons   Tank capacity 30 l

8   Jacques is travelling at 100 km/h on a UK motorway. Is he breaking the speed limit?

30 mph ≈ 50 km/h

70

9   Approximately, what do you weigh:

a) in kilograms   b) in pounds   c) in stones and pounds?

10 Roughly, how tall are you: a) in centimetres, b) in metres and centimetres, c) in feet and inches?

11 Roughly, how far away from school do you live:

a) in miles   b) in kilometres?

BARNSLEY
12 miles 19 km

1 km ≈ 0.6 miles
1 mile ≈ 1.6 km

12 You need some $1\frac{3}{4}''$ nails, but all the sizes in the shop are metric. Which of these should you buy?

4 cm   4.5 cm   3 cm   3.5 cm

13 You need 5 yd² of material to make some curtains. The shop sells the material by the square metre. It has 5 m² in stock. Is this sufficient?

14 Ron can buy two 1 pint bottles of cider for £0.80 each or one 1 litre bottle for £0.80. Which is the better buy?

15 Roughly, how many pounds of tomatoes are there in this can:

A $\frac{1}{4}$lb   B $\frac{1}{2}$lb   C $\frac{3}{4}$lb   D 1lb?

Tomatoes
e 114g

16 Which farm is larger, Home Farm or Burnside Farm?

Home Farm
200 acres

Burnside Farm
110 hectares

B7

17 Visit your local **DIY** store to find the price of wood, concrete, paint, etc., or you may be able to find prices in your local newspaper.
Estimate how much it would cost to build a fence to surround this paddock.

The sizes of wood you need are given below. Write out a full estimate for the fence. Include the cost of all the materials you use, and don't forget to include the cost of labour.

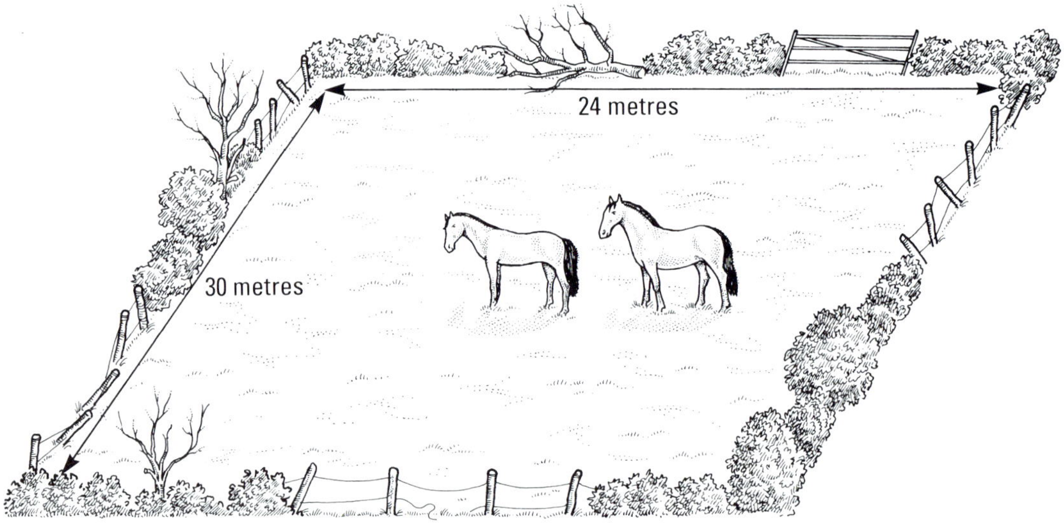

24 metres

30 metres

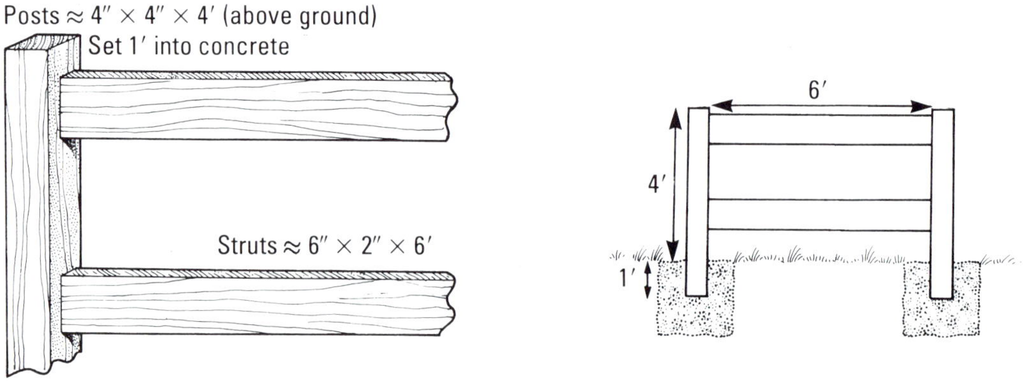

Posts ≈ 4″ × 4″ × 4′ (above ground)
Set 1′ into concrete

Struts ≈ 6″ × 2″ × 6′

6′

4′

1′

B7

CORE

1 a) A is a hexagon. It has 6 sides and 6 interior angles. C is an 'exploded' hexagon. It is made up of 4 triangles. Use the exploded hexagon to help you to explain why the interior angles of a hexagon add up to $4 \times 180°$.

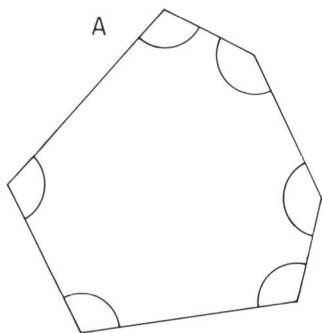

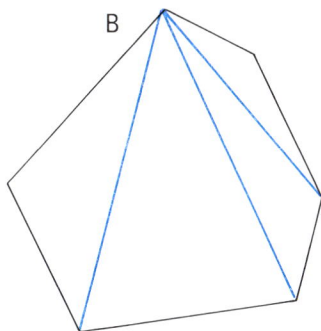

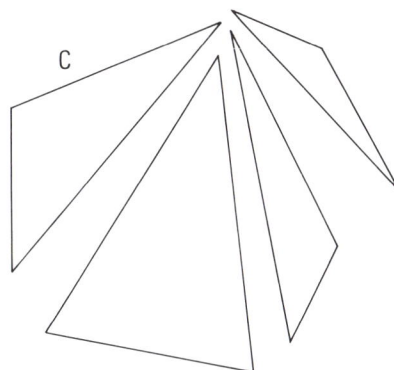

A   B   C

b) This is a pentagon. Use the exploded diagram to explain why the interior angles of a pentagon add up to $3 \times 180°$.

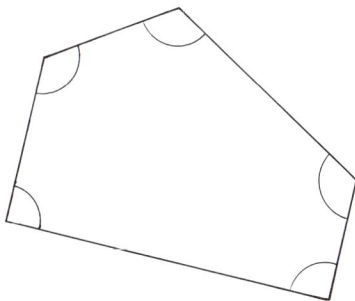

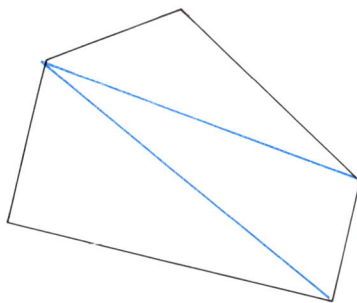

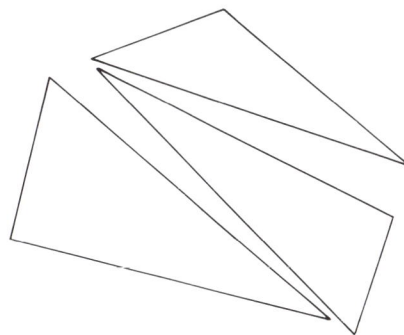

c) The *angle sum* of a hexagon is $4 \times 180° = 720°$.
The *angle sum* of a pentagon is $3 \times 180° = 540°$.
What is the angle sum of (i) an octagon, (ii) a 10-gon?

d) Find a general rule which tells you the angle sum of any polygon.

e) Use the result in c).
Check that each interior angle of
a regular pentagon measures $108°$.

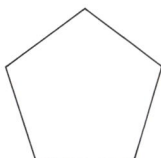

f) Find the size of each interior angle of (i) a regular octagon, (ii) a regular 10-gon.

2   Here are three ways of making a regular hexagon from triangles.

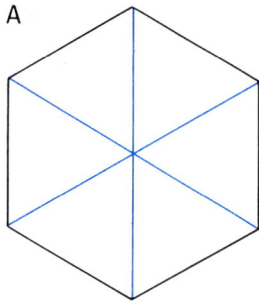

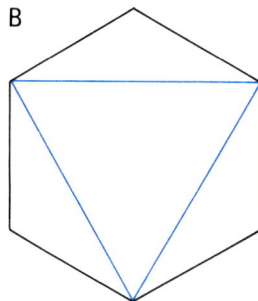

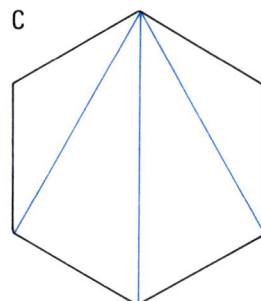

a) Sketch the different types of triangles used to make each shape (that is, one triangle for A, 2 triangles for B, 2 triangles for C).

b) Calculate the size of each angle in each triangle. Write the angle sizes on your sketches.

3   a) Repeat question 2 for these regular octagons ...

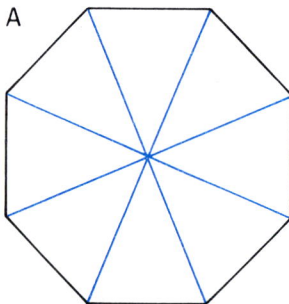

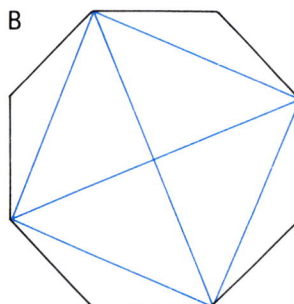

b) and for these regular pentagons ...

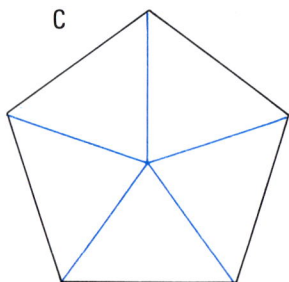

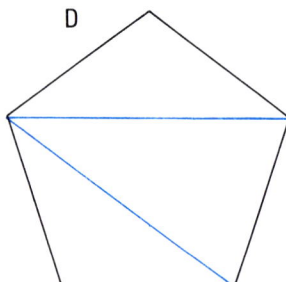

c) and for this regular 10-gon.

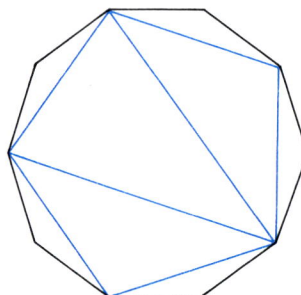

4 a) Follow these instructions.

> Move forward 3 cm.
> Turn anticlockwise 60°.
> Repeat until you return to START.

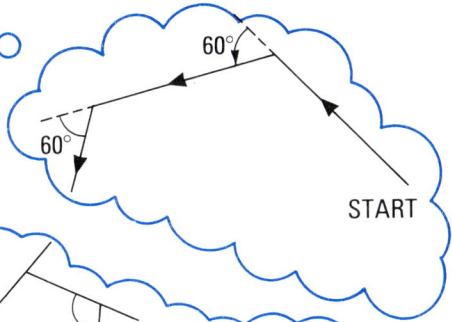

Sketch the completed shape.

60°

60°

START

b) What kind of shape do the instructions produce?

c) How many degrees is each:
   (i) exterior angle
   (ii) interior angle in the shape you drew?

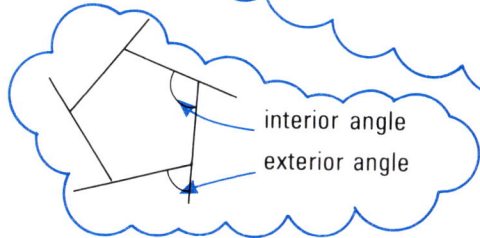

interior angle

exterior angle

d) Describe the symmetries of the shape you have drawn. (How many lines of symmetry? What is the order of turn symmetry?)

e) These instructions will give a regular 12-sided figure.

> Move forward 4 cm.
> Turn clockwise $k°$.
> Repeat until you return to START.

Find out what number k stands for.

f) Investigate what happens when k is replaced by other numbers.

Choose some of these.

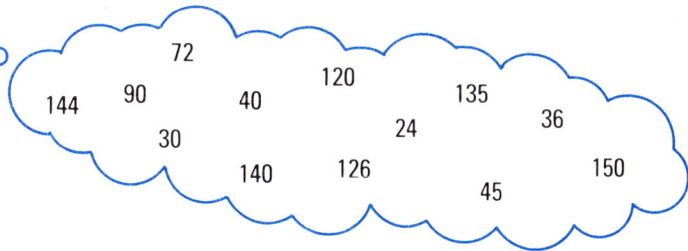

144 90 72 120 135
      30 40   24   36
   140 126 150
         45

Then try some of these.

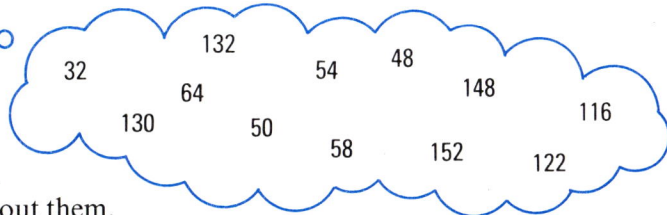

32 132 54 48
   64   148
130 50   116
   58 152 122

g) Think about these questions.
   Write down what you can about them.

   A Do you always end up at START no matter what angle you choose?
   B What different kinds of shapes are produced?
   C Is there any way to predict the shape if you know the angle?

B8

5   Triangles like A will fit together to make the 12-sided regular polygon B.

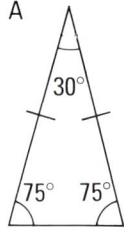

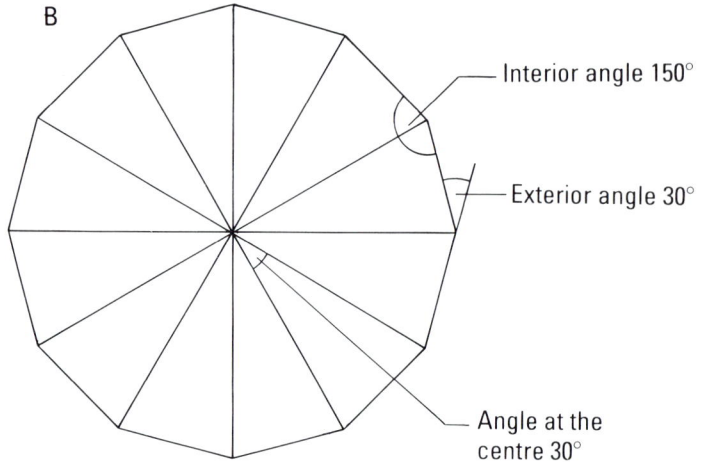

A

B

Interior angle 150°

Exterior angle 30°

Angle at the centre 30°

30°

75°   75°

a) Make a sketch of the triangles which fit together to make:

  (i) a regular 9-sided figure
  (ii) a regular 3-sided figure.

b) Sketch each regular figure and mark the size of the interior angle, exterior angle, and angle at the centre.

6   In the star, the centre is a regular pentagon.

Calculate

a) k b) m c) t

$m°$

$t°$

$k°$

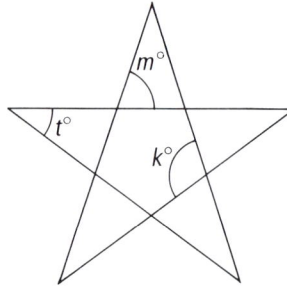

7   Calculate the size of the angle at each point of these regular 8-pointed stars.

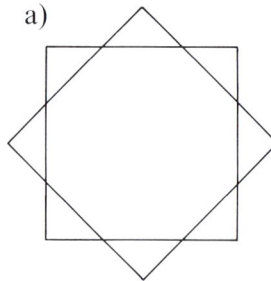

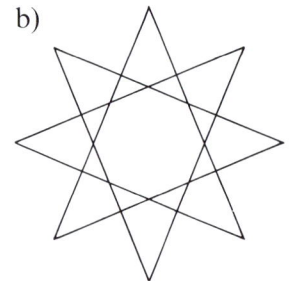

a)

b)

EXPLORATION

8   How many different types of 12-pointed stars are there? Sketch them all.

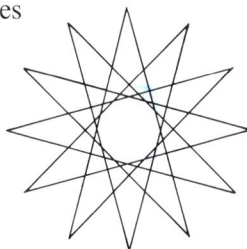

(Draw a regular 12-gon [as in question 5]. Trace it each time to help you.)

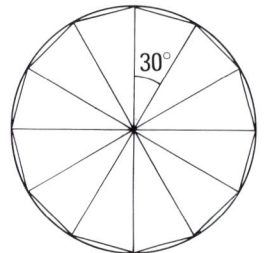

30°

B8

ENRICHMENT

EXPLORATION: TILING PATTERNS

1   a) Copy and complete this table about regular polygons.

Number of sides	3	4	5	6	7	8	9	10	11	12
Interior angles	60°		72°		$128\frac{4}{7}°$				$147\frac{3}{11}°$	

b) Together, squares and regular octagons can be used to make a tiling pattern,

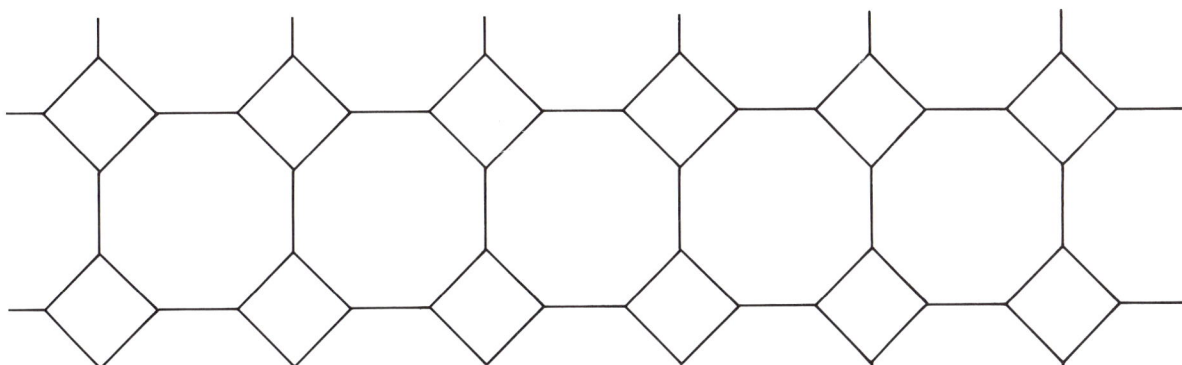

but squares and hexagons cannot.

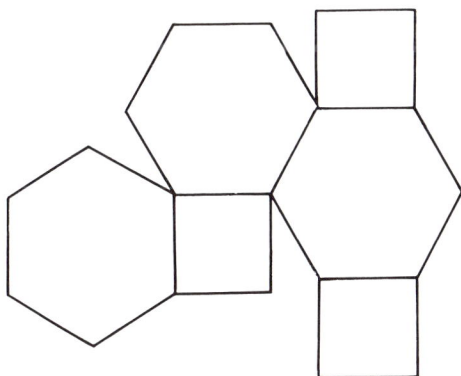

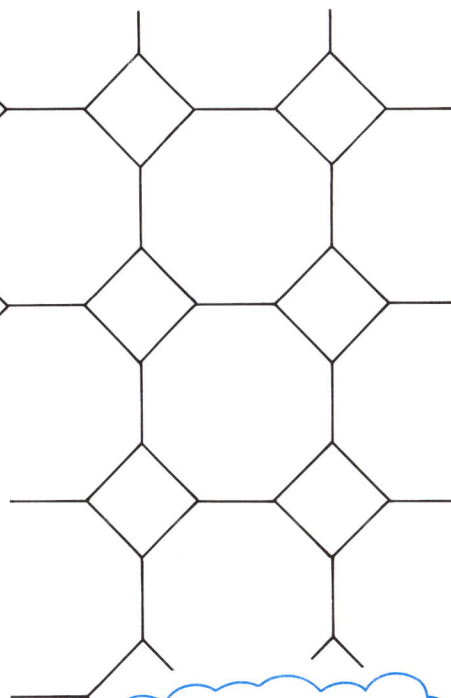

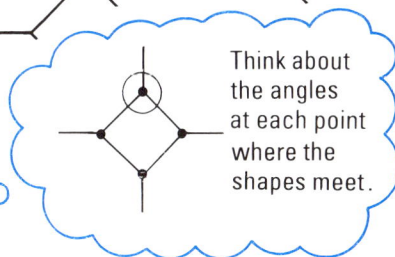

Use your table in a) to explain why in each case.

Think about
the angles
at each point
where the
shapes meet.

c) Investigate which other pairs of regular polygons make tiling patterns (tessellations). Record what you discover. Sketch a tiling pattern for each pair you find.

d) Can more than two kinds of regular polygon fit together to make a tiling pattern?
    If you say 'Yes', record which ones and make sketches.
    If you say 'No' explain why not.

B8

B9 WORKING WITH FRACTIONS

CORE

1 Do these in your head. Write down only the answers.

a) The flour is separated into $\frac{1}{2}$ kg piles. How many piles is this?

b) The flour is separated into 15 equal piles. How many kilograms is this in each pile?

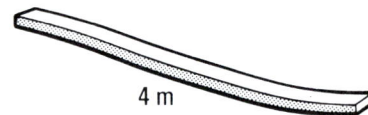

c) The plastic strip is cut into $\frac{1}{3}$ m lengths. How many pieces is this?

d) The canoe travels 10 km at $2\frac{1}{2}$ km per hour. How long does the journey take?

e) Every $\frac{1}{2}$ hour Greg walks $\frac{2}{3}$ km. How far does he walk in 3 hours?

f) A robot shuffles $\frac{1}{2}$ km at $\frac{1}{4}$ km per hour. How many hours does this take?

2 Write the correct sign, '+', '÷', '×' or '−', in each of these sentences. Your results in question 1 will help you.

a) $3 \,?\, \frac{1}{2} = 6$ b) $5 \,?\, 15 = \frac{1}{3}$ c) $4 \,?\, \frac{1}{3} = 12$

d) $10 \,?\, 2\frac{1}{2} = 4$ e) $\frac{2}{3} \,?\, 6 = 4$ f) $\frac{1}{2} \,?\, \frac{1}{4} = 2$

3 Copy and complete:

a) $12 \times \frac{1}{2} = \square$ b) $4 \div \frac{1}{2} = \square$ c) $7 \div \frac{1}{4} = \square$

d) $8 \times \frac{1}{4} = \square$ e) $8 \div \frac{1}{4} = \square$ f) $5 \times \frac{1}{4} = \square$

g) $2 \div \frac{1}{5} = \square$ h) $2\frac{1}{5} \div \frac{1}{5} = \square$

4 This diagram shows that $2 \div \frac{2}{3} = 3$, and $3 \times \frac{2}{3} = 2$.

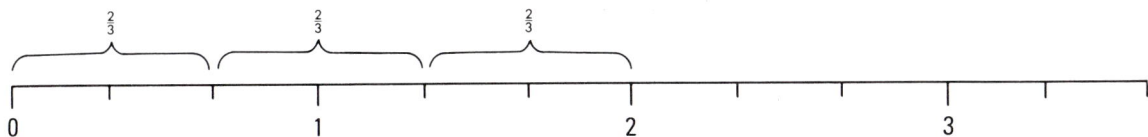

Use it to help you to do these calculations:

a) $4 \div \frac{2}{3} = \square$ b) $\frac{2}{3} \times 6 = \square$ c) $3\frac{1}{3} \div \frac{2}{3} = \square$

THINK IT THROUGH

5 Draw a diagram to help you to do these calculations:

a) $4 \div \frac{4}{5} = \square$ b) $6 \div \frac{3}{8} = \square$ c) $3\frac{3}{5} \div 9 = \square$

6 Pepper is doing this division: $5\frac{3}{5} \div 7$

This is her working. She is not very neat, and she doesn't make things very clear. Study carefully what Pepper has done, then write an explanation for each of these:

$$5\frac{3}{5} \qquad 5 \times 5 \rightarrow \begin{array}{r} 25 \\ + \ \ 3 \\ \hline 28 \end{array}$$
$$28 \div 7 = 4$$
$$\text{Ans}: \ \frac{4}{5}$$

a) Why did she multiply 5×5, then add 3? What does the result tell her?

b) Why did she divide 28 by 7 to get 4, then write $\frac{4}{5}$?

7 Use Pepper's method to do this: $9\frac{4}{5} \div 7$

8 Use any method you like to do this: $7\frac{1}{12} \div 5$

9 This is another example of Pepper's divisions, but this time she has made mistakes. Find out what they are and rewrite her working correctly.

$$4\frac{4}{9} \div 5$$
$$4 \times 4 \rightarrow \begin{array}{r} 16 \\ + \ \ 9 \\ \hline 25 \end{array}$$
$$25 \div 5 = 5 \qquad \text{Ans } \frac{5}{9}$$

10 Here are some different ways, using '$\div$', of getting $\frac{3}{8}$ litre.

$\frac{3}{8}$ litre $\div 1$ $1\frac{1}{8}$ litre $\div 2$ $1\frac{4}{8}$ litres $\div 3$ 6 litres $\div 10$

Write four different ways, using '$\div$', of getting $\frac{2}{3}$ kg.

11 Here are some different ways, using '$\times$', of getting $\frac{3}{8}$ km.

$\frac{3}{8}$ km $\times 1$ $\frac{3}{40}$ km $\times 5$ $\frac{3}{16}$ km $\times 2$ $\frac{1}{2} \times \frac{3}{4}$ km

Write four different ways, using '$\times$', of getting $\frac{5}{7}$ cm.

B9

Multiplying fractions

1   $1\frac{1}{2}$ is 'three halves' or $\frac{3}{2}$.

Two halves and one half

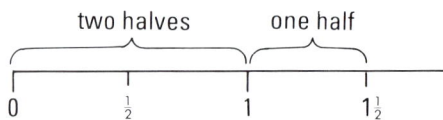

two halves   one half

0    $\frac{1}{2}$ 1 $1\frac{1}{2}$

$2\frac{2}{3}$ is 'eight thirds' or $\frac{8}{3}$.

Six thirds and two thirds

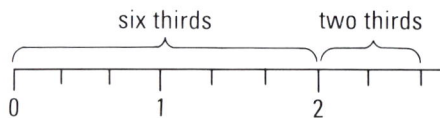

six thirds   two thirds

0   1   2

a) Write each of these in the form $\frac{\square}{\square}$: (i) $2\frac{1}{4}$ (ii) $3\frac{2}{3}$ (iii) $4\frac{7}{8}$ (iv) $5\frac{9}{10}$ (v) $11\frac{5}{9}$

b) Write each of these in the form $\square\frac{\square}{\square}$: (i) $\frac{7}{3}$ (ii) $\frac{9}{5}$ (iii) $\frac{13}{7}$ (iv) $\frac{63}{13}$ (v) $\frac{100}{19}$

2   a) Do these calculations:

   (i) $(4 \div 2) \times (6 \div 3)$ (ii) $(12 \div 6) \times (9 \div 3)$ (iii) $(25 \div 5) \times (8 \div 4)$

b) Write down a similar calculation of your own, and find the result.

c) Calculate:

   (i) $(4 \times 6) \div (2 \times 3)$ (ii) $(12 \times 9) \div (6 \times 3)$ (iii) $(25 \times 8) \div (5 \times 4)$

d) Compare your results in a) and c). Write down what you notice.

e) Copy and complete:

   $(8 \div 4) \times (9 \div 3) = (\square \times \square) \div (\square \times \square)$

f) Try part e) using some numbers of your own. (Start with your calculation in b).)
   Is what you found in d) always true?
   If you say 'No', give an example where it is not true.

g)   (i) Write this multiplication as two divisions multiplied together (as in a)):

      $\frac{2}{3} \times \frac{5}{7}$

   (ii) Now write it as one multiplication divided by another multiplication (as in c)).

   (iii) Copy and complete: $\frac{2}{3} \times \frac{5}{7} = \frac{\square \times \square}{\square \times \square} = \frac{\square}{\square}$

h) Do these calculations:

   (i) $\frac{4}{5} \times \frac{2}{3}$ (ii) $\frac{3}{8} \times \frac{5}{9}$ (iii) $1\frac{1}{2} \times \frac{3}{5}$ (Remember $1\frac{1}{2} = \frac{3}{2}$.)   (iv) $2\frac{1}{2} \times \frac{7}{8}$

   (v) $2\frac{1}{3} \times 3\frac{1}{3}$ (vi) $\frac{5}{8} \times 3\frac{1}{2}$ (vii) $1\frac{1}{2} \times \frac{2}{3} \times \frac{5}{8}$

═══ TAKE NOTE ═══

$\dfrac{2}{3} \times \dfrac{4}{5} = \dfrac{2 \times 4}{3 \times 5} = \dfrac{8}{15}$ $\dfrac{7}{3} \times 4 = \dfrac{7}{3} \times \dfrac{4}{1} = \dfrac{7 \times 4}{3 \times 1} = \dfrac{28}{3}$

3   a) After how many multiplications is $\frac{2}{3} \times \frac{2}{3} \times \frac{2}{3} \times \frac{2}{3} \ldots$ less than $\frac{1}{10}$?

   b) After how many multiplications is $1\frac{1}{3} \times 1\frac{1}{3} \times 1\frac{1}{3} \times 1\frac{1}{3} \ldots$ greater than 10?

4   A rectangular lawn measures $8\frac{1}{2}$ m by $3\frac{1}{3}$ m. What is its area?

5   Find three different pairs of fractions whose product is $\frac{24}{35}$.
   (The *product* is the result when the two fractions in a pair are multiplied together.)

6   $\frac{2}{3} \times \frac{a}{b} = \frac{8}{9}$. What is $\frac{a}{b}$?

7   The letter ⊤ is enlarged using a scale
   factor $\times 1\frac{1}{2}$ by a photocopier. The resulting
   ⊤ is enlarged again using a scale factor $\times 2\frac{1}{3}$.
   How tall and how wide is the final ⊤ ?

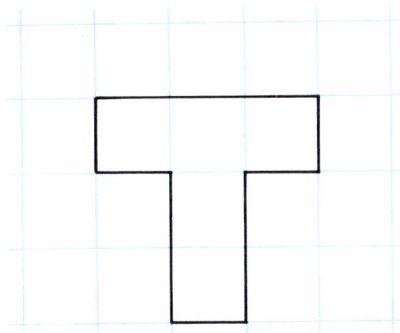

8   Calculate the area of glass needed for each
   section of the window. Check that your four
   areas amount to $1\,m^2$ of glass.

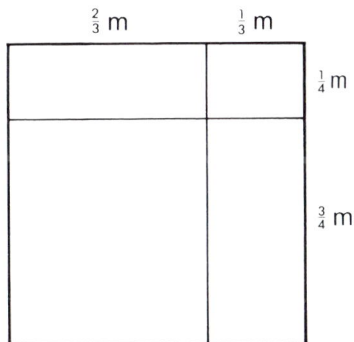

9   a) Copy and complete
       the pattern:

       ⟶ means 'multiply by $\frac{2}{3}$'
       ⟶ means 'multiply by $\frac{3}{4}$'

       Write each fraction as
       simply as you can.
       (For example, $6 \div 12$ is
       the same as $1 \div 2$,
       so $\frac{1}{2} = \frac{6}{12}$.)

   b) What does ⤏ mean?

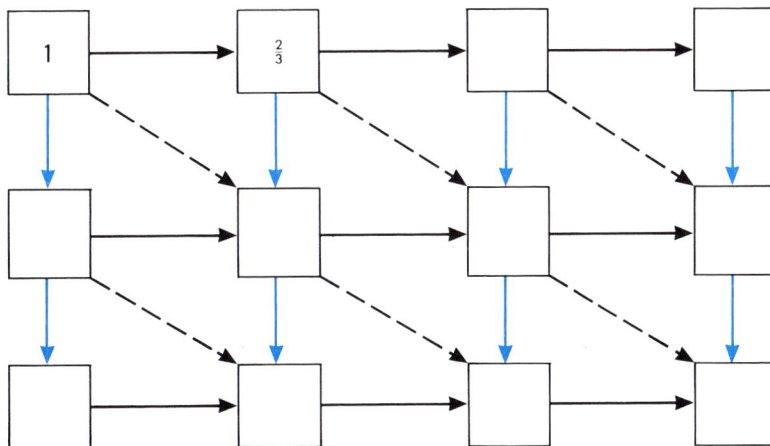

Dividing by a fraction

1 a) Do these calculations:

(i) $(8 \div 4) \div (12 \div 6)$ (ii) $(24 \div 4) \div (4 \div 2)$ (iii) $(30 \div 6) \div (4 \div 8)$

b) Write down a similar calculation of your own, and find the result.

c) Calculate:

(i) $(8 \div 4) \times (6 \div 12)$ (ii) $(24 \div 4) \times (2 \div 4)$ (iii) $(30 \div 6) \times (8 \div 4)$

d) Compare your results in a) and c). Write down what you notice.

e) Copy and complete:

$$(8 \div 5) \div (9 \div 4) = (\square \div \square) \times (\square \div \square)$$

f) Try some more examples of your own like the divisions and multiplications in a) and c). (Start with your calculation in b).)
Is what you found in d) always true? If you say 'No', give an example where it is not true.

g) (i) Write the division as a division of two whole number divisions (see a):

$$\tfrac{2}{3} \div \tfrac{3}{5}$$

(ii) Now write the division as a multiplication of two whole number divisions, as in c).
(iii) Copy and complete:

$$\tfrac{2}{3} \div \tfrac{3}{5} = \tfrac{\square}{\square} \times \tfrac{\square}{\square} = \tfrac{\square}{\square}$$

h) Copy and complete:

(i) $\tfrac{3}{4} \div \tfrac{2}{5} = \tfrac{3}{4} \times \tfrac{\square}{\square}$ (ii) $1\tfrac{1}{2} \div 2\tfrac{1}{4} = \tfrac{\square}{\square} \div \tfrac{\square}{\square}$

$\ = \tfrac{3 \times \square}{4 \times \square}$ $\ = \tfrac{\square}{\square} \times \tfrac{\square}{\square}$

$\ = \tfrac{\square}{\square}$ $\ = \tfrac{2}{\square}$

i) Do these calculations:

(i) $\tfrac{5}{6} \div \tfrac{8}{9}$ (ii) $1\tfrac{1}{3} \div 2\tfrac{1}{3}$ (iii) $4\tfrac{1}{2} \div \tfrac{3}{4}$

j) Write down a division of fractions of your own which gives the result:

(i) $\tfrac{2}{3}$ (ii) $\tfrac{4}{7}$

TAKE NOTE

$\dfrac{2}{3} \div \dfrac{4}{7}$ gives the same result as $\dfrac{2}{3} \times \dfrac{7}{4}$, which is $\dfrac{14}{12}$ $\left(\text{or } \dfrac{7}{6}\right)$.

2 Find three different pairs of fractions whose quotient is $\tfrac{2}{3}$. (The quotient is the result when one fraction is divided by the other.)

3 $\dfrac{2}{3} \div \dfrac{c}{d} = \dfrac{6}{21}$.   What is $\dfrac{c}{d}$?

4   Calculate:

a) $\frac{2}{3} \div (\frac{4}{7} \times \frac{1}{2})$ b) $(\frac{2}{3} \div \frac{4}{7}) \times \frac{1}{2}$

5   a) How many divisions are needed before the result is greater than 10?

$\frac{2}{3} \div \frac{2}{3} \div \frac{2}{3} \div \frac{2}{3} \div \frac{2}{3} \cdots$

b) How many divisions are needed before the result is less than 0.1?

$1\frac{1}{2} \div 1\frac{1}{2} \div 1\frac{1}{2} \div 1\frac{1}{2} \div 1\frac{1}{2} \cdots$

6   How many $7\frac{5}{8}''$ floorboard widths are needed to fit across a room measuring $9'6''$ ($114''$)? ($''$ means inches, $'$ means foot. 1 foot $=$ 12 inches.)

7   Imagine dividing up the 1 m square like this:

a) What is the area of:

   (i) the largest blue square
   (ii) the next largest blue square
   (iii) the fifth largest blue square?

b) If the pattern is continued, what fraction of the whole square will eventually be blue?

c) What result does this 'infinite' addition give?

$\frac{1}{2}^2 + \frac{1}{4}^2 + \frac{1}{8}^2 + \frac{1}{16}^2 + \cdots$
(Remember that $\frac{1}{2}^2 = \frac{1}{2} \times \frac{1}{2}$.)

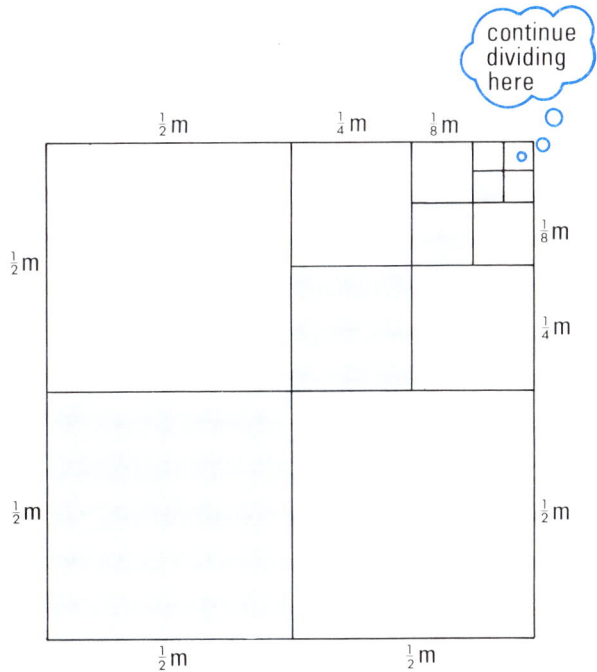

continue dividing here

$\frac{1}{2}$ m $\frac{1}{4}$ m $\frac{1}{8}$ m $\frac{1}{8}$ m $\frac{1}{4}$ m $\frac{1}{2}$ m $\frac{1}{2}$ m $\frac{1}{2}$ m $\frac{1}{2}$ m

---

CHALLENGE

8   An approximation which is often used for π is $3\frac{1}{7}$.

a) Roughly, what is:

   (i) the circumference ($C = \pi \times D$)
   (ii) the area ($A = \pi \times r^2$)
      of a circle whose radius is $1\frac{1}{3}$m?

   Write your results using mixed numbers (for example, $7\frac{1}{3}, 19\frac{1}{4} \ldots$).

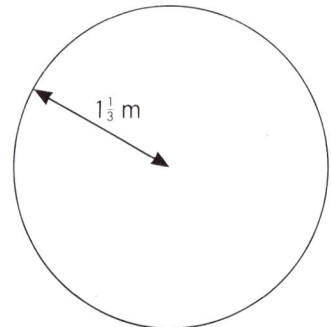

b) The circumference of a circle is $12\frac{1}{2}$cm. Roughly, what is its diameter? Write your result as a fraction. Use $\pi \approx 3\frac{1}{7}$.

$1\frac{1}{3}$ m

ENRICHMENT

Fractions using letters

1 $\dfrac{k}{3} \times \dfrac{4}{m} = \dfrac{5}{6}$. Find a pair of values for k and m.

2 $\dfrac{p}{q}$ and $\dfrac{r}{s}$ are two fractions. Copy and complete each of these statements:

a) $\dfrac{p}{q} \times \dfrac{r}{s} = \dfrac{\square \times \square}{\square \times \square}$ b) $\dfrac{p}{q} \div \dfrac{r}{s} = \dfrac{\square \times \square}{\square \times \square}$

3 $\dfrac{a}{b}$ and $\dfrac{c}{d}$ are two fractions.

a) Use $a = 4$, $b = 2$, $c = 6$, $d = 3$, to decide which of these are definitely **not** correct statements.

A $\dfrac{a}{b} + \dfrac{c}{d} = \dfrac{a+c}{b+d}$ B $\dfrac{a}{b} + \dfrac{c}{d} = \dfrac{a+c}{b \times d}$

C $\dfrac{a}{b} + \dfrac{c}{d} = \dfrac{a \times c}{b+d}$ D $\dfrac{a}{b} + \dfrac{c}{d} = \dfrac{(a \times d)+(c \times b)}{(b \times d)}$

b) For the statement you did not reject in a), try these values for a, b, c and d:

$a = 8$, $b = 2$, $c = 12$, $d = 3$.

Does this also give a correct statement?

c) Try some more values of a, b, c and d in a) D. Do you always get a correct statement?

d) Use what you found in c) to write down the result of this addition: $\frac{2}{7} + \frac{1}{4}$

e)   (i) Explain why $\dfrac{a}{b} \times \dfrac{d}{d} = \dfrac{a}{b}$

   (ii) Explain why $\dfrac{c}{d} \times \dfrac{b}{b} = \dfrac{c}{d}$

   (iii) Use your results in (i) and (ii) to explain why $\dfrac{a}{b} + \dfrac{c}{d} = \dfrac{a \times d}{b \times d} + \dfrac{c \times b}{d \times b}$

f) Use the expression in e) (iii) to write these additions as additions of two fractions with the same denominators:

(i) $\frac{2}{7} + \frac{1}{4}$ (ii) $\frac{3}{8} + \frac{2}{9}$

REVIEW

To help us solve problems we can write number sentences to represent them. The number sentence (or *equation*) below represents this problem:

$(2 \times w) + w + 10 = 31$

A triangle has one side twice as long as another side. The third side is 10 cm long. Its perimeter is 31 cm. How long is each side of the triangle?

w cm
$2 \times w$ cm
10 cm
Perimeter = 31 cm

We can try different possibilities for *w*:

Does *w* represent 3?	$(2 \times 3) + 3 + 10$	is 19	No.
Does *w* represent 5?	$(2 \times 5) + 5 + 10$	is 25	No.
Does *w* represent 10?	$(2 \times 10) + 10 + 10$	is 40	No.
Does *w* represent 7?	$(2 \times 7) + 7 + 10$	is 31	Yes. We say 7 is a *solution* of the *equation*.

The answer to the problem is that the lengths of the sides of the triangle are 10 cm, 7 cm, and 14 cm.

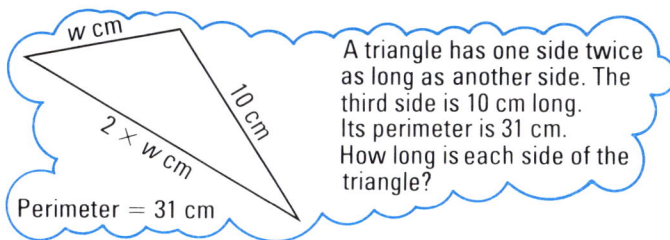

■ Write a number sentence for this problem.

a) Find the number which *k* represents.
b) How long is each side of the triangle?

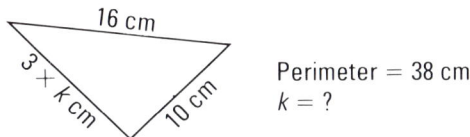

16 cm
$3 + k$ cm
10 cm
Perimeter = 38 cm
k = ?

CONSOLIDATION

1 For each of the situations a) to e):

(i) form an equation
(ii) solve the equation
(iii) write down the information asked for in the situation.

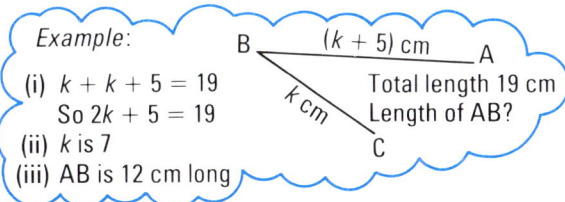

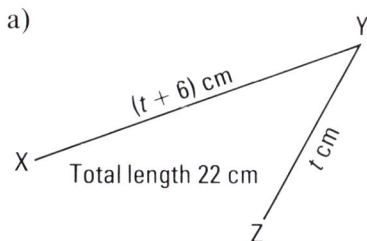

Example:

(i) $k + k + 5 = 19$
So $2k + 5 = 19$
(ii) *k* is 7
(iii) AB is 12 cm long

B
$(k + 5)$ cm
A
Total length 19 cm
Length of AB?
k cm
C

a)

Y
$(t + 6)$ cm
X
Total length 22 cm
t cm
Z

Length of XY = ?

b)

$(k + 48)°$
How many degrees is each angle?
93°
k°

c)

$(S + 7)$ cm
Perimeter = 40 cm
Length of each side?
S cm

d)

| $(2 \times k)$ kg | *k* kg |
| 3 kg | 5 kg |

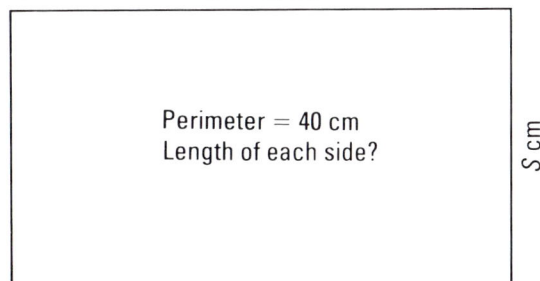

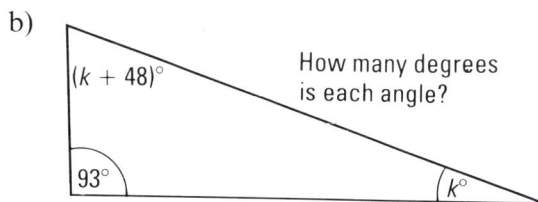

Mass of each weight?

e)

| 12 kg | *m* kg |
| | $(m + 5)$ kg |

Mass of each weight?

2   Here are some number sentences. Think and
    test numbers for each one until you find out
    which number the letter represents. Use your
    calculator if it helps.
    Check your solution each time.
    Write your solutions like this:

    Example: $\frac{p}{3} = 9$

    So p is 27

    Check: $\frac{27}{3} = 9$

    Think: what number divided by 3 gives 9?

a) $k - 5 = 10$

b) $(2 \times m) + 3 = 11$

c) $8 \times t = 56$

d) $\frac{p}{10} = 1$

e) $m + 7 = 2 \times m$

f) $10 - n = n$

g) $t + 3 = 23 - t$

h) $m - 3 = 11 - m$

i) $\frac{16}{k} = 8$

j) $\frac{4}{n} = n + 3$

k) $3 \times p = p + 8$

l) $w \times 4 = 0.4$

m) $(2 \times w) + 4 = 16$

n) $\frac{v}{0.5} = 10$

o) $s - 7 = 12 - s$

p) $\frac{10}{n} = 2.5$

q) $18 - (3 \times d) = 3$

r) $\frac{2}{b} = \frac{b}{8}$

3   a) Write down an addition equation like this
        whose solution is 5.

    $y + 8 = 12$ Solution $y = 4$

    b) Write down a subtraction equation like this
        whose solution is 9.

    $p - 7 = 12$

    c) Write down a subtraction equation like this
        whose solution is 3.

    $15 - t = 9$

    d) Write down a multiplication equation like this
        whose solution is 7.

    $4 \times k = 20$

    e) Write down a division equation like this
        whose solution is 5.

    $\frac{10}{n} = 5$

    f) Write down an equation with letters on each side, like this
        whose solution is 5.

    $4 + m = 2 \times m$

4   The weights on these scales balance.

    a) Copy and complete this
        sentence: $10 + m = \square + ($ $)$

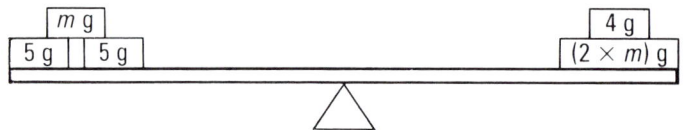

    b) What number does m represent?

5 a) Draw a set of scales to represent this equation: $k + 7 = 1 + 3k$

   b) What number does k represent?

Remember:
3k means 3 × k

6 Angus wants to make a picture frame from this strip of wood.
   He wants the length to be 10 cm
   longer than the width.

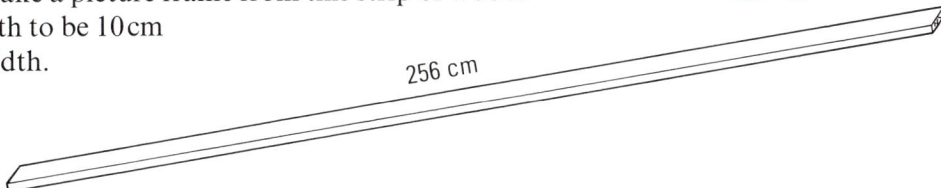

256 cm

   a) Use p cm to represent the width of the picture frame. Write an expression using p to represent
       the length.

   b) Copy and complete this equation: $(4 \times p) + \square = 256$

   c) From your equation find the number which p represents.

   d) How wide and how long should Angus make the picture frame?

THINK IT THROUGH

   e) Now Angus wants to make a frame from this strip of wood. He wants the length to be 12 cm
       longer than the width. Write an equation to represent the situation.
       Use your equation to solve the problem.

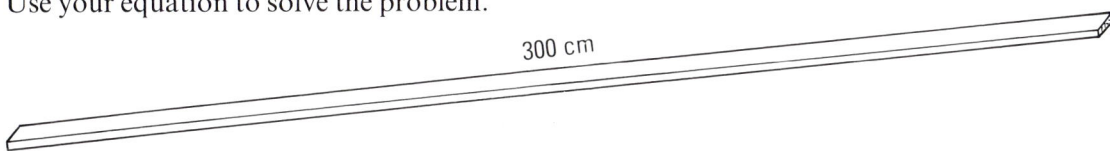

300 cm

7 The 400 g of silver is to be made into a 40 g
   brooch and some 15 g charms.

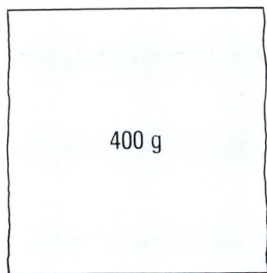

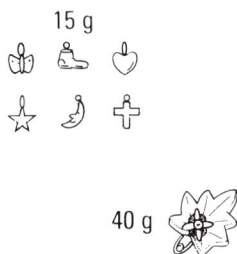

400 g

15 g

40 g

   a) Use k to represent the number of charms.
       Write an equation for the situation.

   b) Find the number which k represents.

   c) Your equation can be written like this: $\square - 15k = \square$. Copy and complete this version.

   d) Write another version of the equation. Check that the value of k satisfies your new version.

The 'balance' method

1 a) The two sides of the scales balance each other. Copy and complete this equation:

$$\square + 4 = T + \square$$

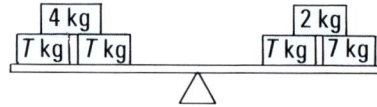

b) Find the value of T.

c) What happens to the scales if we remove 4 kg from each side? Do they tip down to the left, tip down to the right, or stay balanced?

4 kg has been removed from each side of the scales.

d) Copy and complete this sentence: $2T = T + \square$

e) The scales now look like this.

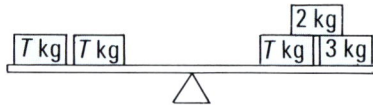

What happens to the scales when we remove T kg from each side? Do they tip down to the left, tip down to the right, or stay balanced?

f) Copy and complete this sentence: $T = \square$

T kg has been removed from each side of the scales in e).

g) Copy and complete this solution to the equation.

h) Explain why -4 and $-T$ are written on the left-hand side of the solution. Check that the solution agrees with your solution in b) and f).

$$
\begin{array}{rl}
 & 2T + 4 \;=\; T + 9 \\
-4 \;:\; & 2T \;\;\;\;\;=\; T + \square \\
-T \;:\; & T \;\;\;\;\;\;\;\;=\; \square
\end{array}
$$

2 a) Write an equation which describes this balance.

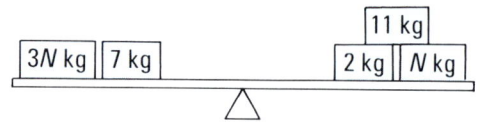

b) Find the solution to your equation.

c) Copy and complete this solution. Write in each box ($\boxed{?}$) what has been subtracted.

$$
\begin{array}{rl}
 & 3N + 7 \;=\; N + 13 \\
\boxed{?} \;:\; & 3N \;\;\;\;\;\;=\; N + 6 \\
\boxed{?} \;:\; & 2N \;\;\;\;\;\;=\; 6 \\
 & \text{So } N \text{ is } \square
\end{array}
$$

═══ THINK IT THROUGH ═══

3 Solve the equation $5M + 9 = 13 + 3M$

a) by thinking and checking   b) by working as in question 2c).

4   Solve each of these equations

   (i) $4+3x = 2x+8$ (ii) $p+22 = 4p+10$

   a) by thinking and checking   b) by using the 'balance' method.

5   a) Try to solve these equations using the 'think and check' method. (If you don't succeed after 5 minutes go on to part b).)

      (i) $8+3k = k+15$ (ii) $5k+7 = 2k+20$

   b) Solve each equation using the 'balance' method.

6   Copy and complete each solution:

a)
$$\begin{aligned}
t + 9 &= 7t + 6 \\
-t: \qquad \square &= \square t + 6 \\
-6: \qquad \square &= 6t \\
\text{So } t &= \square
\end{aligned}$$

b)
$$\begin{aligned}
m + 4 &= 5m + 1 \\
-m: \qquad \square &= \square + 1 \\
-1: \qquad \square &= 4m \\
\text{So } m \text{ is } \square
\end{aligned}$$

7   Copy and complete each solution. Check that your solution fits the original equation.

a)
$$\begin{aligned}
3m - 4 &= 2m + 6 \\
+4: \qquad 3m &= 2m + \square \\
-2m: \qquad \square &= \square
\end{aligned}$$

b)
$$\begin{aligned}
7 - 2k &= 12 - 3k \\
+3k: \qquad 7 + \square &= \square \\
-7: \qquad k &= \square
\end{aligned}$$

8   Solve these equations by using the 'balance' method. (You might like to 'think and check' first.) Don't forget to check that your solution fits the equation.

   a) $3t-4 = t+6$ b) $13-t = t-1$ c) $b+7 = 19-2b$ d) $m-2 = 2-5m$

   ═══════════════════════ CHALLENGE ═══════════════════════

9   Write an equation for each problem and solve it.

   a) How long is each side of the quadrilateral?

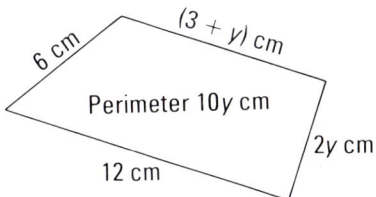

   6 cm
   $(3 + y)$ cm
   Perimeter 10y cm
   2y cm
   12 cm

   b) The length of AB is twice the length of AF. FE is 7cm longer than AF. How long is AB?

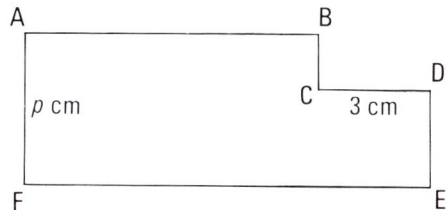

   A   B

   p cm   C   D
              3 cm
   F   E

REVIEW

The incomplete *addition* number sentence goes with this problem:

Number sentence   Problem

$^-3 + {}^+12 = ?$

> The temperature in the greenhouse is $^-3°C$.
> It rises by $12°C$. What is the temperature now?

■ Solve the problem.

■ Copy and complete the number sentence.

This story also fits the number sentence:

A box of cutlery contains three fewer knives than forks. Another box contains twelve more knives than forks. Altogether, the boxes contain ... ... knives than forks.

         a number   'more' or 'fewer'

■ Write down the missing words from the story.

The incomplete *subtraction* number sentence goes with this problem:

Number sentence   Problem

$^+5 - {}^-2 = ?$

> The temperature in the greenhouse is $5°C$. Outside the greenhouse the temperature is $^-2°C$. How much warmer is it inside than outside the greenhouse?

■ Solve the problem.

■ Copy and complete the number sentence.

We often write   4 for $^+4$,   3 for $^+3$, etc.
             -4 for $- {}^+4$, -3 for $- {}^+3$, etc.
$- {}^-4$ is equivalent to $^+4$
$- {}^+4$ and $+ {}^-4$ are equivalent to $^-4$.

■ Write down the result of each of these.

    a) $^+3 + {}^-10$ b) $^+3 + {}^-6$ c) $^+3 + {}^-2$ d) $^+3 + {}^+2$ e) $^+3 + {}^+6$
    f) $^+3 - {}^-10$ g) $^+3 - {}^-6$ h) $^+3 - {}^-2$ i) $^+3 - {}^+2$ j) $^+3 - {}^+6$

CORE

Using addition and subtraction

1 In a photographic competition two judges give $+$ and $-$ scores for photographs. $^+2$ means '2 points above average'. $^-2$ means '2 points below average'. These are the scores for nine photographs.

Photograph	1	2	3	4	5	6	7	8	9
Judge A	$^-1$	0	$^-1$	$^-2$	$^+3$	$^+2$	$^+3$	$^-3$	$^-1$
Judge B	$^-2$	$^-1$	$^-1$	0	$^+1$	$^+4$	$^+2$	$^-4$	$^+1$
Total	$^-3$	$^-1$	0	$^-2$	4	6	5	7	0

a) (i) How many more points did judge A give than judge B for photograph 1?
    (ii) Copy and complete: $^-1-^-2=?$

b) Look at the scores for photograph 9. Explain what this sentence tells us about the judges' scores: $^+1-^-1=2$.

c) What does this tell us about the judges' scores for photograph 8: $^-4-^-3=^-1$?

d) Two of the total scores in the table are incorrect. Which ones, and what is the correct total in each case?

2 On a spreadsheet for a weather forecasting station the temperature in Calgary (Canada) is represented by $K\,°C$. The temperature in Miami (Florida, USA) is represented by $M\,°C$.

spreadsheet		
Calgary ($K\,°$C)	Miami ($M\,°$C)	(M–K)$°$
$^-1$	27	28

a) What does $(M-K)°$ tell you about the temperatures in Calgary and Florida?

b) It is $^-4°C$ in Calgary and $29°C$ in Miami. What is $(M-K)$?

c) What does $(K-M)°$ tell you about the temperatures in Miami and Calgary?

d) Find $(K-M)°$ for the temperatures on the spreadsheet.

3 Copy and complete each sentence.

a) $1-2\quad=\square$

b) $^+4-^+5=\square$

c) $^-1-^-2=\square$

d) $^+2-^+1=\square$

e) $0-^-8\quad=\square$

f) $^-8-0\quad=\square$

g) $^+5-^+9=\square$

h) $7-11\quad=\square$

i) $^+9-^-4=\square$

j) $^-3-6\quad=\square$

k) $8-11\quad=\square$

l) $^-5-7\quad=\square$

m) $\square-^-3=4$

n) $\square-9\quad=^-1$

o) $8-\square=10$

p) $^-3-\square=^-12$

4   Find the value of x in each sentence:

a) $1+x=2$
b) $^-1+x=2$
c) $^-1+x=\ ^-2$
d) $1+x=\ ^-2$
e) $1-x=2$
f) $^-1-x=2$
g) $^-1-x=\ ^-2$
h) $1-x=2$
i) $x-1=2$
j) $x-\ ^-1=2$
k) $x-\ ^-1=\ ^-2$
l) $x-1=\ ^-2$
m) $x+1=2$
n) $x+\ ^-1=2$
o) $x+\ ^-2=\ ^-1$
p) $x+1=\ ^-2$
q) $x+3=\ ^-4$
r) $3-x\ =4$
s) $4-x\ =3$
t) $^-4-x=\ ^-3$

━━━━━━━━━ CHALLENGE ━━━━━━━━━

5   a) Complete this number sentence $(\square+7)-9=\ ^-5$

   b) Write a short story to represent the sentence. (Use bank balances or temperature differences, etc.)

B11 Multiplying directed numbers

1   a) Check that $2\times7=2\times(5+2)=(2\times5)+(2\times2)=14$.

   b) Copy and complete:

   (i) $2\times8=2\times(5+3)\ \ \ =(2\times5)+(2\times?)\ \ =?$
   (ii) $2\times8=2\times(9-1)\ \ \ =(2\times9)-(2\times?)\ \ =?$
   (iii) $2\times8=2\times(9+\ ^-1)\ =(2\times9)+(2\times\ ^-1)=?$
   (iv) $2\times8=2\times(10+\ ^-2)=(2\times10)+(2\times?)\ \ =?$

   c) Question b) (iii) is:

   $2\times8=2\times(9+\ ^-1)=(2\times9)+(2\times\ ^-1)=16$
   $\qquad\qquad\qquad\qquad\ \ \ \uparrow\qquad\ \ \uparrow$
   $\qquad\qquad\qquad\ \ \ 18\ \ +\ \ \ ?\ \ \ =16$

   What does this suggest about the value of $2\times\ ^-1$? Is it $^+2$ or $^-2$?

   d) Look at b) (iv). What does your result suggest about the value of $2\times\ ^-2$?

   e) Copy and complete:

   (i) $3\times9\ \ =\ \ 3\times(10+\ ^-1)\ \ =\ \ (3\times?)+\ (3\times?)=\ \ 30+?=?$
   (ii) $^-3\times9\ =\ ^-3\times(6+3)\ \ \ \ \ \ =(^-3\times?)+(^-3\times?)=\ ^-18+?=?$
   (iii) $^+3\times\ ^+9=\ ^+3\times(^+3+\ ^+6)\ =(^+3\times?)+(^+3\times?)=\ ^+9+?\ \ =?$
   (iv) $^+3\times\ ^-9=\ ^+3\times(^+1+\ ^-10)=(^+3\times?)+(^+3\times?)=\ ^+3+?\ \ =\ ^-27$
   (v) $^+3\times\ ^-9=\ ^+3\times(^-4+\ ^-5)\ =(^+3\times?)+(^+3\times?)=\ ^-12+?=?$

━━━━━━━━━ TAKE NOTE ━━━━━━━━━

$^-2\times3$ (or, $^-2\times\ ^+3$) $=\ ^-6$; $^-3\times4=\ ^-12$; $^+4\times\ ^-6=\ ^-24$;   etc.

2 a) Copy and complete each sentence:

    (i) $4 \times 3 = 4 \times (5-2) = (4 \times 5) - (4 \times ?) = 12$
    (ii) $4 \times 3 = 4 \times (5 + {}^-2) = (4 \times 5) + (4 \times ?) = ?$

  b) Compare your sentences.
    What do they suggest about ${}^-4 \times 2$ and $4 \times {}^-2$?

  c) Copy and complete:

    (i) ${}^-3 \times 2 = 3 \times ?$ (ii) ${}^-3 \times 2 = 2 \times ?$ (iii) ${}^-5 \times 4 = 2 \times ?$

TAKE NOTE

$${}^-8 \times 3 = 8 \times {}^-3 = {}^-8 \times 3 = {}^-24$$

3 a) Copy each of these sentences, and write down the numbers which replace ? each time.

    (i) $2 \times 11 \quad = 2 \times (7+4) \quad = (2 \times 7) + (2 \times 4) \quad = ?$
    (ii) $2 \times 5 \quad = 2 \times (7-2) \quad = (2 \times 7) - (2 \times ?) \quad = ?$
    (iii) $2 \times 5 \quad = 2 \times (7 + {}^-2) \quad = (2 \times ?) + (2 \times {}^-2) = ?$
    (iv) ${}^-2 \times 11 = {}^-2 \times (7+4) \quad = ({}^-2 \times 7) + ({}^-2 \times ?) = ?$
    (v) ${}^-2 \times 5 \quad = {}^-2 \times (3+2) \quad = ({}^-2 \times ?) + ({}^-2 \times 2) = ?$
    (vi) ${}^-2 \times 5 \quad = {}^-2 \times (7-2) \quad = ({}^-2 \times 7) - ({}^-2 \times ?) = ?$
    (vii) ${}^-2 \times 5 \quad = {}^-2 \times (7 + {}^-2) = ({}^-2 \times 7) + ({}^-2 \times ?) = ?$

THINK IT THROUGH

  b) Question a) (vii) is $\quad {}^-2 \times 5 = {}^-2 \times (7 + {}^-2) = ({}^-2 \times 7) + ({}^-2 \times {}^-2) = {}^-10$

$$\uparrow \qquad\qquad \uparrow$$
$${}^-14 \;+\; \quad ? \quad = {}^-10$$

  What does the result suggest about the value of ${}^-2 \times {}^-2$? Is it ${}^+4$, or ${}^-4$?

  c) Copy and complete:
    (i) $\quad 3 \times 9 = \quad 3 \times (10-1) \quad = (3 \times 10) - \quad (3 \times ?) = ?$
    (ii) ${}^-3 \times 9 = {}^-3 \times (10-1) \quad = ({}^-3 \times 10) - ({}^-3 \times ?) = ?$
    (iii) ${}^-3 \times 9 = {}^-3 \times (10 + {}^-1) = ({}^-3 \times 10) + ({}^-3 \times ?) = ?$

  d) What does sentence c) (iii) suggest for the value of ${}^-3 \times {}^-1$?

  e) Copy and complete:

    (i) ${}^-1 \times 5 \quad = {}^-1 \times (6 + {}^-1) \quad = ({}^-1 \times 6) + (? \times ?) \quad = ?$
    (ii) ${}^-8 \times 1 \quad = {}^-8 \times (2 + {}^-1) \quad = ({}^-8 \times ?) + ({}^-8 \times ?) \quad = ?$
    (iii) ${}^-10 \times 4 \quad = {}^-10 \times (5 + ?) \quad = ({}^-10 \times ?) + ({}^-10 \times ?) = ?$
    (iv) ${}^-10 \times {}^-4 = {}^-10 \times (1 + ?) \quad = ({}^-10 \times ?) + ({}^-10 \times ?) = ?$
    (v) ${}^-3 \times {}^-2 \quad = {}^-3 \times ({}^-1 + ?) \quad = ({}^-3 \times ?) + ({}^-3 \times ?) \quad = ?$

B11

■■■■■■■■■■■■■ TAKE NOTE ■■■■■■■■■■■■■

$^-1 \times {}^-1 = {}^+1$ $^-3 \times {}^-5 = {}^+15$ $^-10 \times {}^-9 = {}^+90$

■■■

4   Copy and complete:

a) $^-4 \times 2 = ?$
b) $^-4 \times {}^-2 = ?$
c) $^-\frac{1}{2} \times 6 = ?$
d) $^-\frac{1}{2} \times {}^-6 = ?$

e) $^-9 \times ? = {}^-27$
f) $^-9 \times ? = 27$
g) $? \times {}^-4 = {}^-12$
h) $? \times {}^-4 = 12$

■■■■■■■■■■■■■ EXPLORATIONS ■■■■■■■■■■■■■

5   Write down an expression like this: $A + (B \times C)$.

For example, $^-10 + ({}^+12 \times {}^-2)$.

Change B in a systematic way, so that it eventually changes sign.

For example, $^-10 + ({}^+12 \times {}^-2)$
$^-10 + ({}^+7 \times {}^-2)$
$^-10 + ({}^+2 \times {}^-2)$
$^-10 + ({}^-3 \times {}^-2)$
$^-10 + ({}^-8 \times {}^-2)$ *etc.*

Work out the values of the changing expression. See whether or not there is a pattern in the results. If so, write about it.
Try different expressions.

6   Write down an equation like this: $(a \times x) + b = c$.

For example, $({}^-2 \times x) + 10 = 4$

Change a, b, or c in a systematic way.

For example, $({}^-2 \times x) + 10 = 4$
$({}^-2 \times x) + 10 = 8$
$({}^-2 \times x) + 10 = 12$
$({}^-2 \times x) + 10 = 16$, *etc.*

Work out the various values of x. Write about any pattern that you find.

■■■■■■■■■■■■■ CHALLENGE ■■■■■■■■■■■■■

7   Find the value of x in each sentence. Think and test until you find the value each time. (In some of the parts there are *two* possible values.)

a) $3x = 9$
b) $3x = {}^-9$
c) $^-2 \times x = 4$
d) $^-2 \times x = {}^-4$

e) $2x + 1 = {}^-5$
f) $16 + 4x = 0$
g) $2x - 1 = {}^-1$
h) $x^2 = 25$

i) $x^2 = 36$
j) $x^2 + 1 = 50$
k) $2 \times x \times {}^-3 = 18$

Dividing directed numbers

1 a) Copy and complete each 'directed-number' factor star

   b) Draw your own 'directed-number' factor stars for:

   (i) 8   (ii) $^-12$

2 The directed-number factors of $^+6$ are
   $^-1, ^-2, ^-3, ^-6, ^+1, ^+2, ^+3, ^+6.$

   a) Write down the directed-number factors of $^-6$.

   b) True or false: the directed-number factors of k
      are the same as the directed-number factors of
      ^-k, no matter what number k represents.
      If you say 'false', give an example to show why.

3 Look at your factor star for $^+4$ in question 1 a).

   What is a) $^+4 \div ^+2$ b) $^+4 \div ^-2$ c) $^+4 \div ^-1$ d) $^+4 \div ^+1$?

4 Look at your factor star for $^-12$ in question 1 b).

   What is a) $\dfrac{^-12}{^+2}$ b) $\dfrac{^-12}{^-6}$ c) $\dfrac{^-12}{^+3}$ d) $^-12 \div ^-4$ e) $^-12 \div ^-3$?

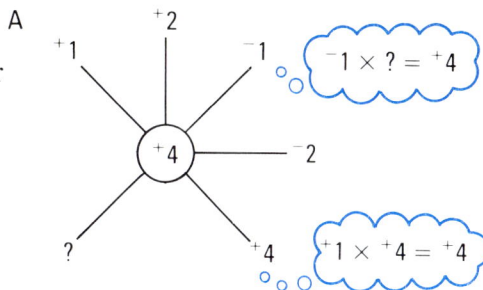

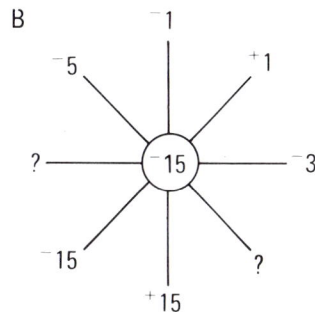

A

$^+1$ $^+2$ $^-1$

$^-1 \times ? = ^+4$

$^+4$ $^-2$

? $^+4$

$^+1 \times ^+4 = ^+4$

B $^-1$

$^-5$ $^+1$

? $^-15$ $^-3$

$^-15$?

$^+15$

B11

TAKE NOTE

$\dfrac{^-4}{^-2} = 2$ $^-8 \div ^-1 = 8$ $^-8 \div ^+1 = ^-8$ $\dfrac{^-4}{^+1} = ^-4$ $\dfrac{^-a}{^-b} = \dfrac{a}{b}$ $^-a \div b = ^-\left(\dfrac{a}{b}\right)$ $a \div ^-b = ^-\left(\dfrac{a}{b}\right)$

---

5 Copy and complete:

   a) $^-1 \times ? = ^+10$ b) $3 \times ? = ^-6$ c) $^-4 \times ? = ^+12$ d) $^-8 \div ? = 4$

   e) $8 \div ? = ^-4$ f) $^-12 \div ? = 4$ g) $? \div ^-4 = 5$ h) $? \div 2 = ^-6$

   i) $\dfrac{^+18}{^+6} = ?$ j) $\dfrac{^-8}{?} = ^-2$ k) $\dfrac{^+10}{?} = ^+2$ l) $\dfrac{^-12}{?} = ^+6$

   m) $0 \times ^-6 = ?$ n) $? \times ^-7 = 0$ o) $? \div 10 = 0$ p) $(^-3)^2 = ?$

   q) $? \times ^-8 = ^-4$ r) $^-7 \times ? = ^-0.7$ s) $(^-1)^2 = ?$ t) $(^-1)^3 = ?$

6 Copy and complete:

   a) $\dfrac{^+1 + ^-4}{^-6} = \dfrac{\square}{^-6} = \dfrac{\square}{2}$ b) $\dfrac{^+4 - ^-6}{^+5} = \dfrac{\square}{^+5} = \square$

   c) $\dfrac{5-9}{1-9} = \dfrac{\square}{^-8} = \dfrac{1}{\square}$ d) $\dfrac{3-11}{^-24} = \dfrac{\square}{^-24} = \dfrac{1}{\square}$

Using a calculator

You need a calculator with a ⌶ key.

1   a) On your calculator press:

(i) [C] [1] [⌶]   (ii) [C] [4] [⌶] [⌶]   Explain what the ⌶ key does.

b) Press [C] [4] [+] [5] [⌶] [=]

Write down the result. Which of these did the calculator calculate?

(i) $^-4 + {}^-5$ (ii) $4 - 5$

c) Press [C] [3] [⌶] [×] [6] [⌶] [=]

Write down the result. Which of these did the calculator calculate?

(i) $^-3 \times 6$ (ii) $^-3 \times {}^-6$

CHALLENGES

2   Solve each equation. Think and test until you find the result.

a) $a \times {}^-3 = {}^-6$ b) $p - 4 = {}^-6$ c) $2p - 4 = {}^-6$ d) $\dfrac{p}{4} = {}^-6$

e) $\dfrac{t-1}{2} = {}^-4$ f) $\dfrac{1-t}{2} = {}^-4$ g) $n - 1 = 2n + 4$ h) $4 - d = d + 10$

i) $0.1 \times t = {}^-0.1$ j) $0.1 \times p = {}^-1$ k) $\frac{1}{2} \times k = {}^-4$ l) $n \times \frac{1}{10} = {}^-2$

m) $^-\frac{1}{2} \times m = 4$ n) $2t \times t = 2$ o) $k^2 = 1.44$ p) $t^3 = {}^-1$
                                      (there are two   (there are two
                                      possible values of t)   possible values of k)

EXPLORATION

3   Choose an integer (a positive or negative whole number or zero), for example, $^+15$. Apply this rule.

If the number is even $\div {}^-2$

If the number is odd $+ {}^+11$

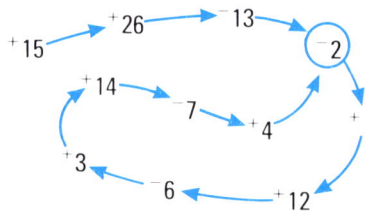

Apply the rule to the answer, and produce a number chain like the one shown above. Does the chain form a loop?

Explore further, by trying different starting numbers, by modifying the rule, etc.

Write a report about what you discover.

B12 SHAPES AND THEIR SYMMETRIES

REVIEW

Parallelograms have:

- 4 sides
- both pairs of opposite sides parallel
- turn symmetry of order 2 or 4

■ Which of these are parallelograms?

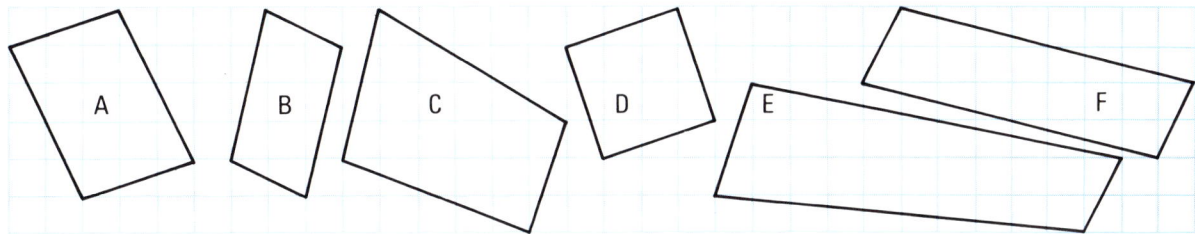

A   B   C   D   E   F

Trapeziums have:

- 4 sides
- at least 2 sides parallel

■ Which of these are trapeziums?

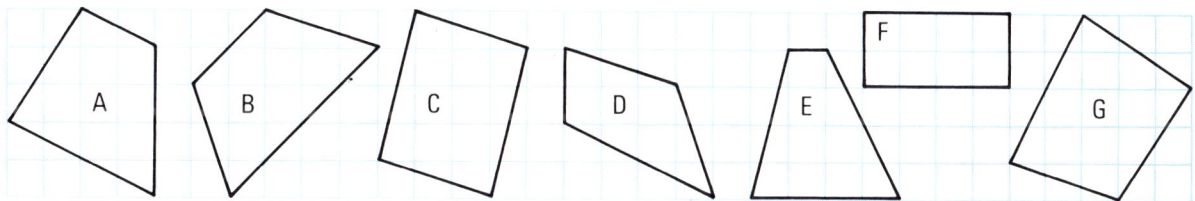

A   B   C   D   E   F   G

Rhombuses have:

- 4 sides all equal
- turn symmetry of order 2 or 4
- two or four lines of symmetry

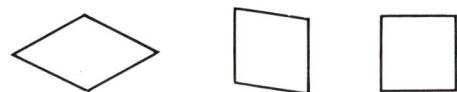

■ Which of these are rhombuses?

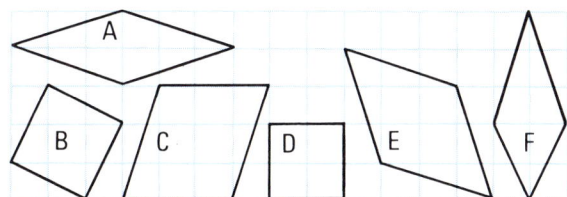

A   B   C   D   E   F

Kites have:

- 4 sides
- at least 1 diagonal as a line of symmetry

Convex kites   Concave kites

■ Which of these are kites?

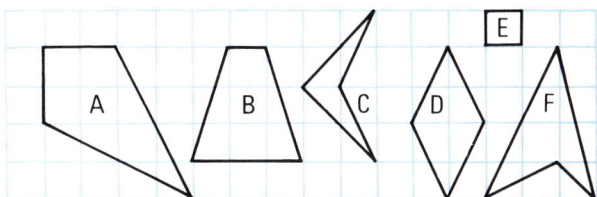

A   B   C   D   E   F

CONSOLIDATION

███ WITH A FRIEND ███

1 a) **EACH OF YOU**
    Draw 3 dots on
    paper – not in a
    straight line.

YOUR FRIEND
Add a fourth dot to your
three dots to try to form a
parallelogram.

YOU
Join the dots
with a ruler.

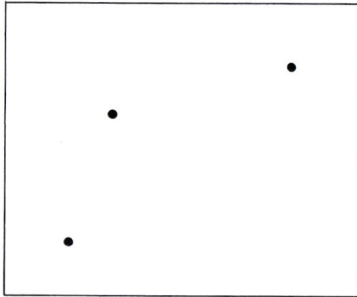   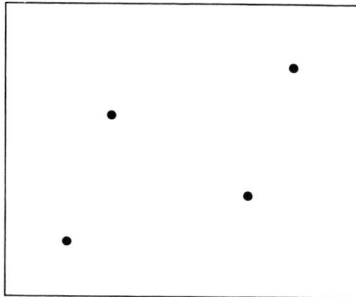   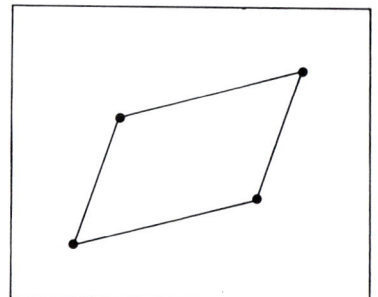

YOU
Measure the size of the angles and sides
of your parallelogram. Write down the
values like this.

BETWEEN YOU
Decide which quadrilateral is nearer to
being a parallelogram. Give reasons
for your choice.

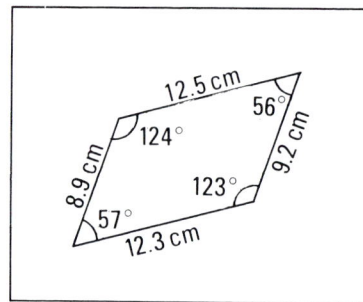

YOU
Write down what changes you would need to make to your quadrilateral (the one for which
you added one dot) to make it into a parallelogram.

b) Repeat the same exercise for a kite.

YOU

YOUR FRIEND

YOU

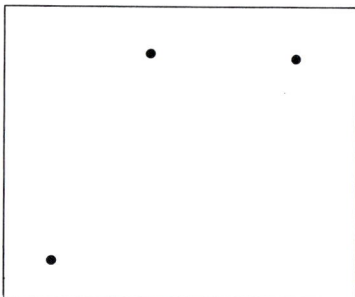   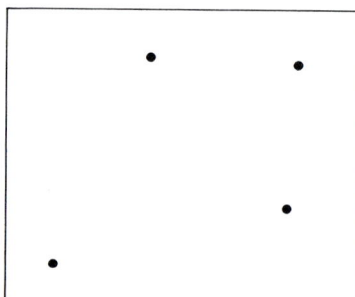   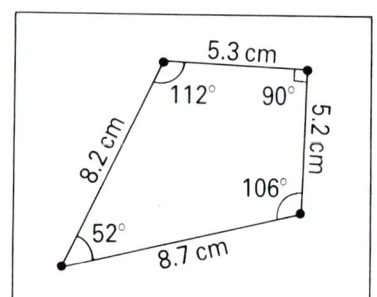

▨▨▨▨▨▨▨▨▨▨▨▨ ACTIVITY ▨▨▨▨▨▨▨▨▨▨▨▨

You need plain paper, thin card, and scissors.

2 a) Draw any triangle on card, like this.

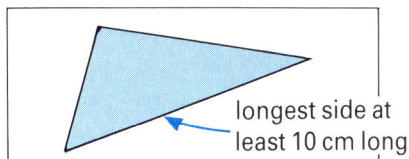

longest side at least 10 cm long

   b) Cut out your triangle. Mark the position
       of its three vertices on plain paper.

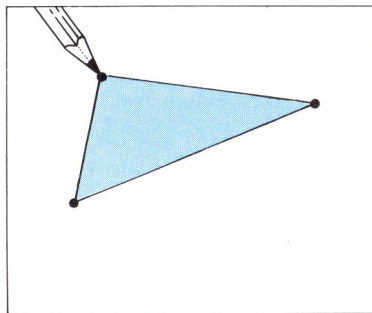

   c) Give your triangle a half turn, and mark a
       fourth point. Join the dots. What kind of
       quadrilateral have you drawn?

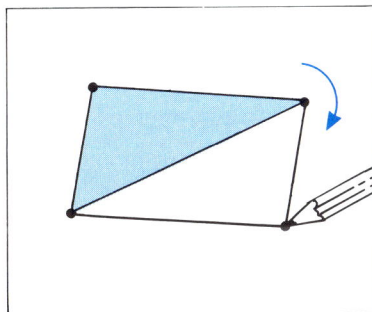

3 Repeat question 2 b) and c), but in c) turn
   your triangle over to make a kite.

4 Repeat question 2 b) and c) again, but in c)
   turn your triangle over to try to form a concave
   kite with the marked angle as large as possible.

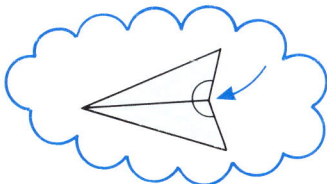

B12

5 Alan makes a parallelogram and a kite using the method in questions 2 and 3. The two shapes are
   identical.

   a) What is special about the triangle?

   b) What special name is given to the quadrilateral?

Angles of parallelograms

1 a) The triangle is
        given a half turn to make
        the parallelogram.
        Sketch the parallelogram;
        mark the size of
        each angle.

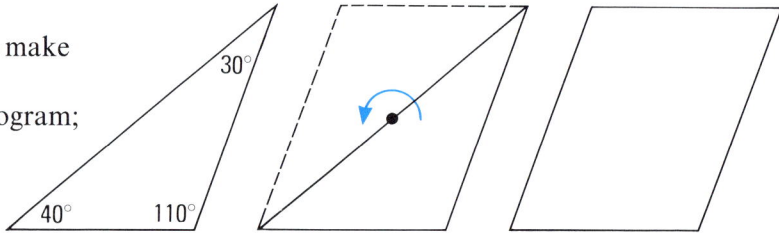

30°

40°   110°

   b) The triangle is again given
        a half turn to make a different
        parallelogram.
        Sketch the parallelogram;
        mark the size of
        each angle.

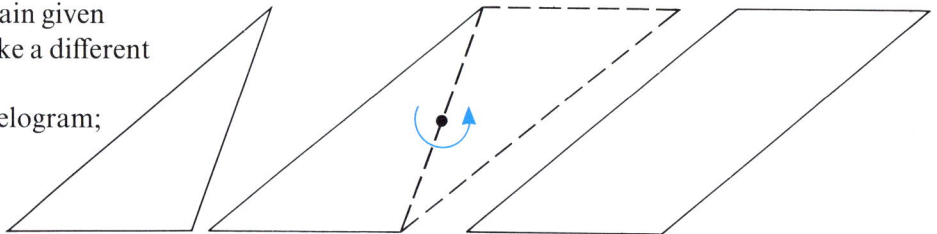

   c) Sketch the third parallelogram that can be made from the triangle; mark the size of each angle.

   d) Look at your three parallelograms; what can be said about:

       (i) any pair of
            opposite angles

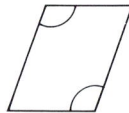

       (ii) any pair of
            adjacent angles?

CHALLENGE

   e) Look at this diagram.
        Use it to explain your
        answers to d) (i)
        and d) (ii).

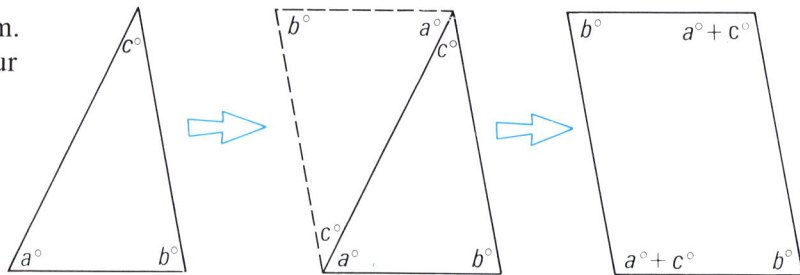

$c°$

$a°$ $b°$

$b°$ $a°$
$c°$
$c°$
$a°$ $b°$

$b°$ $a° + c°$

$a° + c°$ $b°$

TAKE NOTE

Opposite angles of a parallelogram are equal.
Adjacent angles add up to 180°.

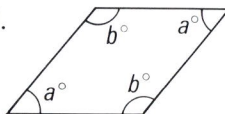

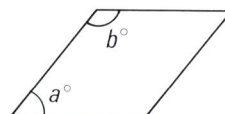

$b°$ $a°$

$a°$ $b°$

$b°$

$a°$

$a° + b° = 180°$

2   The triangle is given a half turn to produce a rhombus. Sketch the rhombus; mark the size of each angle.

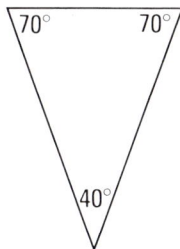

3   Make a sketch of this parallelogram; mark the size of each angle.

4   a) Which of these are parallelograms?

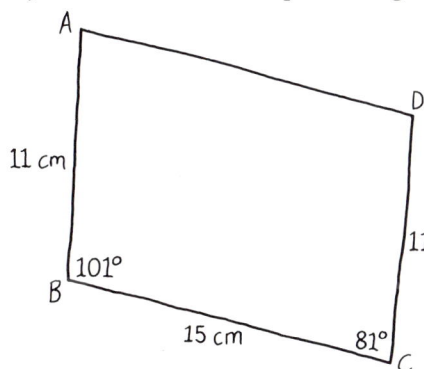

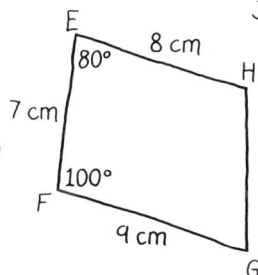

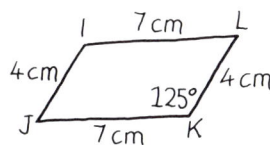

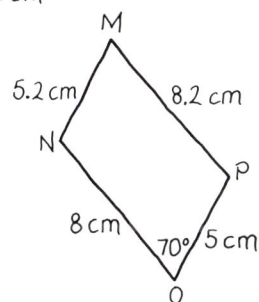

b) Is side AD 15 cm long, or is it shorter or longer than 15 cm?

c) What can you say about:

(i) side GH   (ii) angle JIL   (iii) angle NMP?

5   Imagine giving parallelogram WXYZ a half turn about point O.

a) Line WX goes exactly onto line YZ, so we can say WX = YZ.
List all the other lengths that are equal to each other.

b) Are the diagonals of a parallelogram the same length?

c) Do the diagonals cut each other in half?

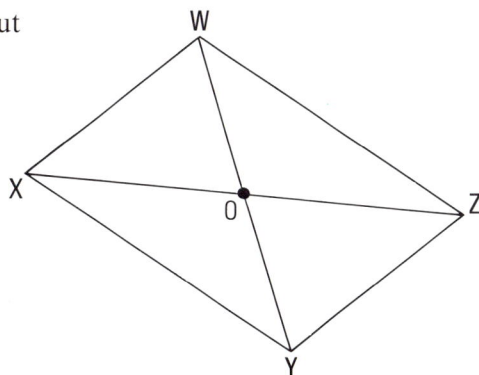

Angles from symmetry

1 a) The triangle is turned over
   (reflected) to make the
   concave kite.
   Sketch the kite; mark the
   size of each angle.

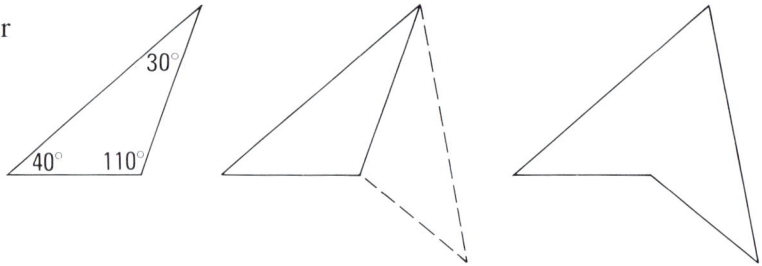

   b) The triangle can be turned over to make another concave kite. Sketch this other kite; mark the
      size of each angle.

   c) The triangle can also be turned over to make a convex
      kite. Sketch the kite; mark the size of each angle.

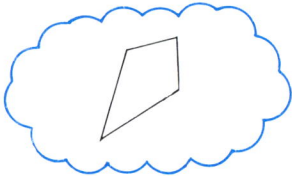

CHALLENGE

   d) Sketch a triangle that can only produce convex kites.
      What can you say about the angles of the triangle?

B12

2 a) What is the size of the third angle
      of this triangle?

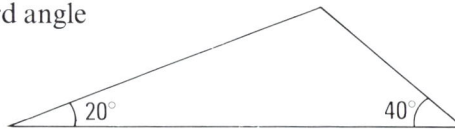

   b) Using two of the triangles we can make different quadrilaterals, for example, A, B, and C. (We
      are allowed to 'flip' the triangle over.)

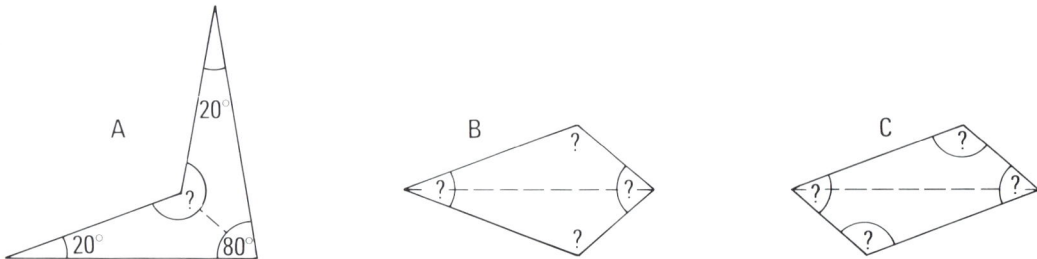

      Make a sketch of *all* the possible quadrilaterals. On your sketches mark in the size of each
      interior angle.

   c) Underneath each of your sketches write down:

      (i) the number of lines of symmetry the shape has

      (ii) the order of turn symmetry of the shape.

      For example
      1 line of symmetry
      Order of turn
      symmetry 1

ENRICHMENT

EXPLORATION: WINDMILLS

1 This shape
has been generated from
the numbers 1, 2, 4.

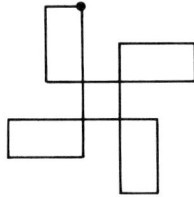

Here are the drawing rules.

Draw a point.	You are facing left. Move 1 unit.	Turn left. Move 2 units.	Turn left. Move 4 units.	Turn left. Move 1 unit again. . . . and so on.

a) Complete this shape for the numbers 1, 2, 5. Use squared paper.

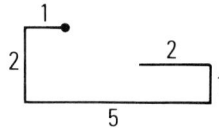

b) For some numbers (for example, 1, 2, 4), the shape has a square in the middle.

For others (for example, 1, 2, 3), the shape has no 'hole' in the middle at all.

For which sets of numbers do the shapes have no 'hole' in the middle? Find a rule.

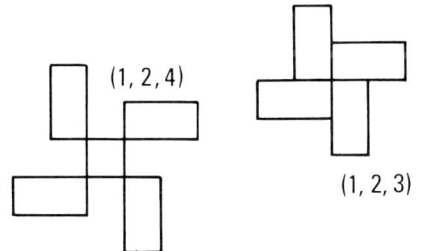

(1, 2, 4)

(1, 2, 3)

c) This shape has turn symmetry.

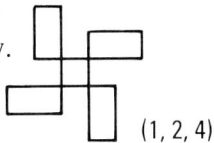

This shape also has four lines of symmetry.

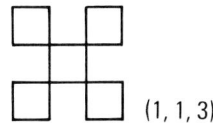

(1, 2, 4)

(1, 1, 3)

Find a rule for shapes to have four lines of symmetry.

B12

REVIEW

A graph is a way of representing information.
The information in the table is also shown
on the graph.

Time since journey began (seconds)	Speed of the steam roller (metres per second)
0	0
5	0.1
10	0.4
15	0.8
20	1.2
25	1.5
30	1.8
35	2.0

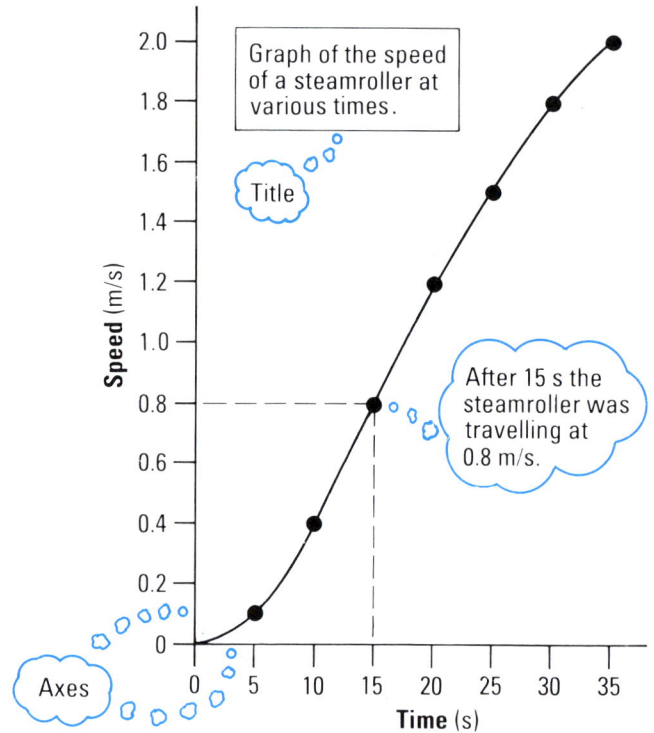

Graph of the speed of a steamroller at various times.

Title

After 15 s the steamroller was travelling at 0.8 m/s.

Axes

■ What are the advantages of having information in a graph rather than in a table? Write down two.

These are two more graphs.

A 'scatter-graph'

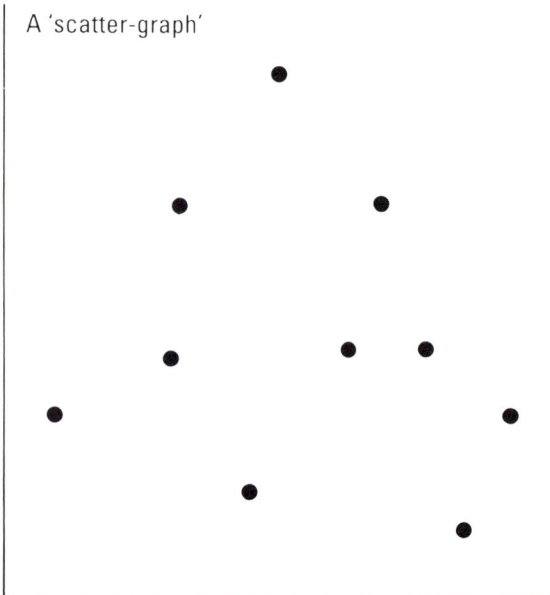

A 'step-graph'

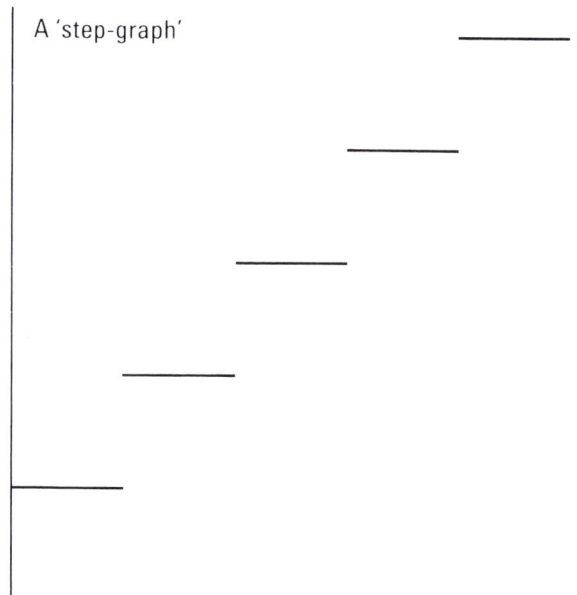

■ What do you think each one might represent? Write down labels for the axes.

CONSOLIDATION

1 On 14 June 1988 £1 was worth $1.8175 (US dollars). Which of these graphs are correct conversion graphs for pounds to US dollars? (Two are.)

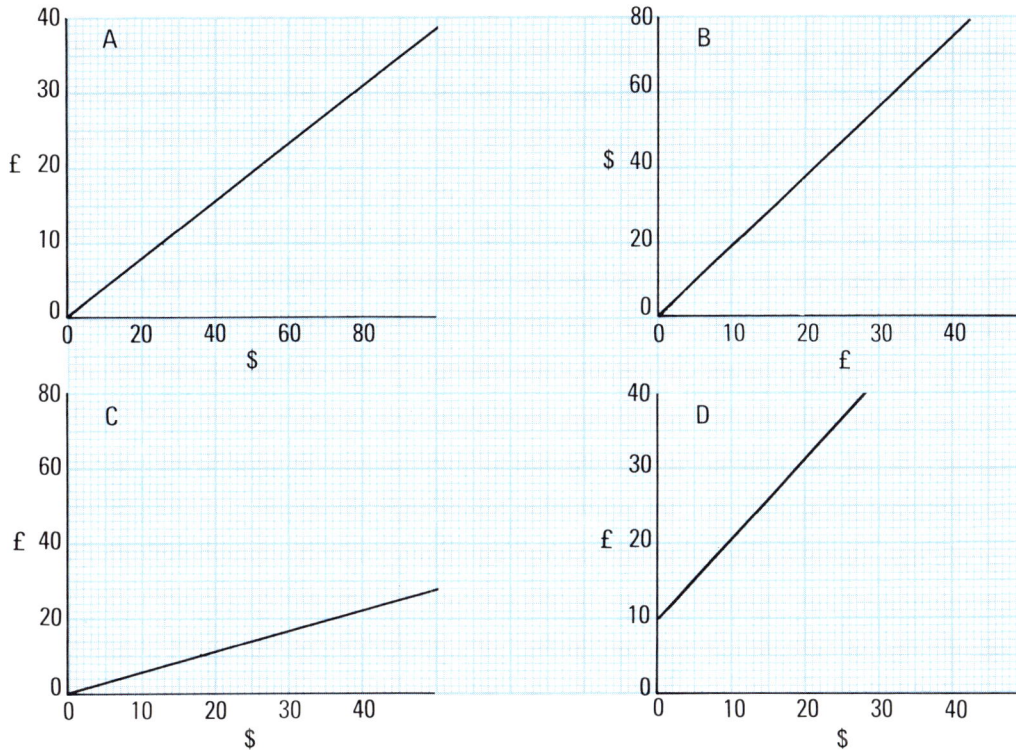

================ ASSIGNMENT ================

B13

2 a) Look in a newspaper for the exchange rates between different currencies and the pound. Choose a currency. Draw a conversion graph for amounts up to £200. Check the exchange rate each week for the next 5 weeks (say each Saturday). Draw an exchange rate graph for each week.

   b) What is (i) the most
                (ii) the least
      that £1000 is worth in your currency during these weeks?

   c) What is the percentage difference between the two values in b)?

3   Temperature can be measured in °F (degrees Fahrenheit) or °C (degrees Celsius).

a) Find out the temperature that water freezes to ice:

(i) in °C   (ii) in °F.

b) This is a conversion graph for °C to °F. Water boils at 100 °C. What is this in °F?

c) The weather report forecasts a temperature of 20 °C. What is this in °F?

d) A washing machine uses water temperatures of 40 °C and 60 °C. What are these in °F?

e) Refrigerators usually maintain a temperature of 38 °F. What is this in °C?

f) Normal body temperature is 98.4 °F. What is this in °C?

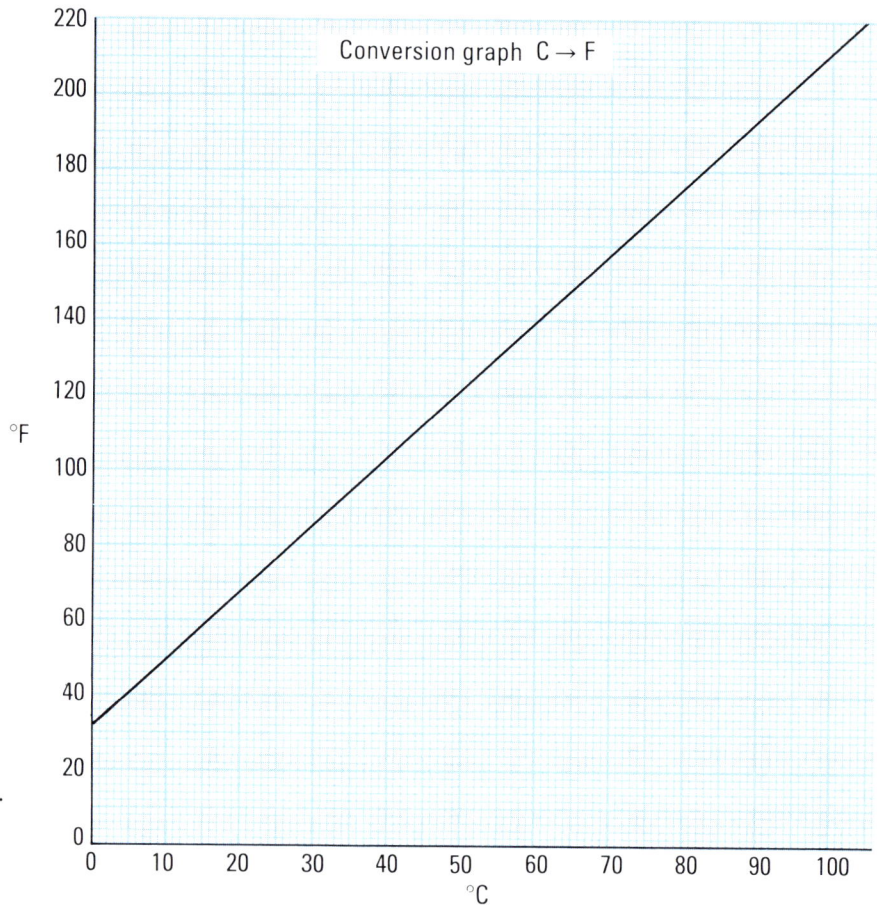

Conversion graph C → F

°F

°C

━━━━━━━━━ WITH A FRIEND ━━━━━━━━━

g) The temperature in a deep freeze is ⁻20 °C. Discuss how you could draw a graph to enable you to find out this temperature in °F. Draw your own graph, and write down your answer.

h) Old cookery books give oven temperatures in °F. Look in a new cookery book to find out these cooking temperatures in °C:

(i) 350 °F   (ii) 375 °F   (iii) 450 °F.

DEEP FREEZE

4   This chart gives information about postal costs in 1987.

Inland Letters and Cards

	60g	100g	150g	200g
1st Class	18p	26p	32p	40p
2nd Class	13p	20p	24p	30p

Overseas Letters (Air)

Zone A (N Africa, Middle East)	29p (10g)
Zone B (Americas, Africa, India, SE Asia)	31p (10g)
Zone C (Australasia, Japan, China)	34p (10g)
Europe (EEC countries)	18p (20g)
Europe (non-EEC) and surface letters	22p (20g)

Please use the postcode and include a return address

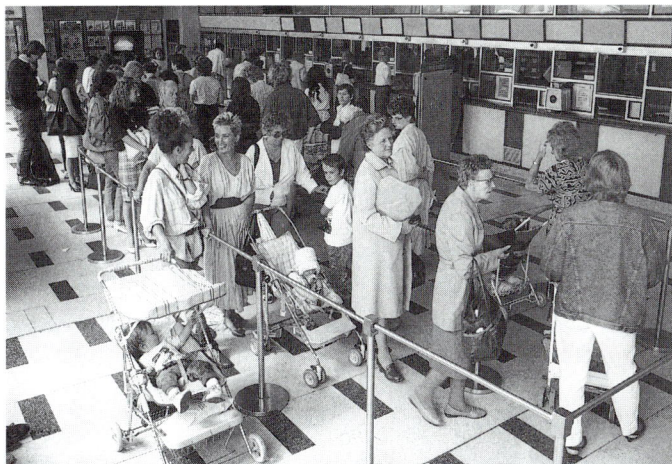

The graph gives the same information for 1st Class letters and cards.

a) Check that the graph shows that an 80 g letter costs 26p.

b) Use 2 mm squared paper. Copy the graph for 1st Class letters and cards (leave out the title – you will need a new one). Use a scale of 1 cm to 5p on the vertical axis and 4 mm to 10 g on the horizontal axis. On the same axes draw the graph for 2nd Class letters and cards. Use a different colour for the 'steps' on this graph. Write a new title for your graph.

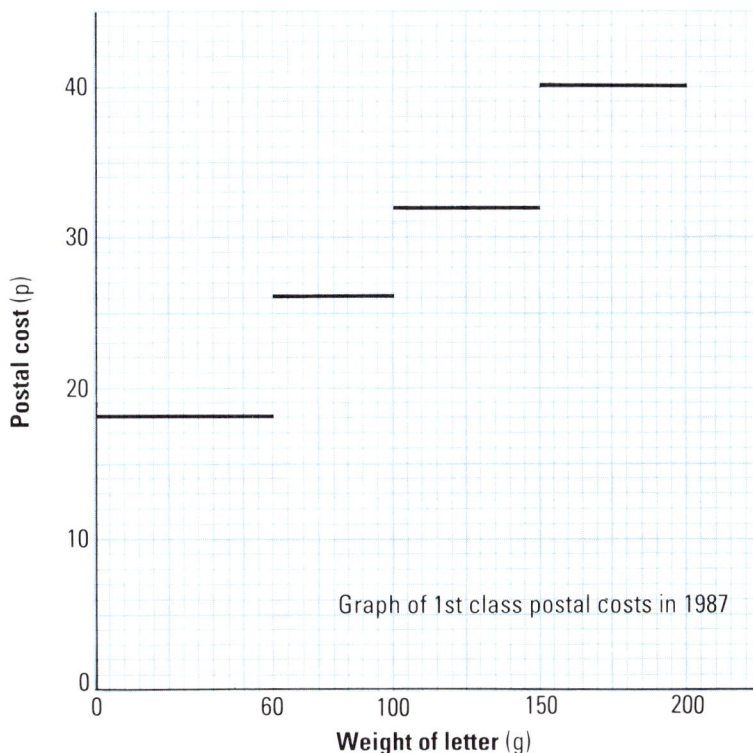

Graph of 1st class postal costs in 1987

c) Use your graph to answer this question:

How much more does it cost to send a 120 g letter by 1st Class rather than 2nd Class post?

Check your result, using the original table of costs.

d) Do you think there are any advantages in having the information about postal costs on a graph rather than in the table? If so, write down how you think this helps.

5 The graph is for the journey of an air-sea rescue helicopter. The newspaper cutting tells you about the rescue. Use the newspaper cutting and the graph to help you to answer the questions below:

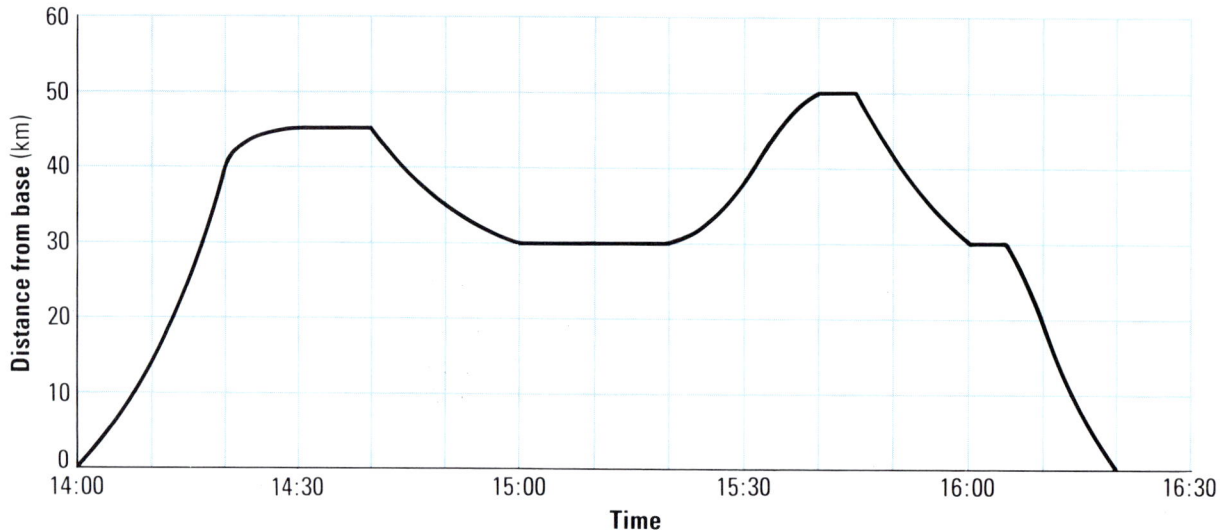

RESCUE AT SEA

The helicopter took off at 2 pm to rescue the crew of a yacht which had sunk. It found two people in a life raft, and airlifted them to the beach. It then returned to search again, found the third crew member, who was wearing a life jacket, and took him to the beach. By 4:20 pm the helicopter was back at base – a job well done.

a) At about what time do you think the first two people were rescued?

b) About how far away from base were they?

c) About how long did it take to pick the two survivors up from the life raft?

d) At what time did they land on the beach?

e) How far away from the place where the first two crew members were picked up was the third survivor?

f) How far did the helicopter fly from the beach back to base?

g) How long did this stage of the flight take?

WITH A FRIEND

6   Here are six distance–time graphs to discuss. Decide what each one might represent. (For
    example, a stone falling to the ground, a car in a traffic jam. ...) Write down a title for each graph.

A

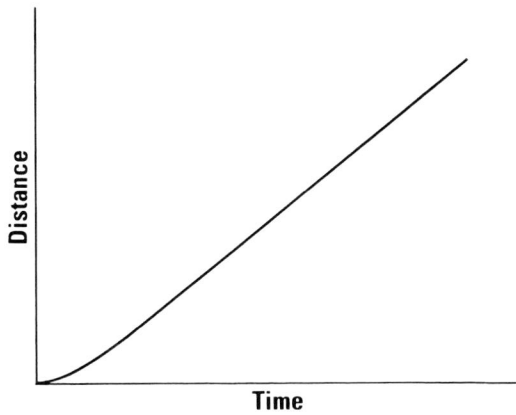

Distance

Time

B

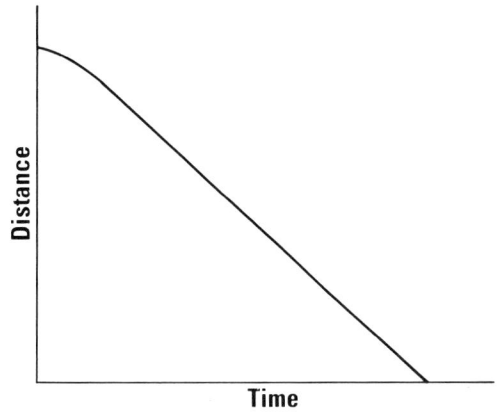

Distance

Time

C

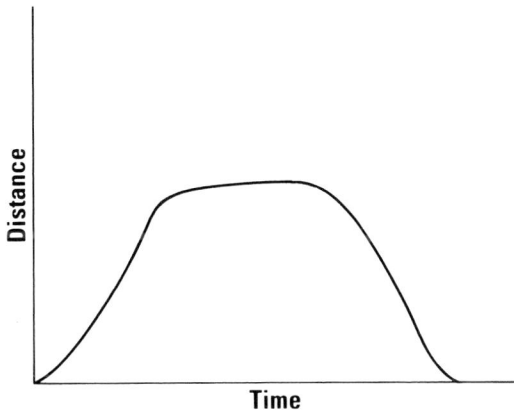

Distance

Time

D

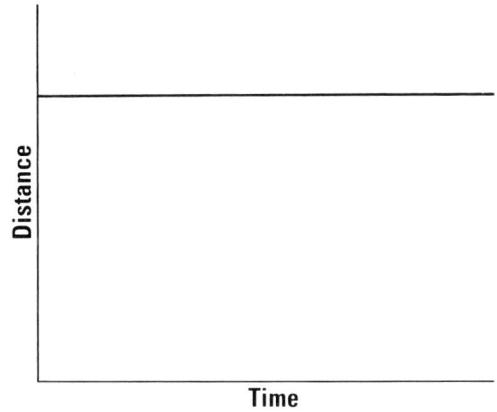

Distance

Time

E

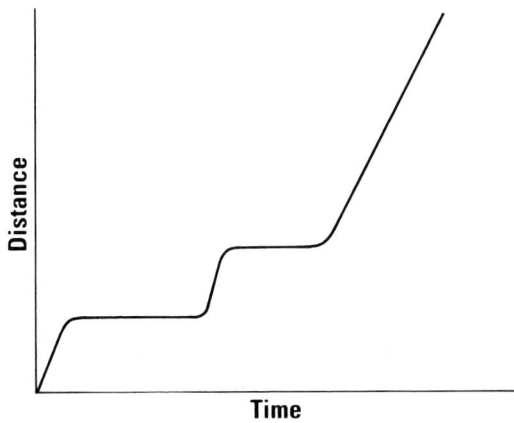

Distance

Time

F

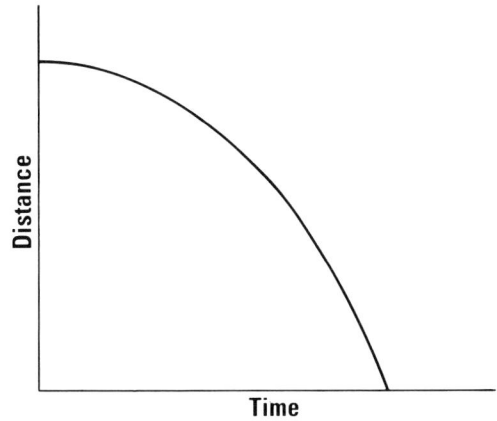

Distance

Time

B13

ENRICHMENT

ASSIGNMENT: PENDULUMS

1 You need a piece of string 1 m long, some weights (four or five different sizes of washers or nuts or bolts will do), and a watch.

a) *Use a different weight each time*. Keep the length of string the same (say about $\frac{1}{2}$ m). Make a pendulum, and find the time for 10 complete swings (from a starting position back to a starting position). Do this for each weight. Draw a graph to show your results and write a short report to explain what you discover.

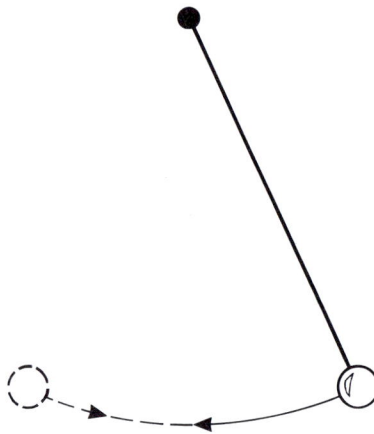

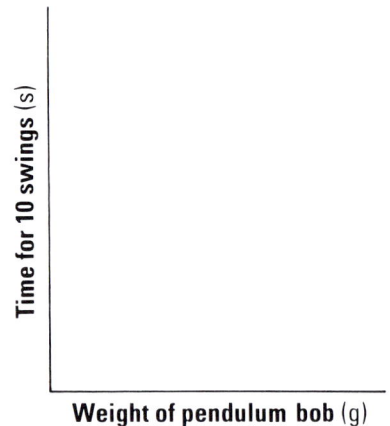

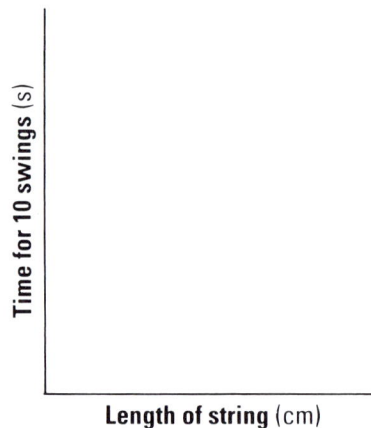

Time for 10 swings (s)

Weight of pendulum bob (g)

b) This time use the *same* pendulum bob each time. Change the length of the string. Find the time for 10 swings for each length of string. Draw a graph to show your results and write a short report to explain what you discover.

Time for 10 swings (s)

Length of string (cm)

c) Think of another way of changing the experiment with the pendulum. Draw a graph of your results and write a short report to explain what you discover.

CORE

1   A small fishing boat is anchored out at sea. This is a map of the area.

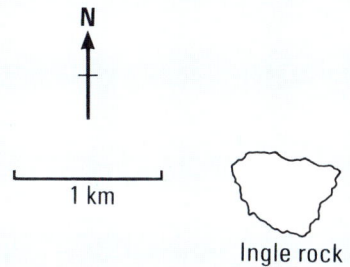

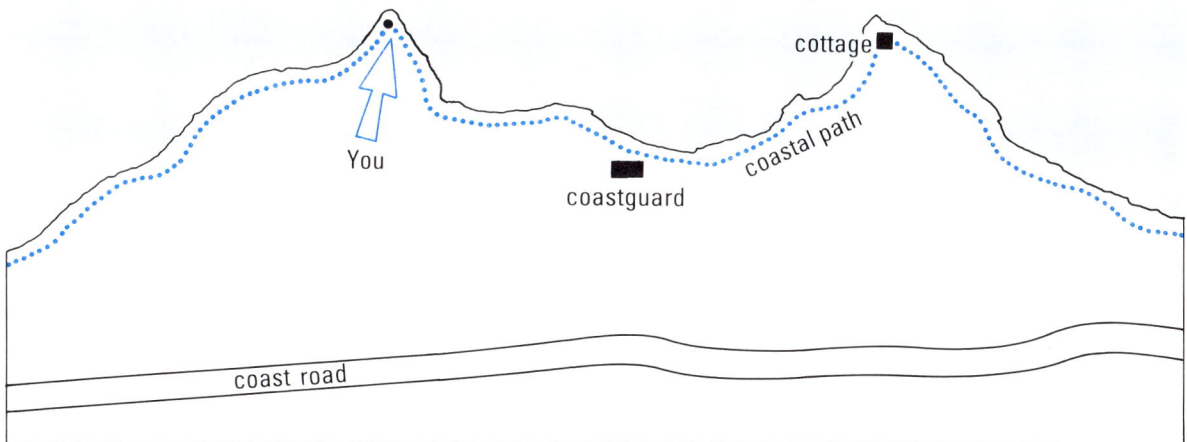

N

1 km

Ingle rock

cottage

You

coastal path

coastguard

coast road

B14

a) Discuss between you how you would locate
and mark the position of the boat on the map.
This is all the equipment you have.

30 cm diameter
protractor

b) Each of you sketch the map.
Show on your sketch how you
would locate the boat.

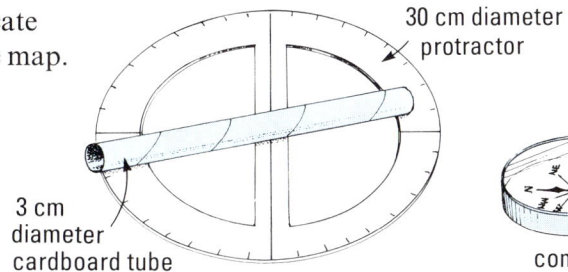

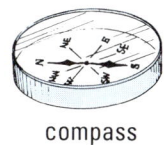

3 cm
diameter
cardboard tube

compass

2   The Loch Ness Society has been searching for 'Nessie' for many
     years. They use instruments like this to fix the exact position
     of 'Nessie' whenever there is a sighting. The drawing below
     shows two of these instruments being used to fix Nessie's position.

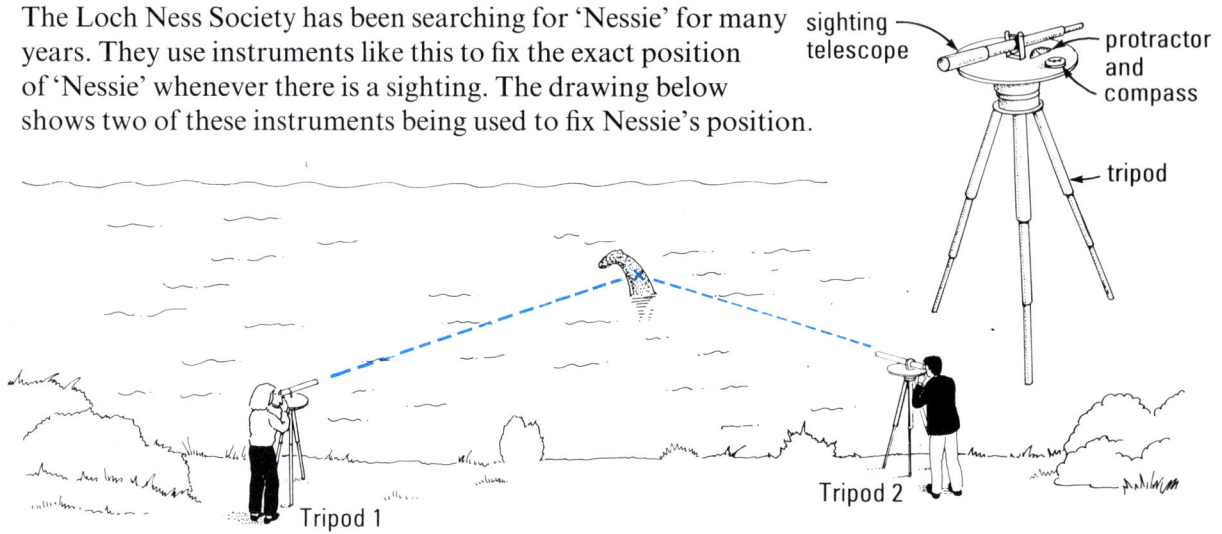

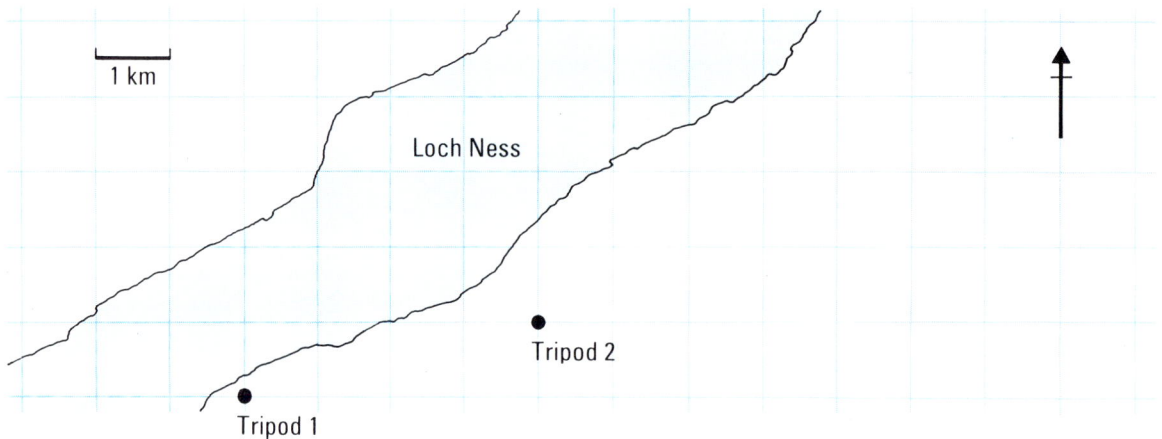

Follow these instructions to fix Nessie's latest position.

a) Copy the sketch map of Loch Ness on
    1 cm squared paper.

b) Tripod 1 measured Nessie's bearing
    as 035°. Draw a line from Tripod 1
    across Loch Ness.

c) Tripod 2 measured Nessie's bearing
    as 284°. Draw a line from Tripod 2
    across Loch Ness.

d) Mark in the position of Nessie.

e) What is the shortest distance from Nessie
    to the shore?

ACTIVITY: ELEVATION AND DECLINATION

3   You need a sheet of A4 paper. Roll up your
    sheet of paper to make a tube.

TUBE OF PAPER

a) Look horizontally through your tube.
   Now raise it to look at a point high up on
   the classroom wall. Estimate the angle
   between your two sightings.

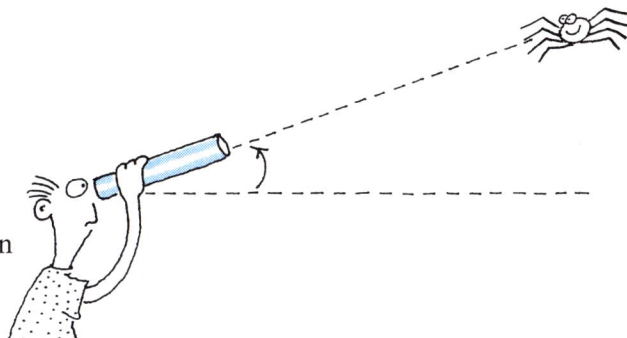

TAKE NOTE

The angle you have estimated is called the angle
of *elevation* of the point you looked at.

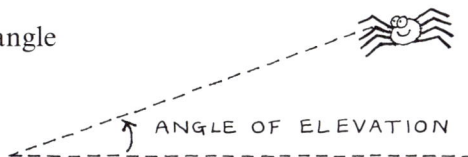

ANGLE OF ELEVATION

b) Choose two more objects above your head level. (For example, the top of the door, a nest in a
   tree outside the window.) Estimate the angle of elevation of each one.

c) Name something whose angle of elevation is about 80°.

d) Look horizontally through your tube.
   Now lower it until you can see a point
   low down on the wall. Estimate the
   angle between your two sightings.

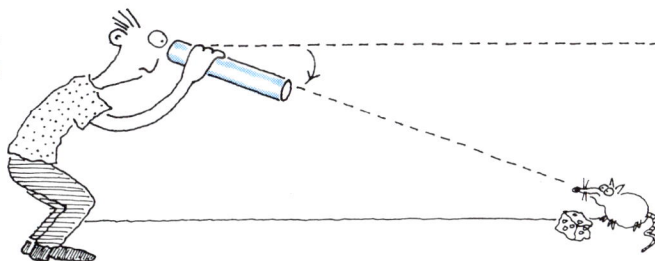

TAKE NOTE

The angle you have estimated is called
the angle of *declination* of the point you looked at.

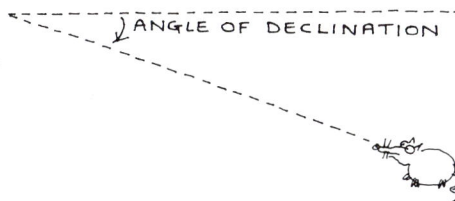

ANGLE OF DECLINATION

B14

e) Choose two more objects below your head level. Estimate the angle of declination of each one.

f) Choose a point on the floor about 1 m away. Roughly, what is its angle of declination?

WITH A FRIEND: LOCATE IT

4 Play this 'Locate it' game. Decide between you which direction is North. Each of you:

- Choose 5 objects you can see outside the window.
- Estimate their bearings and angles of elevation or declination. (Use your tube from question 3.)
- Write down your estimates like this

    (040°, 70°E),   (300°, 20°D)

    ↑   ↑   ↑   ↑

  bearing elevation (E)   bearing declination (D)

Guess what each other's objects are. The one with the larger number of correct guesses wins.

5 Surveyors use *theodolites* to measure bearings, angles of elevation, and angles of declination.

elevation and declination measurement

bearing measurement

a) The surveyor is measuring the angle of elevation of the chimney top. (She needs to find its height.) Use squared paper, with a scale of 1 cm to 1 m. Make a scale drawing to find the height of the chimney for the surveyor.

b) Another surveyor needs to find how high the top of the tower is above the moat. He measures the angle of elevation from two different places. Make a scale drawing to find the height of the tower.

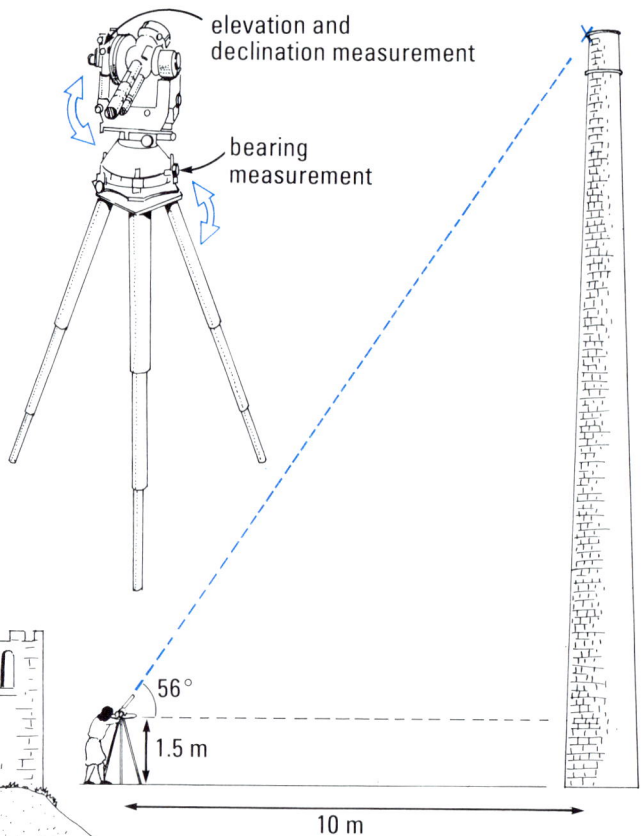

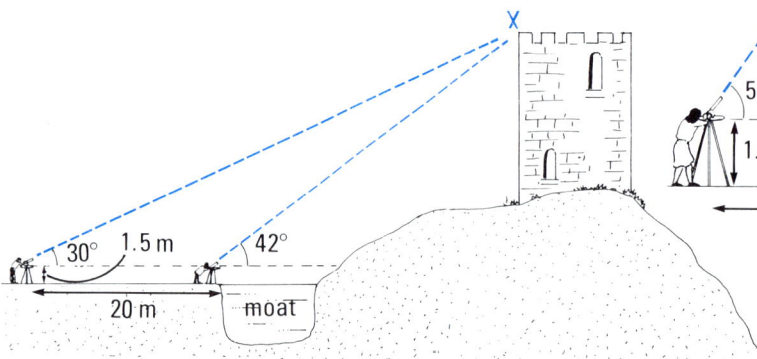

56°

1.5 m

10 m

30° 1.5 m   42°

20 m   moat

THINK IT THROUGH

6 How could you find the width of a river without crossing it? You may use a theodolite. Draw a diagram to explain what you decide.

ENRICHMENT

ASSIGNMENT: A PLOT OF LAND

1   Choose a plot of land close to your school or home (such as a park, a wood, or a city square).
    Make your own instruments to measure bearings and angles of elevation.

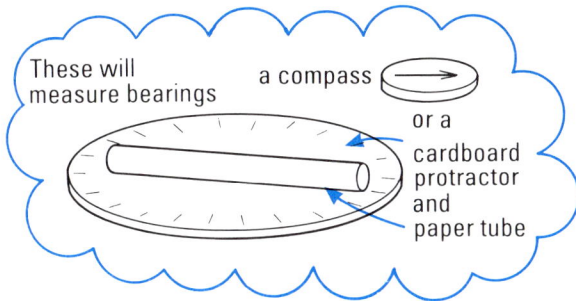

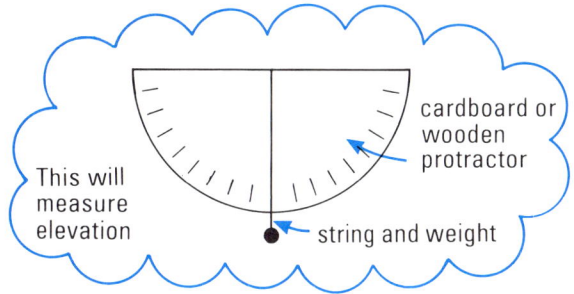

These will measure bearings — a compass — or a cardboard protractor and paper tube

This will measure elevation — cardboard or wooden protractor — string and weight

Make a sketch of the area you have chosen
   a) looking North,
   b) looking South.

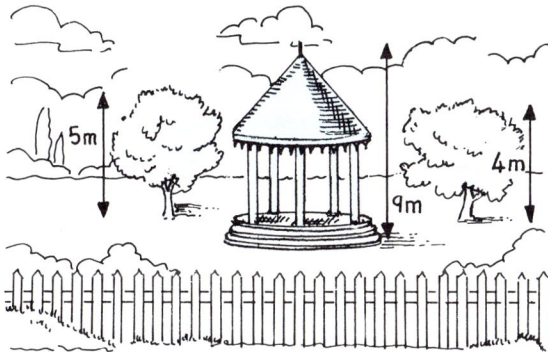

24m

5m

4m

9m

24m

12 m

BANK

For each view measure the heights of at least two things. Mark the heights on your sketches.

c) Draw a plan of the area you chose. Mark the objects whose heights you measured. On your plan show the distances between your objects.

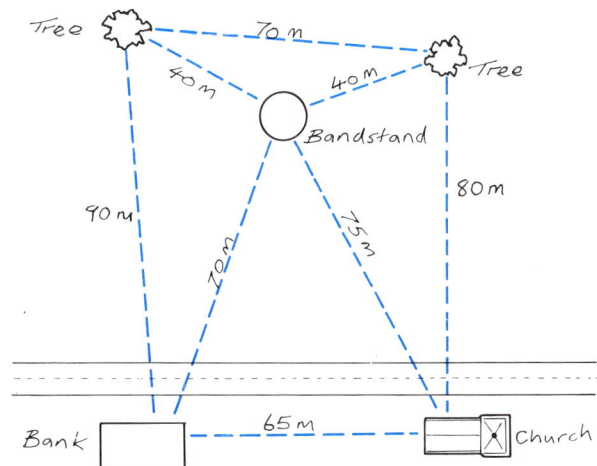

Tree

70 m

40 m

40 m

Tree

Bandstand

80 m

90 m

70 m

75 m

65 m

Bank

Church

B14

CORE

1 a) Six friends are trying to decide which of these mixtures will make the darker blue.

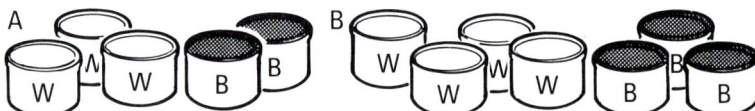

They all have different explanations. Discuss them. Do you agree that each one is a sensible explanation? Which one do you find most convincing?

Winston
If I add 1 tin of white and 1 tin of blue to A I get B. So they're the same.

Tariq
I start with A and add half as much paint again. This gives me 3 blue as in B but 4½ white instead of 4. So B is darker.

Marigold
In A, for every blue tin there are 1½ white tins. In B, for every blue tin there are 1⅓ white tins. That means A has more white and so is lighter.

Sara
B has more blue paint than A so it is darker.

Midge
If I had 12 white tins in A I'd have to have 8 blue tins. If I had 12 white tins in B I'd have to have 9 blue tins. So B is darker.

Moira
If I increased the amount of paint in A by a third I'd get 4 white tins, but only 2⅔ blue tins, so B is darker.

b) Do you have a different explanation? If so, write it down and explain it to your friend.

2 You usually have to dilute household bleach before using it.
For sinks the ratio is 8 ml of bleach : 5 litres of water.
For floors it is 12 ml of bleach : 4 litres of water.
Which is the stronger mix? Write an explanation like the ones in question 1 to explain your choice.

3 a) The green paint was made from a mixture of 3 tins of
       yellow and 5 tins of blue paint. Write the ratio of yellow to
       blue in five different ways,
       for example,

           3:5
           6:10.

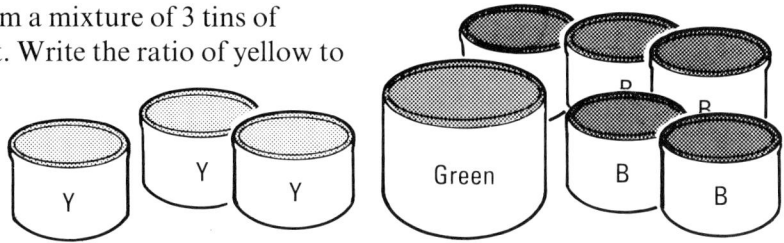

   b) Another green mixture is made from 5 tins of yellow and 8 tins of blue. Write the ratio of
       yellow to blue for this green in five different ways.

   c) Write each ratio in a) and b) like this: 15: ☐.

   d) Which is the lighter green, the 3:5 mixture or the 5:8 mixture?

   e) A certain colour was made by mixing 6 tins of green paint with 15 tins of white. You want the
       same colour and have 2 tins of green paint. How many tins of white should you mix with it?

   f) Copy and complete: 6:15 = 3:☐ = 2:☐ = ☐:30

4 Write the ratio   white paint:blue paint   in A and B like this ☐:1. Which mixture is lighter blue?

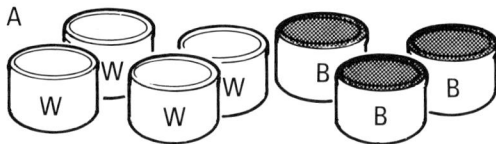

5 Mix 5 tins of white paint with 2 tins of yellow. The ratio is 5:2, which is 2.5:1.

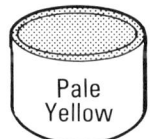

   Add one tin of each colour to the mixture. The ratio is now 6:3, which is 2:1.

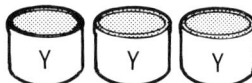

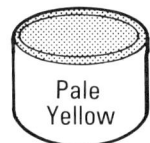

   Add another tin of each colour. Now the ratio is 7:4, which is 1.75:1.

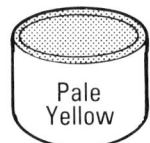

   Is the shade of yellow getting lighter, darker, or staying the same?

TAKE NOTE

We can compare ratios such as 6:15 (A) and 5:13 (B) more easily:

   a) by writing them as 1:☐ or ☐:1
       For example, A is 1:2.5 and B is 1:2.6.
                                              or
   b) by writing them with one number the same.
       For example, A is 30:75 and B is 30:78.

B15

THINK IT THROUGH

6 a) Builders use different ratios of cement to
       sand for different jobs. Which of these will
       give the stronger mix? Explain how you
       decided.

   b) Write down a ratio for a mix which is:

       (i) stronger than **A**
       (ii) weaker than **B**
       (iii) weaker than **A** but stronger than **B**.

	CEMENT		SAND
Ⓐ	3	:	7
Ⓑ	2	:	5

7 The ratio of the length to the width of the table top is 5:2.

A

B

C

E

D

   a) Which of A to E are correct scale drawings?

   b) The top of a bedside table has length to width ratio of 1.2:1.
       It is 40 cm wide. How long is it?

8 18 carat gold is 75% pure gold and 25% other metals.
   So the ratio   pure gold:other metals   is   75:25   or   3:1.
   What is the ratio for these golds?
   (Write each ratio as a ratio of two whole numbers.)

   22 carat gold ⟹ 91⅔ % pure gold

   14 carat gold ⟹ 58⅓ % pure gold

   9 carat gold ⟹ 37.5% pure gold

Making comparisons

1   The table shows how children's pocket money changed between 1980 and 1987.

   a) Look at the ratios of

   14–16 year olds'   11–13 year olds'
   pocket money   ·   pocket money

   for each year. (*The last two columns*)
   Is this what you would expect? Why?

   b) Check that in 1980 the ratio

   14–16 year olds'   11–13 year olds'
   pocket money   ·   pocket money

   was about 1.4 : 1.

   c) Check that in 1987 the same ratio was
   nearer 1.5 : 1.

   d) When was the ratio (i) at its highest,
   (ii) at its lowest?

AVERAGE WEEKLY POCKET MONEY (pence)							
Year	All boys and girls	All boys	All girls	Age 5–7	Age 8–10	Age 11–13	Age 14–16
1980	99	99	99	59	66	109	152
1981	113	117	108	55	87	132	173
1982	95	93	96	64	74	114	128
1983	112	124	115	90	103	141	178
1984	105	101	109	42	73	113	187
1985	109	110	107	50	72	128	188
1986	117	120	114	54	77	142	198
1987	116	112	120	42	76	143	212

Source : Wall's pocket money Monitor by Gallup

   e) Look at the ratios of boys' pocket money : girls' pocket money for each year. (*The second and third columns*)
   Write one or two sentences to describe how the ratio changed during the 1980–1987 period.

   f) Study some ratios of your own choice from the table. Write about what you discover.

2   The table gives some information about
   mothers in a city who have only one child. It
   shows how many go out to work and how
   many do not.

	Working mothers	Non-working mothers
0–4-year-olds	15	38
5–9-year-olds	28	90

   a) Write each of the ratios

   A   Working mothers   Total number of mothers
        of 0–4-year-olds   ·   of 0–4-year-olds

   B   Working mothers   Total number of mothers
        of 5–9-year-olds   ·   of 5–9-year-olds

   like this: ☐ : 1.

   b) Which group of mothers tends to go out to work more, mothers of 0–4-year-olds or mothers
   of 5–9-year-olds? Explain how you decided.

B15

Ratio division

1   9 carat gold is a 3:5 mixture of pure gold and other metals.
    (Other metals are added to harden the gold.)

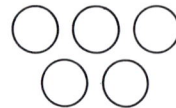

3 parts
pure gold

5 parts
other metals

   a) This is an 8 ounce
       bar of 9 carat gold.
       How many ounces
       of it is pure gold?

   b) A 32 gram bar of 9 carat gold will have 12 grams of pure gold in it. Draw a diagram to help
       you explain why this is true.

   c) This is a 9 carat gold necklace. If only the pure gold was used, how long would the necklace be?

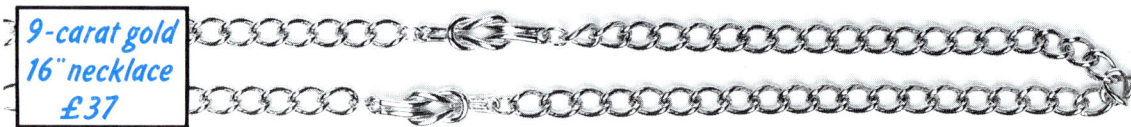

9-carat gold
16" necklace
£37

   d) The necklace weighs 60 g. What weight of pure gold is there in the necklace?

2   Rajiv and Jane put their money together to buy a car. Rajiv puts in £300; Jane puts in £500.
    (So the ratio   Rajiv's share:Jane's share   is 3:5.) Two years later they sell the car for £560.

   a) Explain why Rajiv should get $\frac{3}{8}$ of the £560.

   b) How much should each of them receive when the car is sold?

3   Brazil nuts and cashew nuts are mixed by weight in the ratio 3:2.

   a) Explain why $\frac{2}{5}$ of the weight should be cashew nuts.

   b) How many grams of each type of nut should there be in the bag?

MIXED
NUTS
Brazil/Cashew
mix
$\frac{1}{2}$ kg

TAKE NOTE

£300 is to be shared
between Rula and Tariq
in the ratio 7:3.

Rula gets 7 'parts'

Tariq gets 3 'parts'

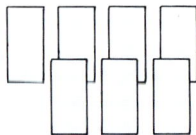

$\frac{7}{10}$ of £300 = £210

$\frac{3}{10}$ of £300 = £90

4   Divide £240 between Rula and Tariq in the ratio 5:7.

ENRICHMENT

Ratio squares

▨▨▨▨▨▨▨▨▨▨ WITH A FRIEND ▨▨▨▨▨▨▨▨▨▨

1 a) These are special arrangements of numbers. Discuss
       between you what is special about them. Each of you write
       down what you decide.

   b) Find two more examples of arrangements like them.

2	12
6	36

5	15
7	21

8	10
28	35

1.2	4.8
2.4	9.6

2 We will call arrangements like those in question 1 *ratio squares*.
   Which of these are ratio squares and which are not?

A
1	3
2	4

B
1	4
$\frac{1}{2}$	2

C
3	12
1	10

D
99	100
9	10

3 Here are two ratio squares. Find the missing numbers.

a)
?	15
4	20

b)
3	12
?	60

4 These are ratio squares. For each one find the pair
   of smallest possible whole numbers which fit.

a)
28	84
?	?

b)
?	100
?	36

5 a) Check that this is a ratio square.

5	15
12	36

   b) Think about rearranging the numbers in different ways.
       Do we always get a ratio square?
       List all the arrangements which are ratio squares.

For example
12	5
36	15

6 a) Check that for this arrangement these sentences are true:

8	40
2	10

$\frac{8}{2} = \frac{40}{10}$, $\frac{8}{40} = \frac{2}{10}$, $\frac{2}{8} = \frac{10}{40}$, $\frac{40}{8} = \frac{10}{2}$, $8 \times 10 = 2 \times 40$.

   b) Write five sentences for this arrangement.

6	18
12	36

B15

━━━━━━━━━ TAKE NOTE ━━━━━━━━━

We will call this kind of arrangement a ratio square.

Notice that $\frac{5}{20} = \frac{3}{12}$, $\frac{5}{3} = \frac{20}{12}$, $\frac{3}{5} = \frac{12}{20}$, $\frac{20}{5} = \frac{12}{3}$, $5 \times 12 = 3 \times 20$

5	20
3	12

---

7 a) Write down a ratio square which goes with this sentence: $\frac{7}{2} = \frac{14}{4}$

   b) Write down four more sentences for the ratio square.

8 a) These are all ratio squares. Find a, b, c, and d.

(i)
a	24
3	12

(ii)
4	28
b	7

(iii)
5	75
2	c

(iv)
1	d
0.5	1.5

   b) Find p, q, r, and s in each sentence. Your results in part a) will help.

   (i) $\dfrac{4}{p} = \dfrac{28}{7}$ (ii) $\dfrac{5}{2} = \dfrac{75}{q}$ (iii) $\dfrac{3}{r} = \dfrac{12}{24}$ (iv) $0.5 = \dfrac{1.5}{s}$

9 a) This is a ratio square.

a	5
b	15

   Copy and complete each sentence.

   (i) $\dfrac{a}{b} = \dfrac{5}{\square}$ (ii) $\dfrac{b}{\square} = \dfrac{15}{5}$ (iii) $\dfrac{a}{\square} = \dfrac{b}{15}$ (iv) $a \times 15 = b \times \square$

   b) Find the smallest whole number for a, and the smallest for b. Check that your numbers fit each number sentence.

10 Use ratio squares to find the number which t replaces in each sentence:

   a) $\dfrac{t}{4} = \dfrac{9}{12}$ b) $\dfrac{7}{t} = \dfrac{49}{63}$ c) $\dfrac{5}{11} = \dfrac{t}{66}$

   d) $\dfrac{15}{t} = \dfrac{3}{8}$ e) $\dfrac{5}{t} = \dfrac{2}{5}$ f) $\dfrac{3}{7} = \dfrac{t}{3.5}$

11 This is a ratio square. Write down five number sentences for the square, for example,

a	b
c	d

   $\dfrac{a}{c} = \dfrac{b}{d}$.

B15

12 a) Check that A′B′C′ is a ×2 enlargement of ABC.

b) Here is one ratio square for the enlargement.

12	15
24	30

Write down another ratio square (use 9, 15, 18, and 30).

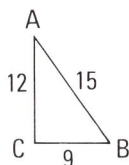

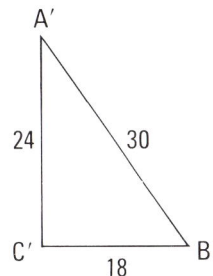

13 This is a ×3 enlargement.

a) Complete these ratio squares for the enlargement.

(i)
3	?
5	?

(ii)
3	6
?	?

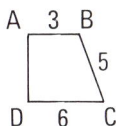

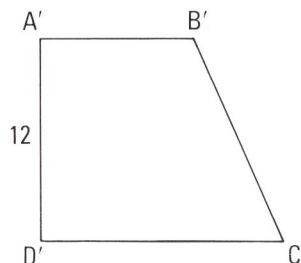

b) Write down a ratio square which uses 5 and 9 and two more numbers.

14 The large rectangle is an enlargement of the small rectangle.

a) Write a ratio square using k, 15, 60, and 28.

b) Write down four different number sentences, using k, for your ratio square.

c) Find k.

d) Check that k fits each of your sentences.

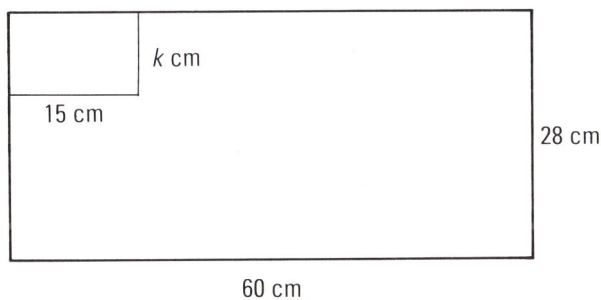

_____ CHALLENGES _____

15 AB′C′ is an enlargement of ABC.
AB = 5 cm, BB′ = 3 cm,
AC = 7 cm, CC′ = k cm. Find k.

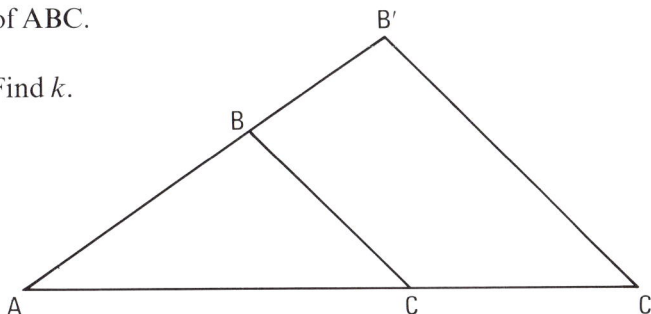

16 Find the number which k represents in each sentence.

a) $\dfrac{k+1}{4} = 2$

b) $\dfrac{8}{k+1} = 4$

c) $\dfrac{4}{5} = \dfrac{12}{k+1}$

d) $\dfrac{15}{18} = \dfrac{k}{k+1}$

CORE

1   Do these in your head. You will need to think hard about some of them. You are not expected to do them quickly!

a) Altogether, is there more than 1 lb or less than 1 lb?

b) Is there more than $\frac{1}{4}$ kg or less than $\frac{1}{4}$ kg left in the bag?

c) The mug is filled from the bottle. Which now contains more water, the bottle or the mug?

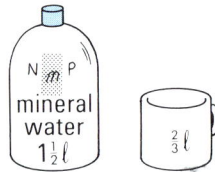

d) $\frac{3}{5}$ m is cut from the pipe. Is the amount left more than $\frac{1}{2}$ m or less than $\frac{1}{2}$ m?

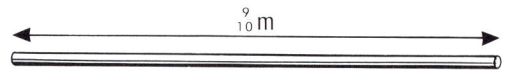

e) The lane is $\frac{3}{8}$ km long. $\frac{1}{4}$ km has been 'tarmac'ed. Is there more than $\frac{1}{4}$ km or less than $\frac{1}{4}$ km left to complete?

f) The flour and icing sugar are mixed together. Altogether, is there more than 1 kg or less than 1 kg?

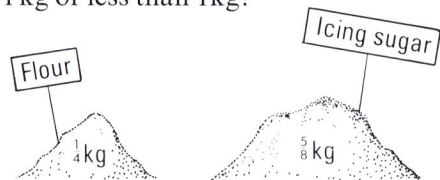

g) Will the screw reach more than halfway or less than halfway into the block?

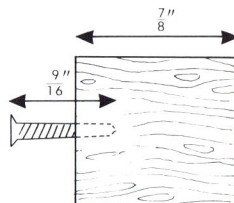

h) Is the width of the washer more than $\frac{1}{4}$ cm or less than $\frac{1}{4}$ cm?

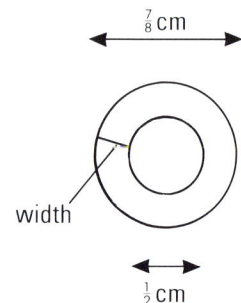

2   Advertisements in the *Daily Listener*
    are sold by the $\frac{1}{2}$-page

$\frac{1}{4}$-page

$\frac{1}{8}$-page

and $\frac{1}{16}$-page.

This grid shows the fractions.
Use it to help you to do these calculations.

a) $\frac{1}{2} + \frac{1}{4} = \frac{\square}{4}$ b) $\frac{1}{2} + \frac{1}{8} = \frac{\square}{8}$

c) $\frac{1}{2} + \frac{1}{16} = \frac{\square}{16}$ d) $\frac{1}{4} + \frac{1}{8} = \frac{\square}{8}$

e) $\frac{1}{4} + \frac{1}{16} = \frac{\square}{16}$ f) $\frac{1}{8} + \frac{1}{16} = \frac{\square}{16}$

g) $1 - \frac{1}{2} = \frac{\square}{2}$ h) $1 - \frac{1}{4} = \frac{\square}{4}$

i) $1 - \frac{1}{8} = \frac{\square}{8}$ j) $1 - \frac{1}{16} = \frac{\square}{16}$

k) $\frac{1}{2} - \frac{1}{4} = \frac{\square}{4}$ l) $\frac{1}{2} - \frac{1}{8} = \frac{\square}{8}$

m) $\frac{1}{4} - \frac{1}{8} = \frac{\square}{8}$ n) $\frac{1}{4} - \frac{1}{16} = \frac{\square}{16}$

o) $\frac{1}{8} - \frac{1}{16} = \frac{\square}{16}$

3   Space in the *Evening Wanderer*
    is sold by the $\frac{1}{3}$-page

$\frac{1}{6}$-page

and $\frac{1}{12}$-page.

Use the grid to help you to do these
calculations:

a) $\frac{2}{3} + \frac{1}{6} = \frac{\square}{6}$ b) $\frac{7}{12} + \frac{1}{4} = \frac{\square}{6}$

c) $\frac{2}{3} - \frac{5}{12} = \frac{\square}{4}$ d) $\frac{5}{6} - \frac{1}{3} = \frac{1}{\square}$

Adding and subtracting fractions

1 a) The wooden bridge can take a 1 tonne load. Jeff and Mutt want to know if their car and trailer can cross safely.

Mutt writes $\frac{1}{3}+\frac{3}{4}=\frac{4}{7}$. (*Mutt adds 'tops and bottoms'.*)

Mutt is obviously wrong. Explain how we can tell simply by looking at the result (compare $\frac{4}{7}$ and $\frac{3}{4}$).

b) Jeff writes two families of fractions: $\left\{\frac{1}{3}, \frac{2}{6}, \frac{3}{9}, \frac{4}{12}, \frac{5}{15}, \cdots\right\}$ $\left\{\frac{3}{4}, \frac{6}{8}, \frac{9}{12}, \frac{12}{16}, \cdots\right\}$

Then he writes $\frac{1}{3}+\frac{3}{4}=\frac{4}{12}+\frac{\square}{12}$ $\frac{3}{4}-\frac{1}{3}=\frac{\square}{12}-\frac{4}{\square}$

$=\frac{\square}{12}$ $=\frac{\square}{12}$

Copy and complete his calculations.

c) Can the car and trailer cross the bridge safely?

d) How much heavier is the car than the trailer?

2 a) Write down four members of the family for $\frac{2}{3}$, including the one with denominator 15.

b) Write down four members of the family for $\frac{1}{5}$, including the one with denominator 15.

c) Use your results in a) and b) to calculate (i) $\frac{2}{3}+\frac{1}{5}$ (ii) $\frac{2}{3}-\frac{1}{5}$

3 a) Write down some members of the family for $\frac{3}{5}$.

b) Write down some members of the family for $\frac{1}{4}$.

c) Calculate   (i) $\frac{3}{5}+\frac{1}{4}$ (ii) $\frac{3}{5}-\frac{1}{4}$

━━━━━━━━━ TAKE NOTE ━━━━━━━━━

To add fractions we DO NOT add 'tops' and 'bottoms' … $\frac{2}{3}\cancel{+}\frac{1}{4}$
To subtract fractions we DO NOT subtract 'tops' and 'bottoms' … $\frac{2}{5}\cancel{-}\frac{1}{4}$

B16

4   Explain in your own words how we should add fractions.
Use this addition to help you to explain: $\frac{1}{3}+\frac{5}{6}$

5 a) Copy the diagram. Shade $\frac{1}{6}$ of the area.

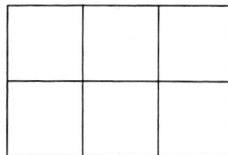

   b) Shade another $\frac{2}{3}$ of the area.

   c) Use your result to complete these:

   (i) $\frac{1}{6} + \frac{2}{3} = \frac{\square}{\square}$ (ii) $\frac{2}{3} - \frac{1}{6} = \frac{\square}{\square}$

6 a) Add any member of family A   A $\left\{\frac{1}{9}, \frac{2}{18}, \frac{3}{27}, \frac{4}{36}, \ldots\right\}$

      to any member of family B   B $\left\{\frac{2}{3}, \frac{4}{6}, \frac{6}{9}, \frac{8}{12}, \frac{10}{15}, \frac{12}{18}, \ldots\right\}$

   b) Subtract any member of family A from any member of family B.

   c) No matter which family members you choose in a) and in b) the result will always be the same. Write one or two sentences to explain why.

7 Here are two fraction families: $\left\{\frac{1}{3}, \frac{2}{6}, \frac{3}{9}, \frac{4}{12}, \ldots\right\}$ $\left\{\frac{1}{4}, \frac{2}{8}, \frac{3}{12}, \frac{4}{16}, \ldots\right\}$

   a) Write down the decimal fraction for each family.

   b) Use the two families to help you to complete these calculations.

   A $\frac{1}{3} + \frac{1}{4}$

   $= \frac{\square}{12} + \frac{\square}{12} = \frac{\square}{12}$

   B $\frac{1}{3} - \frac{1}{4}$

   $\frac{\square}{12} - \frac{\square}{12} = \frac{\square}{12}$

8 a) Write down several members of the fraction family to which:

   (i) $\frac{3}{5}$ belongs   (ii) $\frac{4}{15}$ belongs.

   b) Use your families to help you to do these calculations: (i) $\frac{3}{5} + \frac{4}{15}$ (ii) $\frac{3}{5} - \frac{4}{15}$

9 a) Rewrite each of these pairs of fractions so that they have the same denominator. Choose your own denominator.

   For example, $\frac{2}{3}$ and $\frac{4}{7}$ can be written as $\frac{14}{21}$ and $\frac{12}{21}$ respectively.

   (i) $\frac{5}{8}$ and $\frac{1}{6}$ (ii) $\frac{2}{3}$ and $\frac{1}{4}$ (iii) $\frac{7}{9}$ and $\frac{1}{3}$ (iv) $\frac{8}{10}$ and $\frac{7}{12}$

   b) Use your results in part a) to help you to do these calculations. Write each result with the lowest possible denominator.

   For example, $\frac{1}{3} + \frac{1}{6} = \frac{9}{18} = \frac{1}{2}$.

   (i) $\frac{2}{3} + \frac{1}{4}$ (ii) $\frac{2}{3} - \frac{1}{4}$ (iii) $\frac{5}{8} + \frac{1}{6}$ (iv) $\frac{5}{8} - \frac{1}{6}$

   (v) $\frac{7}{9} + \frac{1}{3}$ (vi) $\frac{7}{9} - \frac{1}{3}$ (vii) $\frac{8}{10} + \frac{7}{12}$ (viii) $\frac{8}{10} - \frac{7}{12}$

B16

━━━━━━━━━━ CHALLENGES ━━━━━━━━━━

10 a) Find two fractions with different denominators whose sum is $\frac{2}{7}$.

   b) Find two fractions with different denominators whose difference is $\frac{2}{5}$.

   c) Find a fraction larger than $\frac{7}{11}$ but smaller than $\frac{8}{11}$.

   d) Which is larger, $\frac{11}{14}$ or $\frac{3}{4}$? Explain why.

11 This is an old rain chart for the week of 21–28 March 1935. Write down the missing totals.

Day	Amount (in inches)	Total so far this week
Monday	$\frac{1}{8}''$	$\frac{1}{8}''$
Tuesday	$\frac{1}{4}''$	$\frac{3}{8}''$
Wednesday	$\frac{1}{8}''$	
Thursday	$\frac{1}{2}''$	
Friday	$\frac{3}{8}''$	
Saturday	$\frac{7}{16}''$	
Sunday	$\frac{5}{16}''$	

12 Copy and complete:

  a) $1\frac{1}{2}+1\frac{1}{3}=2+\frac{1}{2}+\frac{\square}{\square}$

          $=2+\frac{3}{6}+\frac{\square}{\square}$

          $=2+\frac{\square}{6}$

          $=2\frac{\square}{6}$

  b) $1\frac{1}{4}+2\frac{1}{5}=\square$

  c) $2\frac{1}{3}+4\frac{5}{8}=\square$

13 Copy and complete:

  a) $2\frac{1}{4}-1\frac{1}{5}=1+(\frac{1}{4}-\frac{\square}{\square})$

          $=1+(\frac{5}{20}-\frac{\square}{20})$

          $=1+\frac{\square}{20}$

          $=1\frac{\square}{20}$

  b) $3\frac{2}{3}-1\frac{1}{2}=\square$

  c) $4\frac{3}{8}-1\frac{1}{5}=\square$

14 Copy and complete:

  a) $2\frac{1}{4}-1\frac{2}{5}=1+(\frac{1}{4}-\frac{2}{5})$

          $=1+(\frac{5}{20}-\frac{\square}{20})$

          $=(\frac{20}{20}+\frac{5}{20})-\frac{\square}{20}$

          $=\frac{\square}{20}$

  b) $3\frac{1}{5}-2\frac{3}{8}=\square$

  c) $4\frac{5}{12}-2\frac{7}{8}=\square$

15 A batten $1\frac{7}{8}''$ thick has to be screwed onto a $\frac{3}{4}''$ floorboard.

  a) The carpenter wants to leave at least $\frac{3}{16}''$ clearance for the screw. What is the longest screw he should use?

  b) He wants the screw to go at least $\frac{1}{2}''$ into the floorboard. What is the shortest screw he should use?

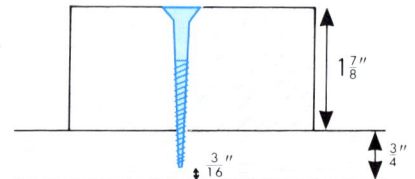

16 The water pipe has an external diameter of $3\frac{5}{32}''$. Its thickness is $\frac{3}{8}''$. What is its internal diameter?

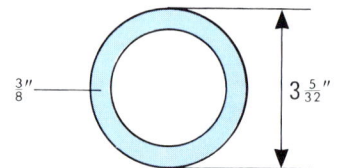

ENRICHMENT

CHALLENGES: FRACTIONS

1   Gwen has a rule for finding fractions between two other fractions. Does it always work? (A calculator might help.) If you say 'No', give an example to explain why.

To find a fraction between $\frac{3}{4}$ and $\frac{7}{10}$

find the average of the numerators → 5

find the average of the denominators → 7

your fraction is $\frac{5}{7}$.

2   How long are the three lines?

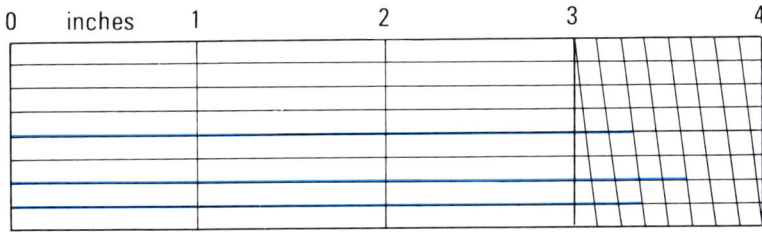

3   Here is a way of dividing these five chocolate bars between six people (A, B, C, D, E, F).

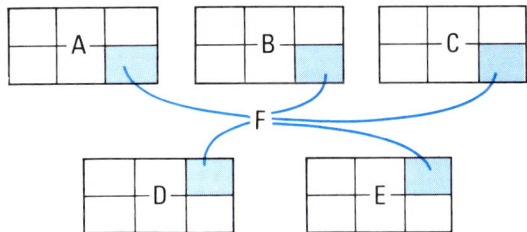

Here is another way.

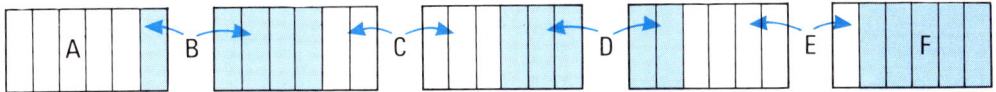

Try to find a different way.

4   Alan needs to drill a hole big enough for a $\frac{3}{10}''$ diameter piece of dowelling.

a) Which of these drill bits could he use?

$\frac{1}{4}''$, $\frac{1}{2}''$, $\frac{3}{8}''$, $\frac{5}{16}''$, $\frac{9}{32}''$.

b) Which one would give the tightest fit?

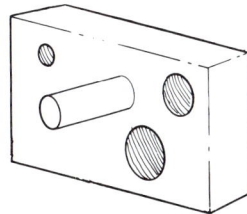

B16

REVIEW

The blue triangle has been *translated* (slid) onto the white triangle. Every point has moved the same distance and direction, as represented by the arrow.

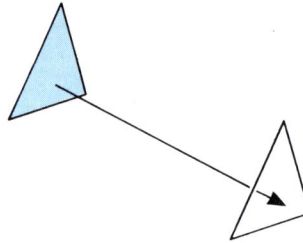

The blue triangle has been *rotated* clockwise onto the white triangle through an angle of 70°. Point C is the centre of rotation.

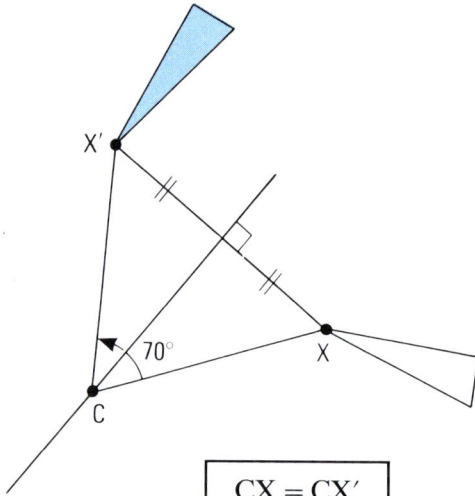

$$CX = CX'$$
$$\angle XCX' = 70$$

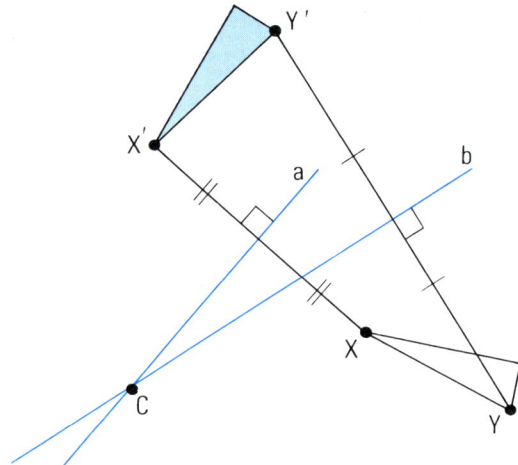

Line a is the perpendicular bisector of XX'.
Line b is the perpendicular bisector of YY'.

The blue triangle has been *reflected* onto the white triangle.
Line m is the mirror line.

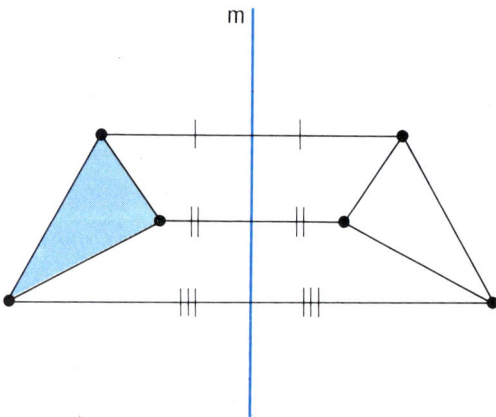

The blue triangle has been *enlarged* onto the white triangle.
C is the centre of enlargement. The scale factor is × 3.

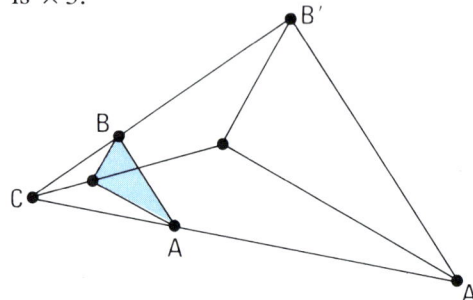

$$CA' = 3 \times CA$$
$$A'B' = 3 \times AB$$

CONSOLIDATION

1   Explain why the blue circle is NOT a reflection of the white circle.

a) in the line m   b) in the line n.

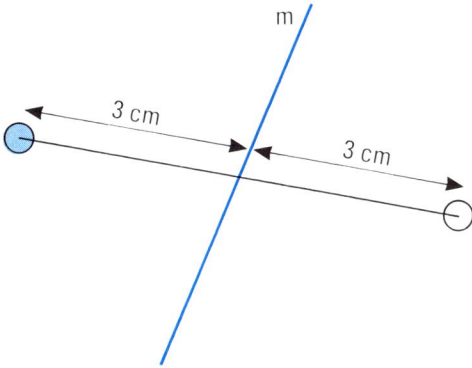

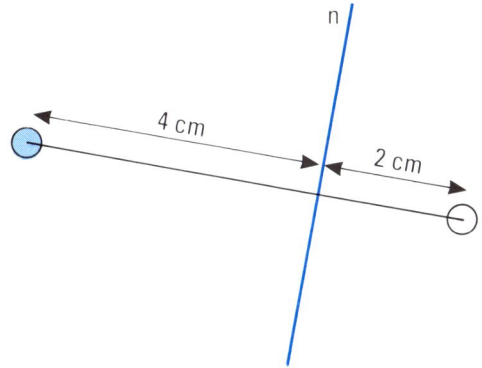

m

3 cm

3 cm

n

4 cm

2 cm

2   The blue triangle has been rotated anticlockwise through 90° onto the white triangle.

a) Explain why
point A is NOT
the centre of
rotation.

b) Explain why
point B is NOT
the centre of
rotation.

c) Explain why
point C IS the
centre of
rotation.

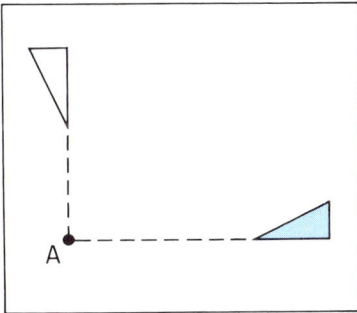

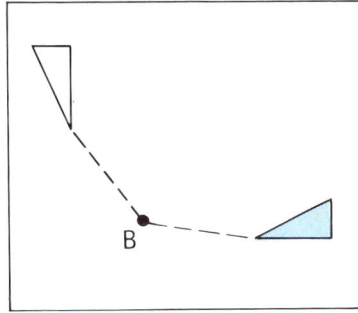

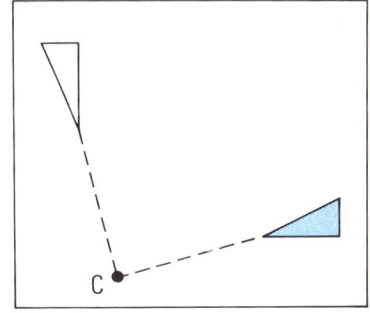

A

B

C

3   The black flag has been rotated anticlockwise through 60°
onto the white flag.
Point A is the same distance from the base point
of both flags.
This is also true for points B, C, D, and E.
Which point is the centre of rotation?
Explain your choice.

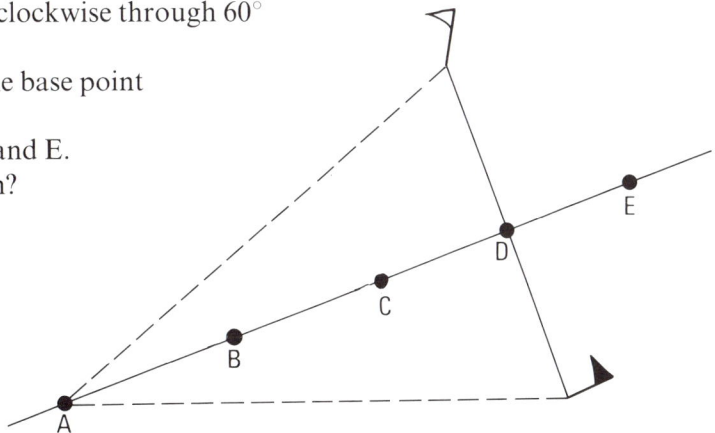

E

D

C

B

A

B17

4   In these drawings the black triangle has been transformed onto the blue triangle. In each case, write down whether the transformation could be a translation, a reflection, a rotation, or an enlargement.

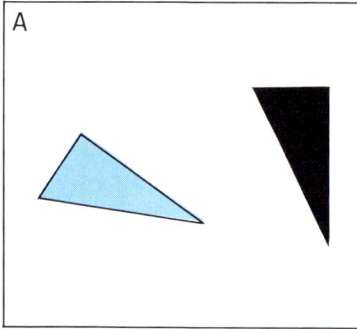

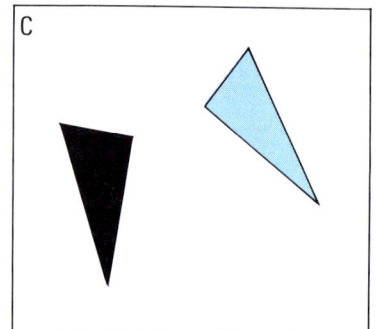

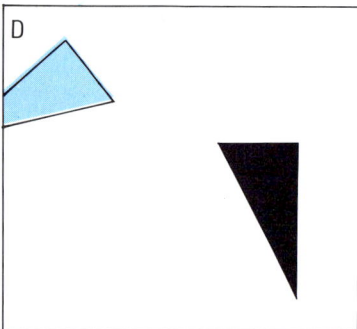

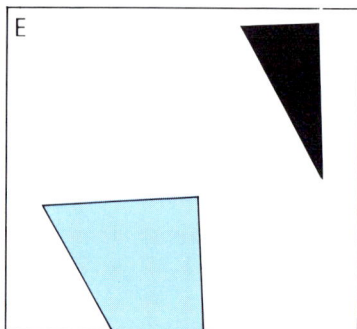

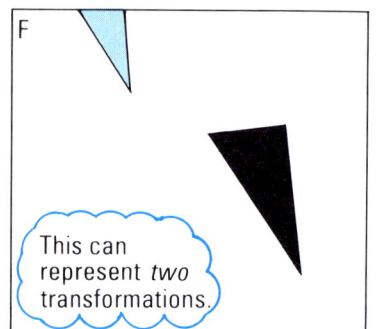

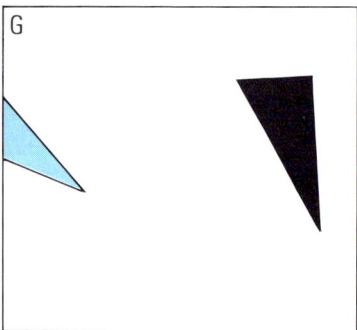

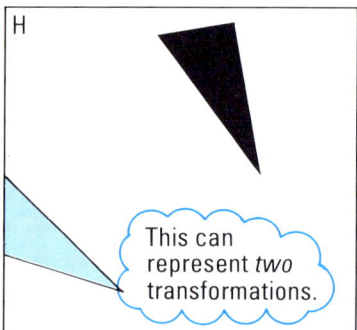

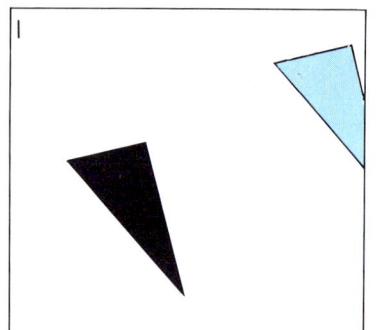

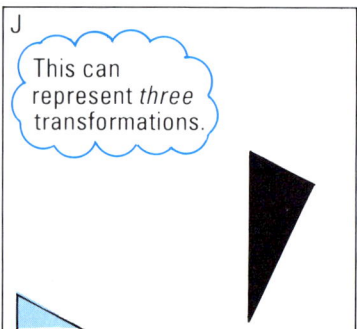

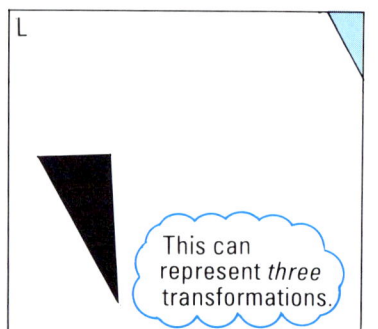

A

B

C

D

E

F

This can represent *two* transformations.

G

H

This can represent *two* transformations.

I

J

This can represent *three* transformations.

K

L

This can represent *three* transformations.

B17

5   Drawing A could represent a translation (B) or an enlargement (C).

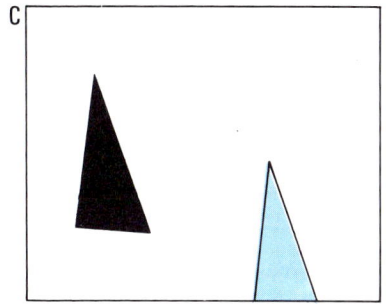

a) Make sketches to show
   how drawing D could
   represent:
   (i) a rotation
   (ii) a reflection.

b) Make sketches
   to show how drawing E
   could represent:
   (i) a translation
   (ii) a reflection.

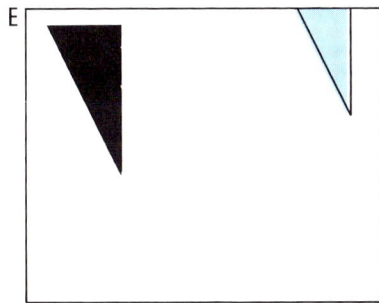

ACTIVITY

6   You will need squared paper and tracing
    paper.
    This drawing could represent a reflection or
    a rotation.

   a) Make a careful copy of the drawing on
      squared paper. Complete it so that it
      represents a reflection. Find the mirror
      line by folding.

   b) Make another copy. Complete it so that it
      represents a rotation. Use tracing paper
      to find the centre of rotation.

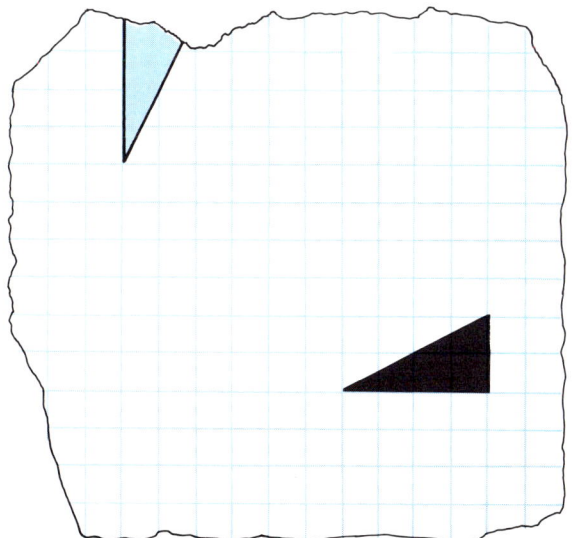

Vectors

1 Triangle A can be translated onto triangle B.
We can describe the displacement in words,
like this: 6 units horizontally to the right, and
2 units vertically upwards.

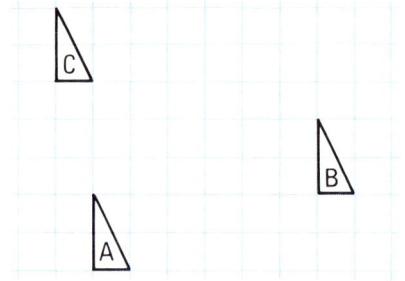

a) Describe the translation from B to A in the same way.

b) Another way of writing the translation from A to B is $\begin{pmatrix} 6 \\ 2 \end{pmatrix}$.

Copy and complete this
description for the translation from B to A: $\begin{pmatrix} -6 \\ ? \end{pmatrix}$.

━━━━━━━━━ TAKE NOTE ━━━━━━━━━

Pairs of numbers like $\begin{pmatrix} 6 \\ 2 \end{pmatrix}$, which represent translations, are called *vectors*.

We often represent vectors by a line and an arrow.

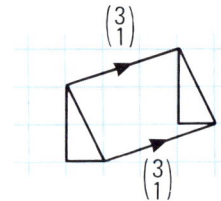

$\begin{pmatrix} 3 \\ 1 \end{pmatrix}$

$\begin{pmatrix} 3 \\ 1 \end{pmatrix}$

c) This vector $\begin{pmatrix} 7 \\ -3 \end{pmatrix}$ describes a translation on the grid at the top of the page.
Which translation? (C to A, or A to C, or B to A, and so on.)

d) Write down the vector which describes the translation: (i) B to C (ii) A to C (iii) C to A.

2 Copy the grid and the triangle.

a) The triangle is translated using the vector $\begin{pmatrix} 2 \\ 3 \end{pmatrix}$.
Draw its new position. Call it T′.

b) T′ is translated using the vector $\begin{pmatrix} 1 \\ -3 \end{pmatrix}$. Draw its new position.
Call it T″.

c) One of these vectors
will take T″ back to T. Which one? $\begin{pmatrix} 0 \\ -3 \end{pmatrix} \begin{pmatrix} -3 \\ 0 \end{pmatrix} \begin{pmatrix} 3 \\ 0 \end{pmatrix} \begin{pmatrix} 0 \\ 3 \end{pmatrix}$

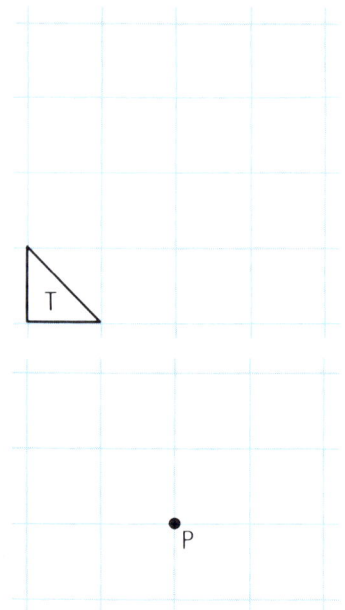

3 The dot P on the grid is translated around
the grid using this sequence of vectors:

$\begin{pmatrix} 1 \\ 0 \end{pmatrix}$ then $\begin{pmatrix} 1 \\ 1 \end{pmatrix}$ then $\begin{pmatrix} -1 \\ 1 \end{pmatrix}$ then $\begin{pmatrix} -1 \\ 0 \end{pmatrix}$ then $\begin{pmatrix} -2 \\ -3 \end{pmatrix}$.

What translation is needed to return it to its starting position?

Transformations on grids

1   You need 1 cm squared paper.

   a) The vertices of triangle A are at the points
      (2,3), (1,), and (2,). Copy and complete
      each coordinate.

   b) Triangle A is reflected in line a. These are
      the coordinates of the vertices of the
      reflected triangle. Copy and complete
      them.

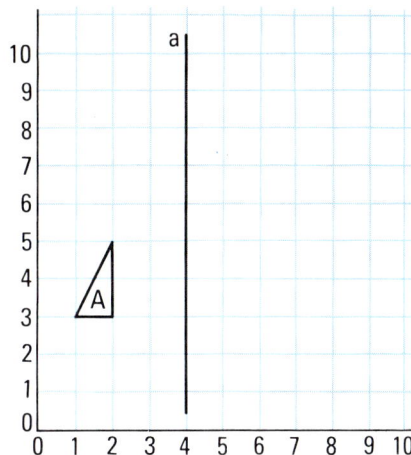

      (6,) (,3) (,5)

   c) Triangle A is translated so that the right-angled vertex goes to the point (9,6).

      (i) Copy and complete the new coordinates of its vertices: (9,6) (,6) (9,).
      (ii) Write down the vector which describes the translation.

   d) Triangle A is enlarged, scale factor 4, centre (0,2).

      (i)   Mark the centre of enlargement on squared paper.
      (ii) Draw the enlarged triangle.
      (iii) Copy and complete the new coordinates of its vertices: (4,) (,6) (8,).

   e) Triangle A is given a quarter turn anticlockwise about the point (5,4).

      (i)   Mark the centre of rotation with an 'R'.
      (ii) Draw the triangle after the rotation.
      (iii) Copy and complete the new coordinates of its vertices: (6,) (,1) (,0).

2   Triangle P can be sent to each of the other
    positions Q, R, S, T by a reflection, rotation,
    translation, or enlargement. For each
    position describe the movement accurately.

    For a *reflection*,
    describe the
    position of its
    line of reflection.

    for example,
    reflection in the line
    which passes through
    (5,0) and (5,9)

    for example, $\frac{1}{4}$ clockwise turn about
    the point (2,3) etc.

    For a *rotation*, state the amount and
    direction of turn, and the turn centre.

    For an *enlargement*, state the centre of enlargement and the scale factor.
    For a *translation*, write down the vector which describes it.

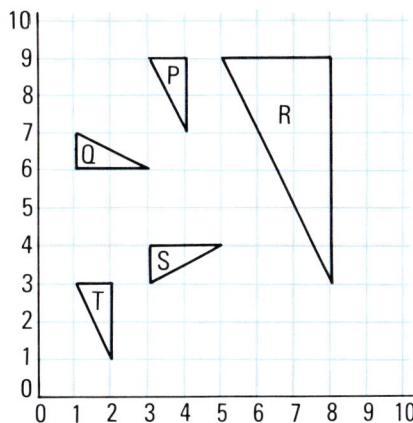

ENRICHMENT

1   The diagram shows the closed and open positions of an 'up and over' garage door. Make a scale drawing of the door in the 'open' and 'closed' positions. Find where the pivot should be located so that the door will swing exactly into its open position.

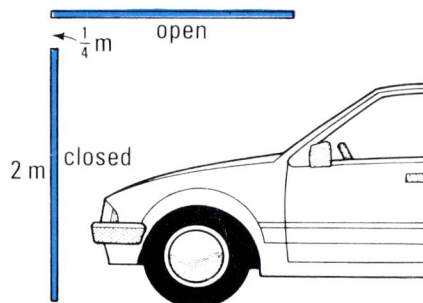

2   We can get the letter F onto its finishing position

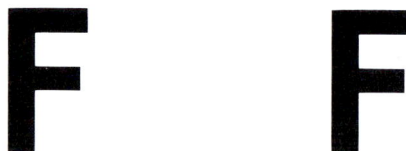

*   by two reflections like this

*   or two half turns like this.

On squared paper draw some pairs of Fs with different start–finish distances between them.

Find out what you can about:

a) where the two mirrors could be placed

b) where the two centres of rotation could be positioned.

Write a report about what you find out.

CORE

1 These are strip patterns made from triangles and circles.

a) Check that each pattern obeys this rule:

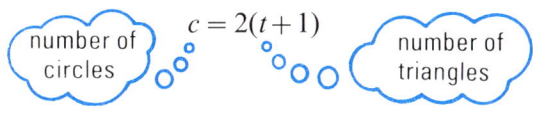

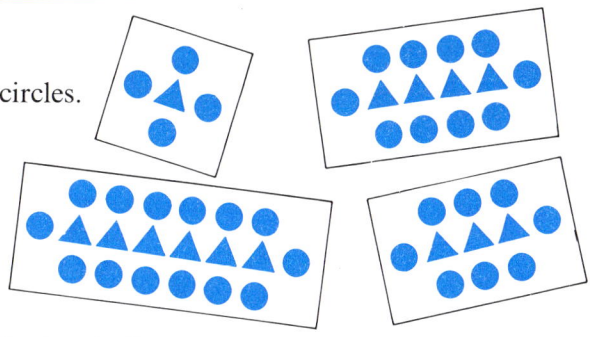

(number of circles) $c = 2(t+1)$ (number of triangles)

b) How many circles are there in a pattern with 13 triangles?

c) How many triangles are there in a pattern with 42 circles?

$\times, \div, +, \text{or} -$

d) This is another way of writing the same rule.
It is incomplete. Use the patterns to help you to complete it. $t = (c-2) \ldots \square$

e) This is a third way of writing the rule. $2t = c \ldots \square$
Use the patterns to help you to complete it.

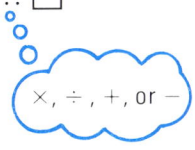

$\times, \div, +, \text{or} -$

f) This is a fourth way of writing the rule.

$c - \square = 2$

Use the patterns to help you to complete it.

═══════ TAKE NOTE ═══════

There are many different ways of writing the same rule. Check that each one of these is correct for the connection between days and weeks.

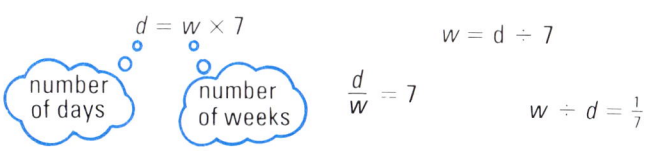

$d = w \times 7$ (number of days) (number of weeks)

$w = d \div 7$

$\dfrac{d}{w} = 7$

$w \div d = \frac{1}{7}$

2 There is a rule which connects the number of blue tiles and white tiles in each of the Ls. These are different ways of writing the rule. (b is the number of blue tiles; w is the number of white tiles.)

a) $b - w = \ldots$
b) $b = \ldots$
c) $w = \ldots$

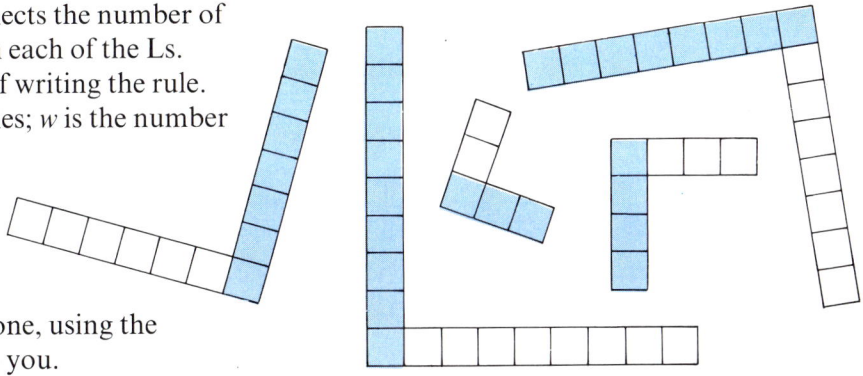

Copy and complete each one, using the drawings of the Ls to help you.

B18

3   'Axis' sells gift sets of pencils and ballpoint pens. This is a rule about the number of pencils (p) and ballpoint pens (b) in different gift sets.

$$p = b + 3$$

These are different ways of writing the rule.

a) $b = p \dots \square$ b) $\square - b = \square$

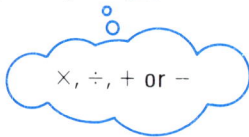

×, ÷, + or −

Copy and complete each one. (You might like to draw some gift sets to help you.)

4   'Axis' also sells gift sets of notebooks and pencils. This is a rule about the number of notebooks (n) and pencils (p) in different gift sets.

$$p = 3n$$

These are different ways of writing the rule. Copy and complete them.

a) $n = p \dots \square$ b) $\dfrac{\square}{\square} = 3$

×, ÷, + or −

THINK IT THROUGH

5   Here are two rules and two different ways of writing each one. Copy and complete them.

a) $p = t - 1$

    (i) $t = \square + \square$

    (ii) $\square - p = \square$

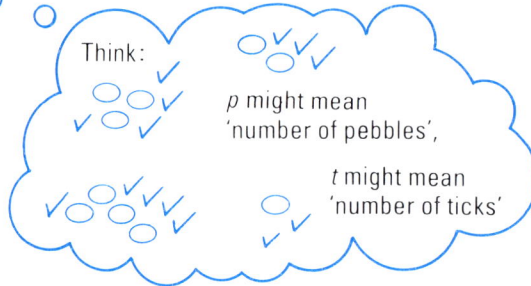

Think: p might mean 'number of pebbles',

t might mean 'number of ticks'

b) $4 \times k = n$

    (i) $k = \dots$

    (ii) $\dfrac{\square}{\square} = 4$

CHALLENGE

6   Here is a set of patterns. These are ways of writing a rule for the patterns (s = number of stars, t = number of triangles).

$$s = t + 1 \qquad s - t = 1 \qquad s - 1 = t$$

✳△✳△✳

Draw a set of patterns of stars and triangles of your own which obey a rule. Write down the rule in as many different ways as you can.

Transforming rules

1 To find the approximate circumference (C)
of a circle we multiply its diameter (D) by 3: $C \approx 3 \times D$

a) Write in words how we
   can find the
   approximate diameter
   of a circle when we
   know its circumference.

b) Write your rule in a)
   like this: $D \approx \ldots$

c) Use your rule in b) to
   check that the radius
   of a water wheel with
   circumference 21 m is
   about 3.5 m.

2 To find the area (A) of a rectangle we multiply its length (l) by its width (w): $A = l \times w$

a) Write in words how we can find the length of a rectangle when we know its area and its width.

b) Write your rule in a) like this: $l = \ldots$

c) Use your rule in b) to check that the length of a rectangular room of area 36 m² and width 5 m
   is 7.2 m.

3 To find the area of a square, we square the length of any of its sides: $A = l^2$

a) Write in words how we can find the length of one side of a square when we know its area.

b) Write your rule in a) like this: $l = \ldots$

c) Use your rule in b) to check that the length of each side of a square meadow of area 280 900 m²
   is 530 m.

4 To find the steady speed (S km/h) for a journey we
  divide the distance travelled (D km) by the time taken (T hours): $S = \dfrac{D}{T}$

a) Write in words how we can find the distance we have travelled when we know the steady speed
   and the time taken.

b) Write your rule in a) like this: $D = \ldots$

c) Write in words how we can find the time taken for a journey when we know the steady speed
   and the distance travelled.

d) Write your rule in c) like this: $T = \ldots$

CHALLENGE

5   To find the area (A) of a circle
we square the radius (r),
and then multiply by π:

$A = r^2 \times \pi = \pi r^2$ ○○○ ($r \times r \times \pi$)

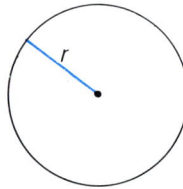

a) The area of a circle is $9 \times \pi$ cm^2. What is its radius?

b) The area of a circle is 42 cm^2.

(i) Explain how we can find r^2 for the circle
(ii) Find r^2
(iii) Find r.

c) Write in words how we can find the radius of a circle when we know the area. (Your work in parts a) and b) will help you.)

d) Write your rule in c) like this: $r = \ldots$

TAKE NOTE

When we write a rule (or formula) in a new way, we say we
have *transformed* the formula. For example, we can start with $C = 2\pi r$, and rewrite it as $r = \dfrac{C}{2\pi}$.

In $C = 2\pi r$ we say that C is the subject of the formula.

In $r = \dfrac{C}{2\pi}$ we say that r is the subject of the formula.

6   a) Check that this rule describes the piles of
nuts and bolts correctly (n = number of
nuts in a pile, b = number of bolts in that pile).

$n = 2b + 1$

b) Use the drawings of the piles of nuts and
bolts to help you to write the rule with b as
the subject:

$b = \ldots$

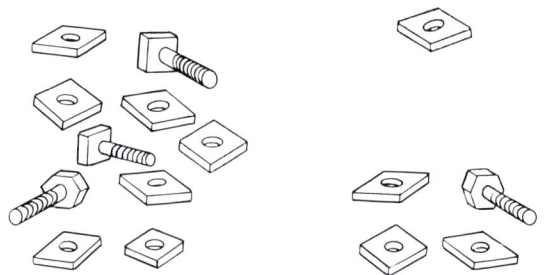

7   This is how Alex transforms the rule $n = 3b + 2$:

a) Copy and complete his solution.

b) Check that $b = 4$, $n = 14$ fits the original rule.

c) Check that $b = 4$, $n = 14$ fits your final rule.
If it does not, check your working.

$$n = 3b + 2$$
$$-2: \quad n - 2 = ?$$
$$\div 3: \quad \frac{n-2}{?} = b$$

8 Copy and complete each of these, to make k the subject of the formula. Each time, choose values for p and k to check that your solution is correct.

a)

$$\begin{aligned} & & p &= 1 - 2k \\ +2k&: & p + ? &= 1 \\ -p&: & ? &= 1 - ? \\ \div ?&: & k &= \frac{1-?}{?} \end{aligned}$$

b)

$$\begin{aligned} & & p &= \frac{2+k}{4} \\ \times 4&: & 4p &= ? \\ -2&: & 4p - ? &= k \end{aligned}$$

c)

$$\begin{aligned} & & p &= 2(k-1) \\ \div 2&: & \frac{p}{2} &= ? \\ +?&: & \frac{p}{2} + ? &= k \end{aligned}$$

d)

$$\begin{aligned} & & p &= \frac{4}{k+1} \\ \times(k+1)&: & p(k+1) &= ? \\ \div p&: & k+1 &= \frac{?}{p} \\ -1&: & k &= \frac{?}{p} - 1 \end{aligned}$$

THINK IT THROUGH

9 Write each of these formulas with m as its subject.

a) $t+m=4$ b) $t-m=3$ c) $\frac{t}{m}=10$ d) $\frac{m}{5}=t$

e) $2m-1=t$ f) $t=3-4m$ g) $\frac{t}{m}+1=3$ h) $t(m-1)=m$

REVIEW

- Triangles with the same sized angles are always *similar*. One is an enlargement of the other.

These triangles are **similar**

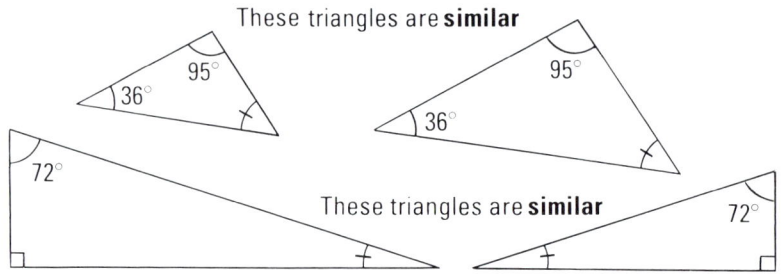

These triangles are **similar**

- *Similar* triangles have pairs of sides in the same ratio.

Similar triangles

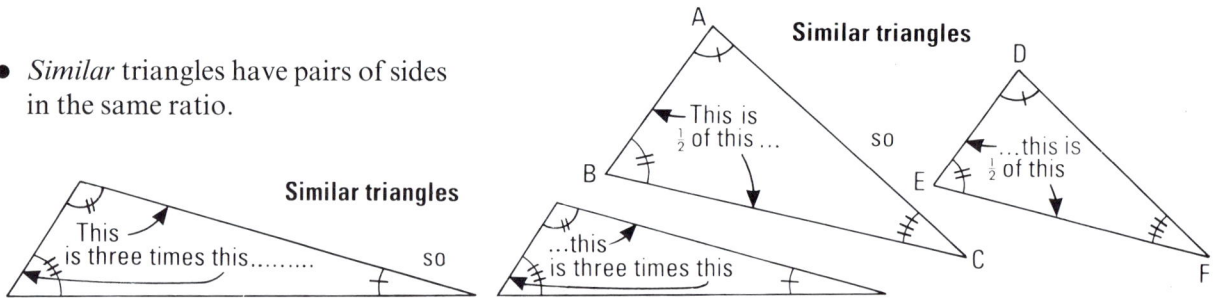

This is ½ of this ...

so

...this is ½ of this

Similar triangles

This is three times this.........

so

...this is three times this

■ Copy and complete these ratios: $\dfrac{AB}{BC} = \dfrac{DE}{\square} = \dfrac{1}{2}$ $\dfrac{AB}{AC} = \dfrac{DE}{\square}$

- Also, in similar triangles the ratio of corresponding sides of the two triangles are equal.

Similar triangles
This is 4 times this so this is 4 times this

Similar triangles
This is 0.6 times as long as this, so this is 0.6 times as long as this

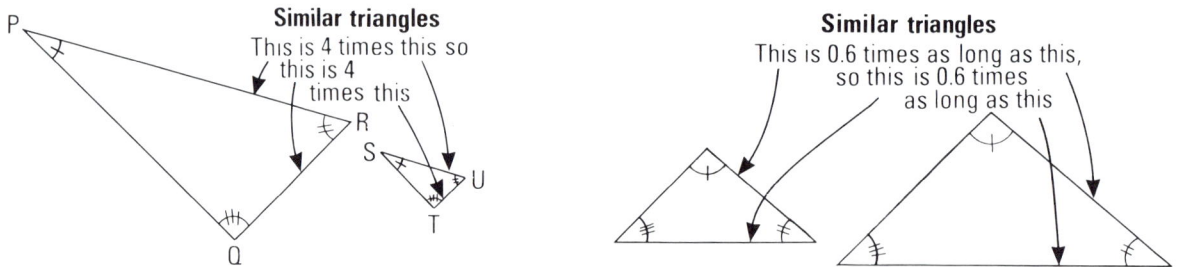

■ Copy and complete these ratios: $\dfrac{PR}{SU} = \dfrac{\square}{TU} = 4$ $\dfrac{TU}{QR} = \dfrac{ST}{\square} = \square$

These triangles are similar.

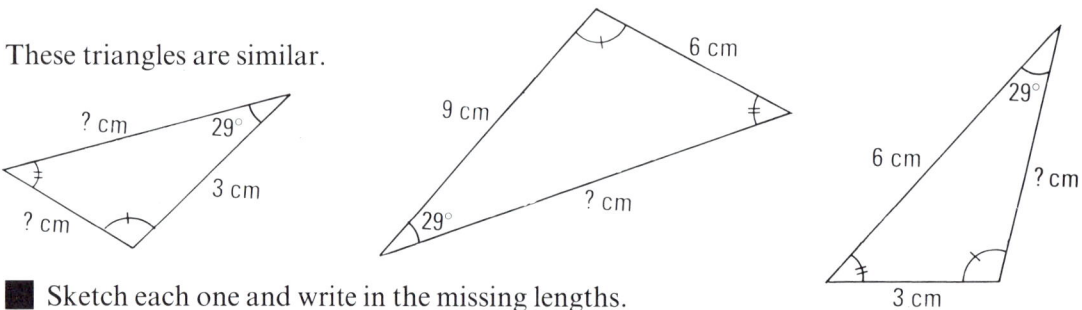

■ Sketch each one and write in the missing lengths.

CONSOLIDATION

1   The two triangles are similar (one is an
    enlargement of the other).

    a) What is the scale factor for enlargement?

    b) How long is the side marked '?cm'?

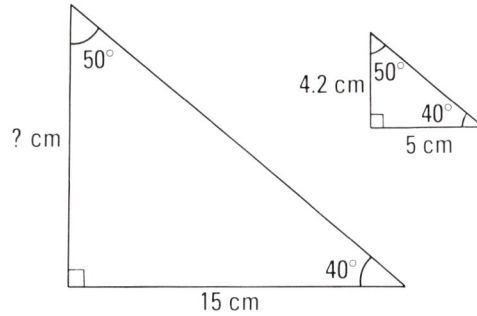

2   The larger triangle is an enlargement of the
    smaller triangle.

    a) Write one or two sentences to explain
        how we can tell.

    b) What is the enlargement scale factor?

    c) How long is the side marked '?cm'?

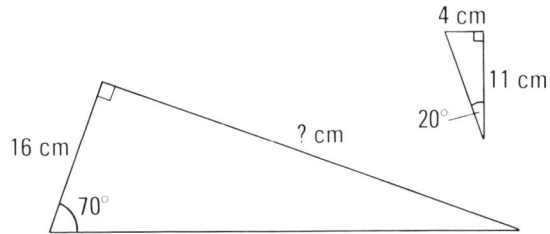

▨▨▨▨▨▨▨▨▨▨ WITH A FRIEND ▨▨▨▨▨▨▨▨▨▨

3   You need 1 cm squared paper.

    a) Both of you draw two *different* triangles.
        Each triangle should have an angle
        of 35° and an angle of 60°.

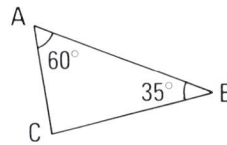

    b) Measure the length of sides of each of your triangles. For each one calculate $\dfrac{AC}{AB}$.
        Compare your results with your friend's results.

    c) Why should the four ratios you have found be equal? Decide between you, then write one or
        two sentences to explain.

    d) Check that the four ratios for another pair of sides are also equal.

4   a) The three triangles are similar. What is the value of:

    (i) $\dfrac{AC}{CB}$ (ii) $\dfrac{AB}{AC}$

    (iii) $\dfrac{DE}{DF}$ (iv) $\dfrac{GI}{HI}$

    (v) $\dfrac{DF}{EF}$ (vi) $\dfrac{HI}{GH}$?

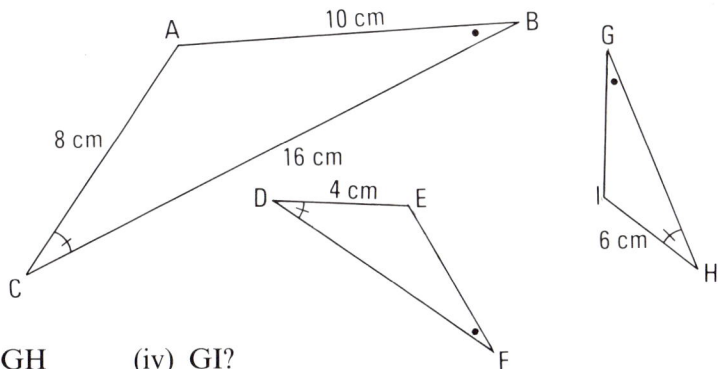

    b) How long is:

    (i) DF   (ii) EF   (iii) GH   (iv) GI?

CHALLENGES

Try to solve these problems without making scale drawings (except for 5b)).

5   a) The shadow of the large flagpole is 7 m long. How long is
the shadow of the smaller flagpole?
Write one or two sentences to
explain how you decided.

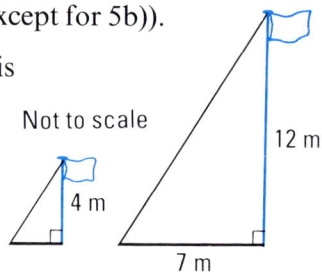

Not to scale

12 m

4 m

7 m

b) Draw accurately a right-angled triangle with one angle 35°.
Choose the other measurements yourself. Use your triangle
to help you to find the height of this flagpole.
Write one or two sentences to explain
how your method works.

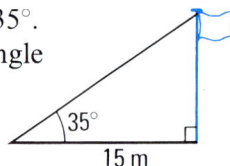

35°

15 m

B19

6   The sketches show some measurements
taken for four pine trees.

a) Which tree is taller, A or B?

b) By looking at the
measurements decide
which of the four trees is

(i) tallest   (ii) shortest.

A

B

37°

12 m

40°

12 m

c) The height of tree B is 10.1 m
(to the nearest 0.1 m).
The height of tree C is 8.3 m
(to the nearest 0.1 m).
Find the heights of the other two trees.

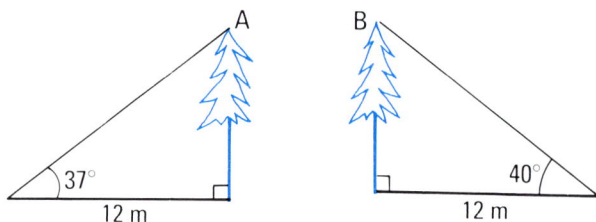

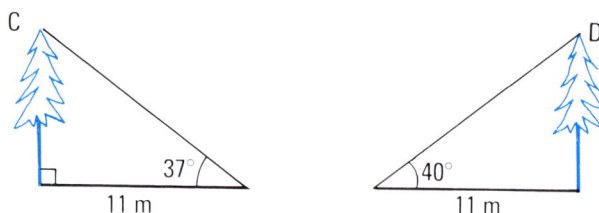

C

D

37°

11 m

40°

11 m

7   The sketches show some measurements
taken for four radio masts.

a) List the masts in order, tallest first.

b) The tallest mast is 46.8 m tall. The
shortest is 30.3 m tall. Find the
heights of the other two masts.

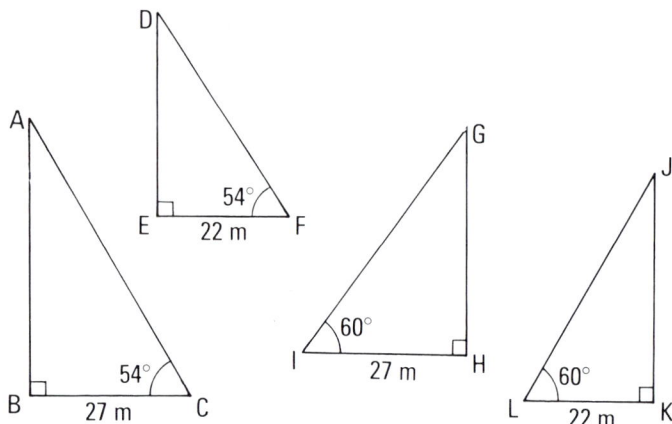

D

A

54°

E   22 m   F

G

J

60°

I   27 m   H

54°

B   27 m   C

60°

L   22 m   K

CORE

Calculating lengths

1   Meg drew these triangles and measured the
    lengths of the sides.

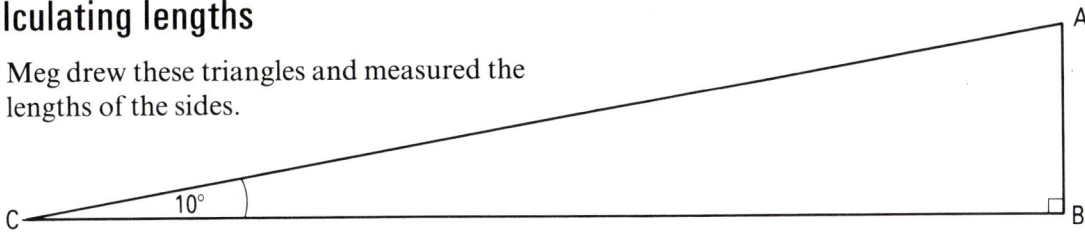

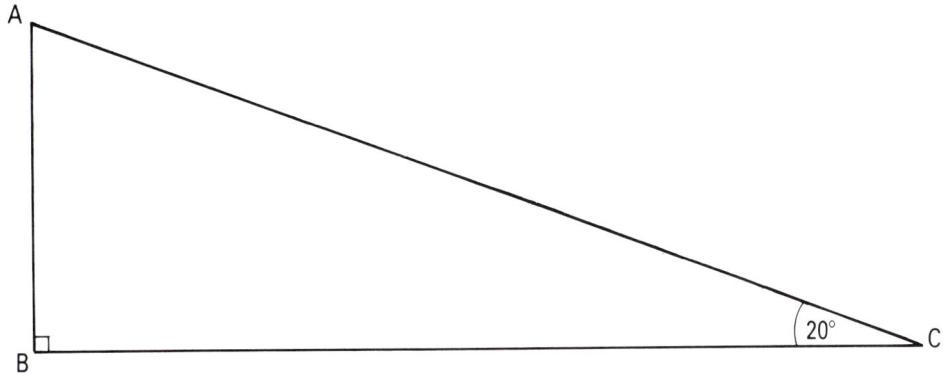

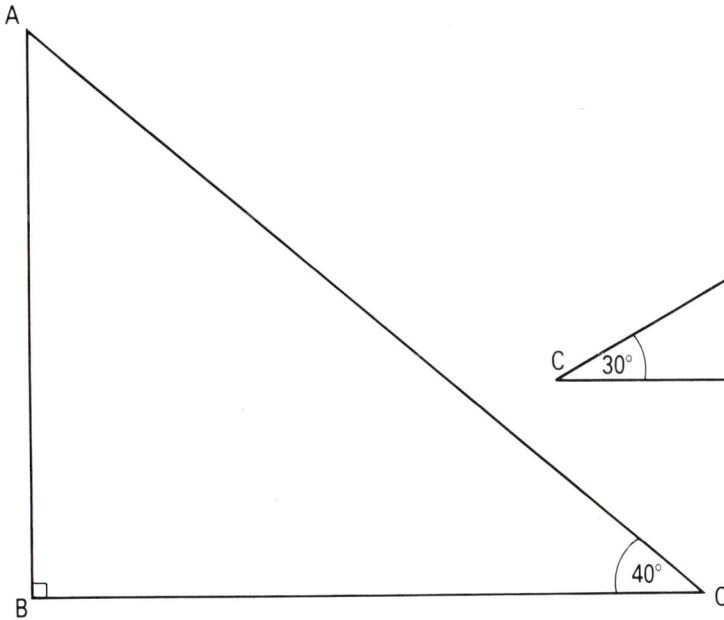

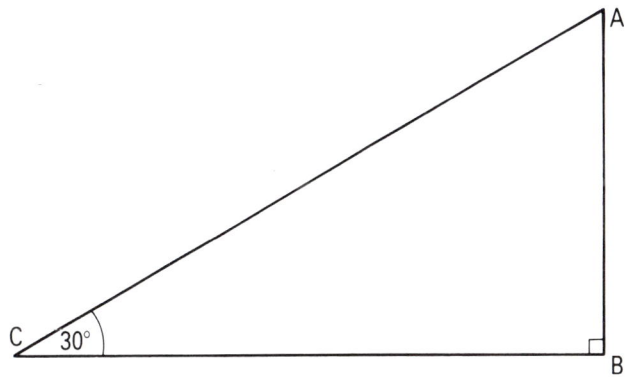

B19

She produced this table.

Angle ACB	10°	20°	30°	40°	50°	60°	70°	80°
AB/BC to 2 DP	0.18	0.36	0.58	0.84				

Copy and complete the table by drawing accurate triangles and making your own measurements.

B19

2   Use your table in question 1. Write down the ratio $\dfrac{AB}{BC}$ for each of these triangles.

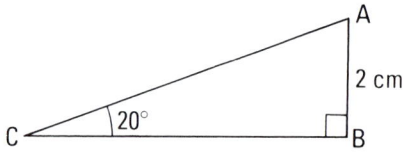
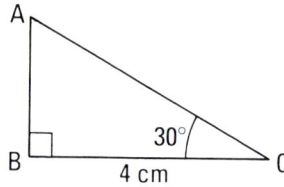
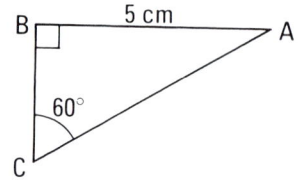

TAKE NOTE

For all right-angled triangles ABC,
if $\angle C = 20°$ and $\angle B = 90°$,

then the ratio $\dfrac{AB}{BC}$ is always

the same (*about 0.36*) because all the triangles are similar.

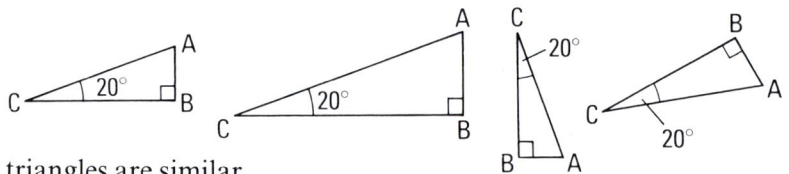

---

3   What is the ratio $\dfrac{AB}{BC}$ for all right-angled triangles with $\angle B = 90°$ and $\angle C$ measuring:

a) 60°   b) 80°   c) 45° (Sketch a triangle and think carefully.)

d) 65° (You will need to draw a triangle and measure.)

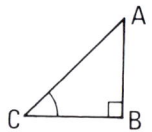

THINK IT THROUGH

4   Use your table from question 1. The angle marked * in each of these situations is 10° or 20° or 30° or ... or 80°. Find the angle in each situation.

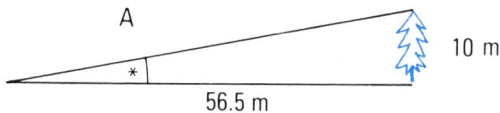

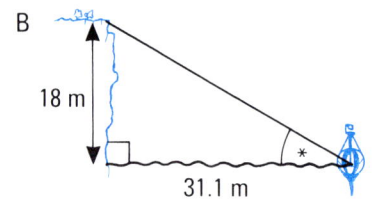

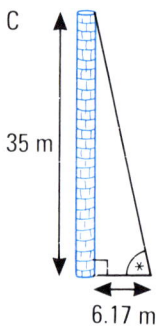

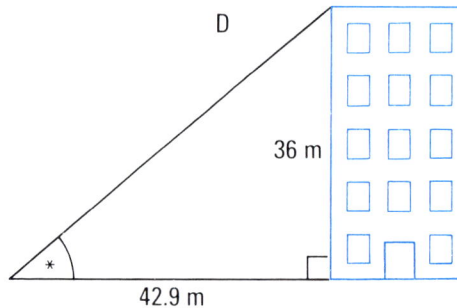

═══════ ACTIVITY ═══════

5   a) Draw some right-angled triangles of your own – accurately. Use them to complete these tables.

Table 1

Angle ACB	10°	20°	30°	40°	50°	60°	70°	80°
Ratio $\dfrac{AB}{AC}$								

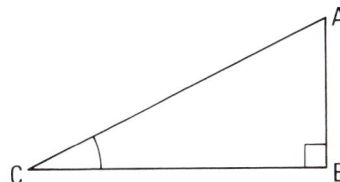

Table 2

Angle ACB	10°	20°	30°	40°	50°	60°	70°	80°
Ratio $\dfrac{BC}{AC}$								

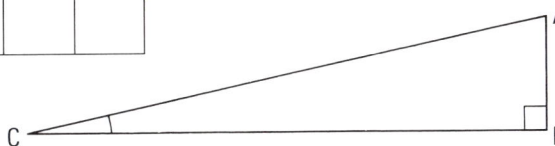

B19

   b) Do you notice any pattern in the two tables? (For example, look at the ratio for 10° in table 1 and for 80° in table 2.)

   Write one or two sentences to explain why the ratios in the two tables are the same.

6   a) This is a ramp. Use your table 1 from question 5 to answer the questions.

   (i) What is the value of $\dfrac{h}{2}$?
   (ii) Calculate h.

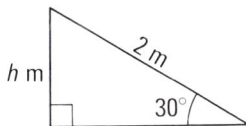

   b) This is another ramp. Use your table 2 from question 5 to answer the questions.

   (i) What is the value of $\dfrac{k}{4}$?
   (ii) Calculate k.

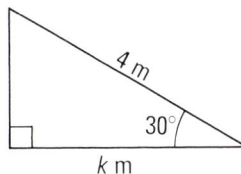

═══════ TAKE NOTE ═══════

For this 24° triangle

$\dfrac{AB}{BC} \approx 0.45\,(2\,DP)$ $\dfrac{AB}{AC} \approx 0.41\,(2\,DP)$ $\dfrac{BC}{AC} \approx 0.91\,(2\,DP)$.

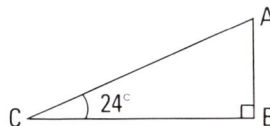

We can use the ratios to help us to find distances for other 24° right-angled triangles.

░░░░░░░░░░ CHALLENGE ░░░░░░░░░░

7   Use only your calculator and your tables from questions 1 and 5.
Find approximately the distance marked ? in each situation.

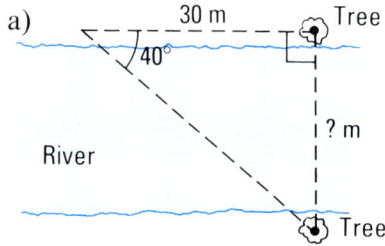

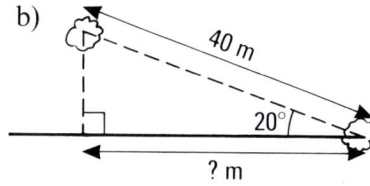

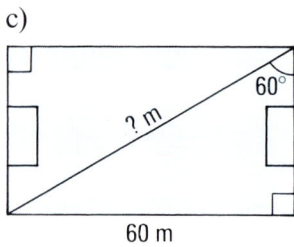

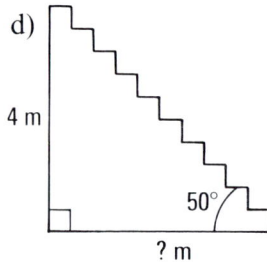

a)

30 m   Tree
40°
? m
River
Tree

b)

40 m
20°
? m

c)

60°
? m
60 m

d)

4 m
50°
? m

░░░░░░░░░░░░░░░░░░░░░░░░░░░░░░░░░░░░░░

This table gives the ratio $\dfrac{AB}{BC}$
for angles measuring
a whole number of degrees
from 0° to 89°.

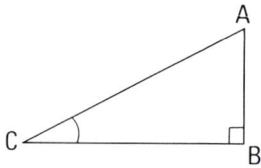

A

C   B

Angle	Ratio $\frac{AB}{BC}$ (to 3 DP)
0	0.000
1	0.017
2	0.035
3	0.052
4	0.070
5	0.087
6	0.105
7	0.123
8	0.141
9	0.158
10	0.176
11	0.194
12	0.213
13	0.231
14	0.249
15	0.268
16	0.287
17	0.306
18	0.325
19	0.344
20	0.364
21	0.384
22	0.404
23	0.424
24	0.445

Angle	Ratio $\frac{AB}{BC}$ (to 3 DP)
25	0.466
26	0.488
27	0.510
28	0.532
29	0.554
30	0.577
31	0.601
32	0.625
33	0.649
34	0.675
35	0.700
36	0.727
37	0.754
38	0.781
39	0.810
40	0.839
41	0.869
42	0.900
43	0.933
44	0.966

Angle	Ratio $\frac{AB}{BC}$ (to 3 DP)
45	1.000
46	1.036
47	1.072
48	1.111
49	1.150
50	1.192
51	1.235
52	1.280
53	1.327
54	1.376
55	1.428
56	1.483
57	1.540
58	1.600
59	1.664
60	1.732
61	1.804
62	1.881
63	1.963
64	2.050
65	2.145
66	2.246
67	2.356
68	2.475
69	2.605

Angle	Ratio $\frac{AB}{BC}$ (to 3 DP)
70	2.747
71	2.904
72	3.078
73	3.271
74	3.487
75	3.732
76	4.011
77	4.331
78	4.705
79	5.145
80	5.671
81	6.134
82	7.115
83	8.144
84	9.514
85	11.43
86	14.30
87	19.08
88	28.64
89	57.29

These are
given to 2DP

8   a) The ratio $\dfrac{AB}{BC}$ for this triangle is 0.510 (to 3 DP).

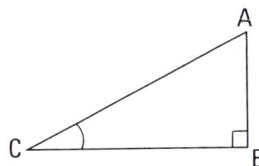

Angle ACB measures a whole number of degrees.

Use the table on page 254 to help you to find out how many degrees.

b) Use the table on page 254 to help you to write down the approximate values of the ratios:

(i) $\dfrac{PR}{RQ}$ (ii) $\dfrac{RQ}{PR}$

in this triangle.

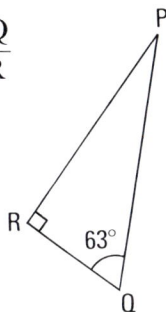

============ CHALLENGE ============

9   a) Wensley is 4.2 km North and 3.7 km East of Fasten.

What is the bearing of Wensley from Fasten? (Calculate a ratio, then use the table on page 254. Do your best to give your answer to the nearest degree.)

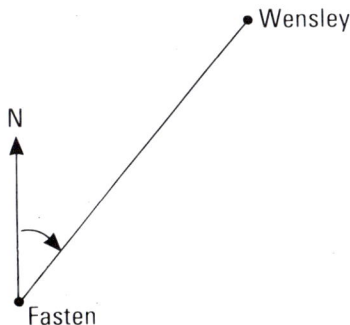

b) An aeroplane flies in the direction shown in the drawing until it is 114.7 km West of the airport. Use the table on page 254 to help you to find how far North of the airport the aeroplane is.

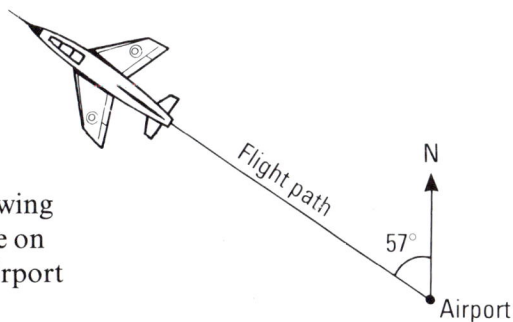

============ FIND OUT YOURSELF ============

10 You need a scientific calculator.

Investigate the [sin] [cos] and [tan] keys; for example, type [C] [5] [5] [tan].

What do you think the display tells us when we press these keys?

CORE

═══ WITH A FRIEND ═══

1   Here are some pairs of
    expressions.
    Decide between you which
    pairs are:

    (i)   *always equal* (equal no
          matter which value
          you choose for the
          letter)
    (ii) *sometimes equal* (equal
          for at least one value
          of the letter, but not
          for all values)
    (iii) *never equal*.

A $t + t$ $2 \times t$ B $m + 4$ $4 \times m$

C $c \times a$ $a \times c$ D $n + n + n$ $3 + n$

E $a + 7$ $7 + a$ F $p \times p$ $p + p$

G a^2 a^3 H $3 \times n$ $n \times 3$

I $a \times 2 \times b$ $2 \times b \times a$ J $a + 3 + b$ $3 + a + b$ K $5 - k$ $k - 5$

Each of you write down what you decide. For those that are 'sometimes equal' write down the
values which make them so.

2   Here are two expressions: $2 + k$, $2 \times k$

    a) Choose k to be 2. Are the two expressions equal for this value of k?
    b) Are the two expressions always equal? Explain your answer.

3   a) Check that the expressions a^2 and $3 \times a$ are equal when a is 3.
    b) Are the expressions always equal? Explain your answer.

═══ TAKE NOTE ═══

To decide if two expressions are *always, sometimes* or *never* equal, replace the letters by numbers.
Always try more than one number. (You will not be able to try every possible number. Try as many
as you need to convince yourself.)

4   a) Check that $n + n$ and $2 \times n$ are always equal.   b) Check that $2 \times c$ and $c \times 2$ are always equal.
    c) Check that $a \times d$ and $d \times a$ are always equal.   d) Check that $4 + k$ and $3 + k$ are never equal.
    e) Check that $4 + n$ and $n + n$ are sometimes (but not always) equal.
    f) Write down another pair of expressions which are never equal.

═══ TAKE NOTE ═══

$2 \times n$ and $n+n$ are always equal (for all values of n we care to choose).
We write $2 \times n \equiv n+n$ to show that they are always equal.
(Notice: we call '$\equiv$' an *identity* sign. It has three lines instead of two.)

5   Which of these are true and which are false?

a) $k+k+k \equiv 3 \times k$ b) $p+p \equiv 2+p$ c) $n \times m \equiv m \times n$

d) $\dfrac{m}{n} \equiv \dfrac{n}{m}$ e) $5-t \equiv t-5$ f) $m+k+1 \equiv 1+k+m$

═══ TAKE NOTE ═══

We often miss out the '$\times$' sign when we are using letters.
For example, instead of $2 \times n$ we write $2n$.
            Instead of $c \times a$ we write ca.

6   Which of these are true and which are false?

a) $pq \equiv qp$ b) $m+2n \equiv n+2m$ c) $5ab \equiv 5+ab$ d) $2 \times a \times 3 \equiv 6a$

7   Explain why this is not true: $m+6 \equiv 6m$.

8   Match each of the expressions a) to e) with one from A to E to which it is always equal:

a) $a+a+a$ d) $5a-1$ A $a+4a-1$ D $3a$
b) $a+a+4$ e) $7a-3a$ B $4a$ E $2a+4$
c) $a \times 2 \times 3$ C $6a$

═══ THINK IT THROUGH ═══

9   a) Three of these are correct expressions for the
        perimeter of the top rectangle. Which three?

$k+k+6$ cm
$2k+6$ cm
$8k$ cm
$2(k+3)$ cm (that is, $2 \times (k+3)$ cm)

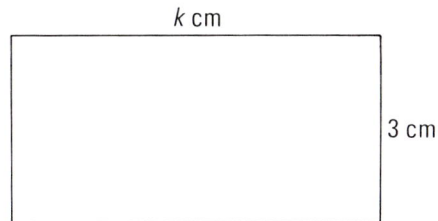

b) Write the perimeter of the second rectangle as
    simply as you can.

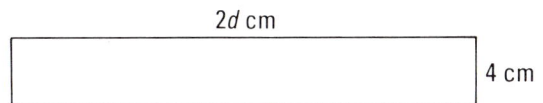

c) Tariq says that $6a+3b$ can be written as $9ab$.
    Choose values for a and b which show that this is wrong.

d) This is an *equation*: $7k+4 = 32$. It is true for one value of k. Which value?

B20

Simplifying expressions

━━━━━━━━━━━━ TAKE NOTE ━━━━━━━━━━━━

This is an expression: $3a+2b+4b+a+b+2a$.

We can write the expression in other ways: $3a+a+2a+4b+2b+b$
$a+b+2b+2a+3a+4b$ and so on.

The simplest ways we can write it are: $6a+7b$ and $7b+6a$.

($6a+7b \equiv 3a+2b+4b+a+b+2a$, that is, the two expressions are equal for all values of a and b.)

Writing the original expression in a simpler way is called *simplifying* the expression.

1   Simplify each of these expressions as much as you can. (Some are already as simple as possible.)

   a) $2a+a+3a$ b) $a+b+a+b+b$ c) $3a-a+b$ d) $7a+2b$
   e) $7a-3a+2b$ f) $k+l+2l+3k$ g) $x+2y+3z+x$ h) $1+5a+b$

2   True or false? $2a+3a \equiv 5a^2$ ($5a^2$ means $5 \times a^2$)
   If you say 'false', choose a value for a to explain why.

3   a) We can write $3k+7m$ like this: $k+2k+7m$
                      or this: $k+2k+3m+4m$
      or ...
   Find three more ways of writing $3k+7m$.

   b) Find four ways of writing $k-7m$. (For example: $k-(3m+4m)$ or $k+m-8m$)

4   Write down an expression for the perimeter of each shape. Write your expression as simply as you can.

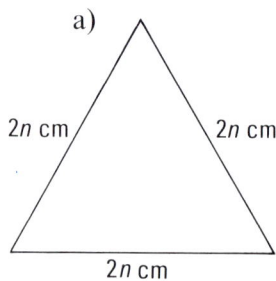

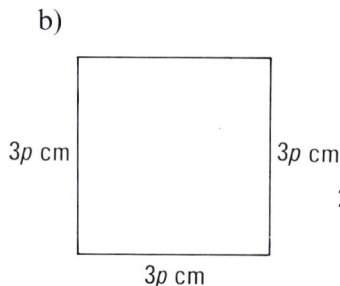

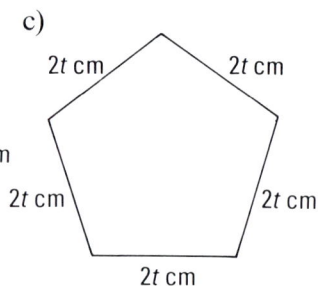

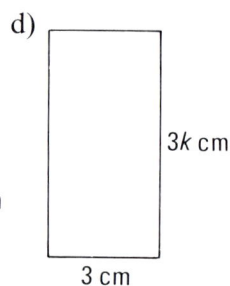

a) $2n$ cm, $2n$ cm, $2n$ cm (triangle)
b) $3p$ cm, $3p$ cm, $3p$ cm (square)
c) $2t$ cm, $2t$ cm, $2t$ cm, $2t$ cm, $2t$ cm (pentagon)
d) $3k$ cm, 3 cm (rectangle)

5   Match each expression on the left with one on the right to which it is always equal.
   (If you are not sure, check different values for k until you *are* sure.)

   a) $\frac{1}{2}k$ A $k \div 2$
   b) $k-\frac{1}{4}k$ B $k+2k$
   c) $k \times 3$ C $3k \div 4$
   d) $k \div 4$ D $\frac{1}{4}k$

6   Here are four different ways of writing the expression 'n':

$\frac{1}{2}n + \frac{1}{2}n$, $2n - n$, $1 \times n$, $2n \div 2$

Find four different ways of writing each of these expressions.
Use '$+$' '$-$' '$\times$' or '$\div$' only once in an expression:

a) p b) $\frac{1}{2}n$ c) $\frac{k}{4}$ d) $\frac{4}{t}$

Expressions using a^2

1   a) $2a^2$ and $(2a)^2$ mean different things.
$2a^2$ means $2 \times a^2$.
$(2a)^2$ means $2a \times 2a$.
For one particular value of a the expressions are equal to 32 and 64. What is the value of a?

b) For another value of a, $2a^2$ is 50. What is $(2a)^2$?

c) Try some more values of a. Write down the values of $2a^2$ and $(2a)^2$ that go with them.

d) Write down what you notice about the values of $2a^2$ and $(2a)^2$.

e) Copy and complete: $(2a)^2 \equiv \Box a^2$.

f) One of these drawings represents $2a^2$.
The other represents $(2a)^2$.
Which is which?

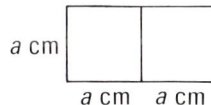

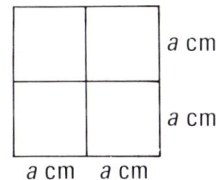

Consider their areas

a cm, a cm a cm, a cm, a cm a cm

g) Make your own drawings to represent $3k^2$ and $(3k)^2$.

h) For one value of k, $3k^2$ is 3. What is the value of $(3k)^2$?

i) For another value of k, $(3k)^2$ is 144. What is $3k^2$?

j) Try some more values of k. Write down the values of $3k^2$ and $(3k)^2$ that go with them.

k) Write down what you notice about the values of $3k^2$ and $(3k)^2$.

l) Copy and complete: $(3k)^2 \equiv \Box k^2$.

2   Are there any values of m for which $5m^2$ and $(5m)^2$ are equal? If so, write them down.

3   a) Explain why this is not true: $4t^2 \equiv (4t)^2$. (*Notice*: $\equiv$, not $=$)

b) Find a value of t for which this is true: $4t^2 = (4t)^2$. (*Notice*: $=$, not $\equiv$)

4 a) Use the drawing of the cuboid to help
you explain why this is true:

$$4k^2 + 3k^2 = 7k^2$$

(Think about volumes.)

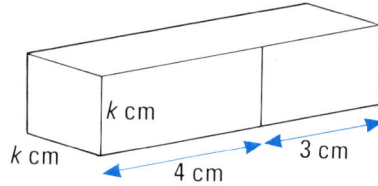

b) Which of these are true?

(i) $4a^2 + a^2 \equiv 5a^2$ (ii) $4a^2 + a \equiv 4a^3$ (iii) $4a^2 + a \equiv 5a$

5 Match each of the expressions a) to e) with one from A to E to which it is always equal:

a) $2a^2 + 3a^2 + a^2$ d) $b + b + a + 2$ A $a^2 + a + a^2$ D $6a^2$
b) $2a^2 + a$ e) $b^2a^2 + b^2a^2$ B $2b + a + 2$ E $2ab^2$
c) $a \times b \times b \times 2$ C $2b^2a^2$

━━━━━━━━━ EXPLORATION ━━━━━━━━━

6 a) By putting $a = 2$ and $b = 10$ explain why $ab + a$ and $2ab$ are not identically equal expressions.

b) By putting $k = 3$ and $m = 10$ explain why $k^2m + k$ and $2k^2m$ are not identically equal.

c) Decide which of these are pairs of identically equal expressions:

A $a^2b + a^2b$, $2a^2b$ B $ab + a^2b$, $2a^2b$ C $a^2b^2 + 2a^2b^2$, $3a^2b^2$
D $4abc - abc$, $3abc$ E $np^2 + pn^2$, $2n^2p^2$ F $2np^2 - np^2$, np^2

d) Investigate some examples of your own like those in c). Try to decide when it is possible to write an addition or a subtraction without $+$ or $-$ signs. For example, $a^2b + ba^2 \equiv 2a^2b$. Write a report to explain what you find out.

7 Write these additions using the number of $+$ signs given in brackets.
(For example, $5mt + 2n$ can be written using four '$+$'s as $mt + 2mt + 2mt + n + n$.)

a) $3kn + 2pt$ (four $+$ signs)   b) $3ab + 5ab^2$ (three $+$ signs)   c) a (one $+$ sign,
d) ax (one $+$ sign)   e) fx^2 (one $+$ sign)   think about fractions)
f) $3gn$ (five $+$ signs)

Saving time

1 In this spiral each section is twice as long as the section before it.
There are 12 sections.

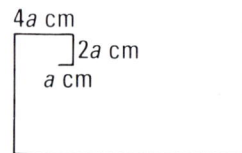

a) If the first section is 3 cm long, how long is the spiral?

b) Copy and complete this expression for the total length of the spiral: $a + 2a + 4a + \ldots$ cm

c) Write your expression in b) as simply as you can.

d) In your expression in c), choose a to be 3. Check that your result agrees with your result in a).

e) What is the total length of the spiral if the first segment is 4 cm long?

Expressions can save us time when we need to make different calculations for the same situation. For example, the length of this spiral is 36$a$cm, so if a is 1 the length is 36cm, if a is 2 the length is 72cm and so on.

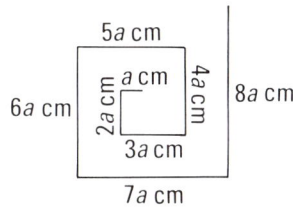

2 a) A, B, and C are regular polygons. Find an expression for the perimeter of each one. Write each of your expressions as simply as you can.

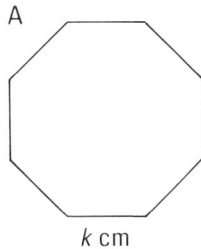

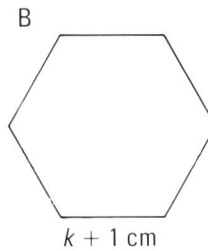

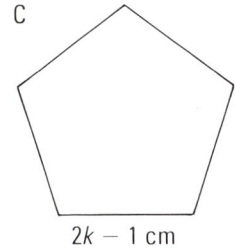

A — k cm

B — $k + 1$ cm

C — $2k - 1$ cm

   b) What is the perimeter of A, B, and C when k is chosen to be 2?

   c) What is the perimeter of each when k is 5?

3 In this sequence of squares the length of side of each square is k cm longer than the length of side of the square on its left. There are 5 squares altogether.

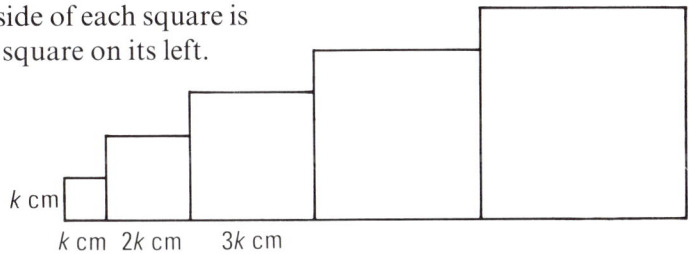

k cm
k cm $2k$ cm $3k$ cm

   a) If k is chosen to be 1, what is the total area covered by the squares?

   b) Copy and complete this expression for the total area covered by the squares:

   $k^2 + 4k^2 + 9k^2 + \ldots \text{cm}^2$

   c) Simplify your expression in b) as much as you can.

   d) Replace k by 1 in your expression in c). Check that your result agrees with your result in a).

   e) What is the total area covered if the first square has area 9 cm²? (Think first! Then use your expression in c).)

4 a) Find an expression for the perimeter of this 'bow tie' shape.

   b) Use your expression to find the perimeter when:

       (i) a is 2 and k is 3
       (ii) a is 5 and k is 7.

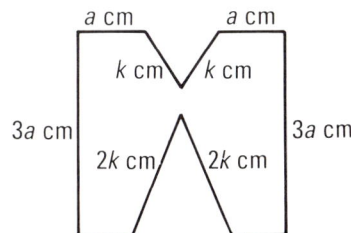

a cm a cm
k cm k cm
$3a$ cm $3a$ cm
$2k$ cm $2k$ cm

B20

Expressions with brackets

1 a) In this pattern, there are b circles in each row of the right-hand rectangle. Which of these gives the total number of circles? (Two of them do. Try some values for b until you are sure.)

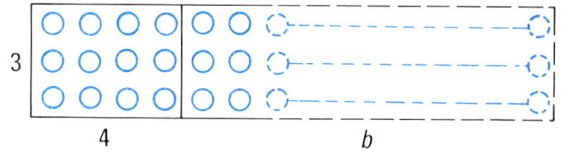

   A $3(4+b)$ (i.e. $3 \times (4+b)$)   B $(3 \times 4)+b$ C $12+b$ D $12+3b$

   b) Sketch a circles diagram which represents $4(3+a)$ circles.

   c) Copy and complete: (i) $3(4+b) \equiv (3 \times \square)+(3 \times \square)$ (ii) $4(3+a) \equiv (4 \times \square)+(4 \times \square)$

2 a) Two of these tell us how many circles there are in the rectangle A. Which two?

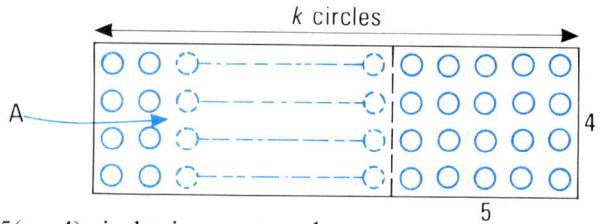

   A $4(k+5)$ B $4(k-5)$
   C $4k+20$ D $4k-20$

   b) Sketch a circles diagram which represents $5(t-4)$ circles in a rectangle.

   c) Copy and complete: (i) $4(k-5) \equiv (4 \times \square)-(4 \times \square)$ (ii) $5(t-4) \equiv (5 \times \square)-(5 \times \square)$

B20

3 Copy and complete: a) $2(t+6) \equiv 2t+\square$ b) $3(6-k) \equiv 18-\square$ c) $4(n+b) \equiv 4n+\square$

━━━━━ CHALLENGE ━━━━━

4 Draw a circles diagram to represent: a) $4(n+b)$ circles   b) $4(n-b)$ circles.

5 Solve these equations.

   a) $2(a+1)=4$ b) $3(t-2)=9$ c) $4(5-p)=12$ d) $\dfrac{4}{t+1}=2$ e) $\dfrac{a+6}{a}=2$ f) $\frac{1}{2}(a-1)=4$

   (Think and test until you find the correct value of the letter.
   For example, $\dfrac{15}{n+2}=5$ when $n=1$ because $\dfrac{15}{1+2}$ is 5.)

6 Which of these is a correct expression for the total area of the rectangle? (Four are.)

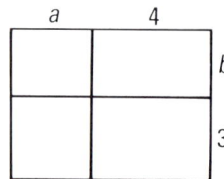

   A $(a+4)(b+3)$ B $(3+a)(4+b)$
   C $3b+4a$ D $ab+4b+12+3a$
   E $b(a+4)+3(a+4)$ F $4(b+3)+a(b+3)$

━━━━━ CHALLENGE ━━━━━

7 Draw a diagram to help you explain why $a(b+c) \equiv ab+ac$

CONSOLIDATION

1 a) This is a 1 cm² square.
      How many mm² is this?

   b) The cost of an
      advertisement is £2.00
      per cm². How much is
      this per mm²?

   c) Calculate the cost of
      each of the
      advertisements.

Farm&Country HOLIDAYS
South West, Yorkshire, Lakes and Borders
Selected cottages and farmhouses providing excellent value holidays. Self-catering and half-board. *Colour Brochure* Farm and Country Holidays, Jennings (N.T.),First Roydon, Bideford,Devon. Tel:24 hrs.
(02372) 95429

NORTH YORKSHIRE MOORS NATIONAL PARK
Elegant Georgian Country House Hotel in celebrated beauty spot at the heart of Herriot/Captain Cook Country. En suite rooms, English Breakfast, Evening Meal. Licensed. 2 night breaks from £17 pp/day.
ELIZABETH MARIE'S HUNTLEY HOUSE Ripon,RP6 7UA Tel(07515) 678

2 a) The area of the window is 1 m².
      How many cm² is this?

   b) Glass costs £35.20 per m².
      How much is this per cm²?
      (to the nearest 0.1p)

   c) How much would you
      expect to pay for the
      glass in this small window?

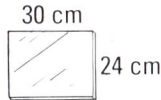

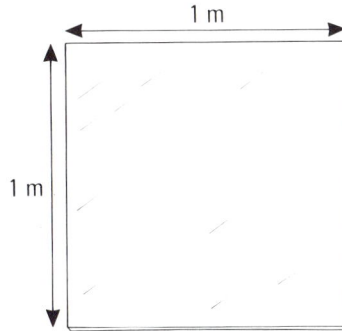

30 cm
24 cm

1 m
1 m

3 Some netball pitches are to
   be marked out as shown
   in the diagram.
   About how many pitches
   will fit into the 1 hectare
   square?

Do you remember?
A 100 m by 100 m square
has an area of 1 hectare.

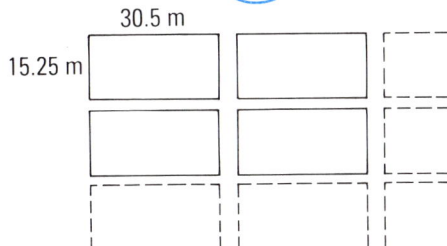

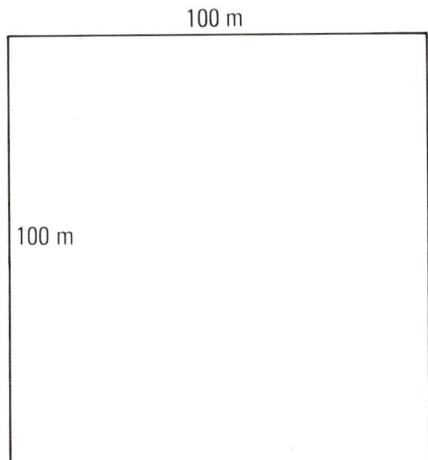

30.5 m
15.25 m

100 m
100 m

ASSIGNMENT

4 Estimate how many litres of emulsion paint it would take to give the inside walls of your school
   gym one coat of paint. In your report explain how you made your estimate.

5 a) Check that each of these fields has an area of 1 hectare.

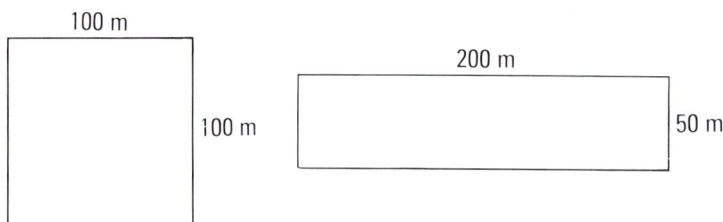

100 m

100 m

200 m

50 m

Sketch two more fields which have an area of 1 hectare.

b) Roughly, what is the area of this L-shaped field:

(i) in hectares   (ii) in m²?

c) Sketch an L-shaped field of your own whose area is about 2 hectares. Don't forget to write in the important measurements.

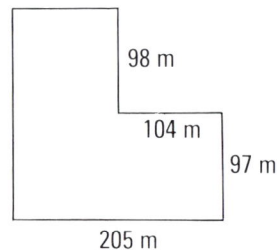

98 m

104 m

97 m

205 m

6 In this aerial photograph the area taken up by the house marked A is about 100 m². Estimate the area of its garden:

a) in hectares   b) in m².

A

B21

════════════ ASSIGNMENT ════════════

7 Carrhouse Builders have bought 2 hectares of land (roughly a rectangle shape, twice as wide as it is long). Estimate how many detached houses, each with a garden of about 500 m², can be built on the plot. Make a sketch of the plot to show your suggested layout.

FOR SALE
2 HECTARES

CORE

Units for volumes

1   The volume of this cube is $8\,cm^3$.

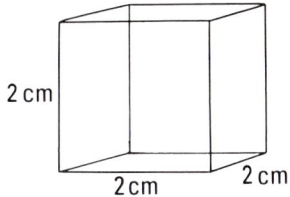

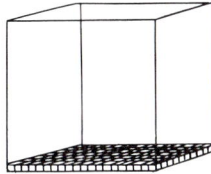

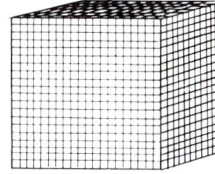

2 cm

2 cm   2 cm

a) One layer of 1 mm cubes.
How many 1 mm cubes?

b) 20 layers of 1 mm cubes.
How many mm^3 is $8\,cm^3$?

c) How many mm^3 is $1\,cm^3$?

2   What is the surface area of a 1 cm cube a) in cm^2, b) in mm^2?

3   a) Design a box whose capacity is 1 million cubic millimetres ($1\,000\,000\,mm^3$).
      Make the box if you wish.

   b) What is the surface area of your box (i) in cm^2, (ii) in mm^2?

4   A plastics firm makes rectangular tubing like this.
   The thickness of the plastic is 1 mm. How
   many cm^3 of plastic is needed for
   each 1 m of tubing?

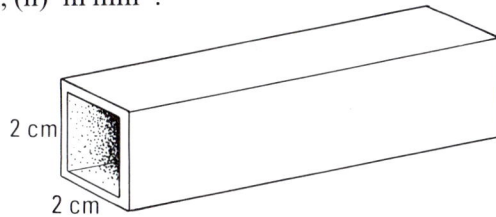

2 cm

2 cm

5   Ready-mix concrete costs £6.70 per m^3.
   Calculate the cost for these foundations.
   (The trench is to be filled with concrete.)

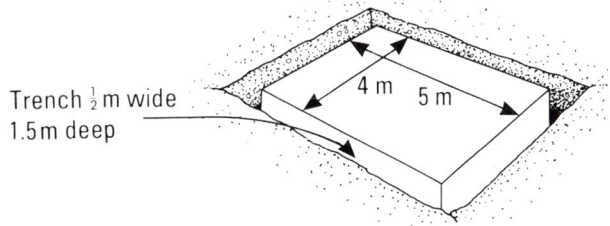

Trench $\frac{1}{2}$ m wide
1.5 m deep

4 m   5 m

6   Each silver trinket on Rupinder's bracelet
   uses $240\,mm^3$ of silver. How many trinkets
   can be made from the silver block?

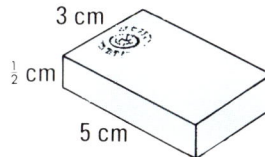

3 cm

$\frac{1}{2}$ cm

5 cm

━━━━━━━━━ CHALLENGE ━━━━━━━━━

7   Roughly, how many ccs (cubic centimetres) of ice cream could you pack into your classroom?
   How many litres is this? How many ice cream cones full is it?

CORE

TAKE NOTE

Another word for 'chance' is probability. We say that the probability that a coin will turn up *heads* is 0.5 or $\frac{1}{2}$.

1   Write down the probability of each of the following (i) as a fraction, (ii) as a decimal.

a) that five will be thrown on a die
b) that a heart will be chosen from a pack of cards (52 cards)
c) that an even number will be thrown on a die
d) that an ace will be selected from a pack of cards.

2   You spin a coin, then spin it again.

a) Which of these do you think is the probability that you will get two *heads* (HH)?

$$\frac{1}{4}, \quad \frac{1}{8}, \quad \frac{1}{2}, \quad \frac{3}{4}$$

b) These are the possible outcomes when you spin a coin twice: that is, there are four possible outcomes. Each one is equally likely. HH is one of the possibilities, so its probability is $\frac{1}{4}$. (Did you choose this in a)?) What is the probability that you will spin:

    H H
    H T
    T H
    T T

(i) TT   (ii) HT   (iii) a *head* and a *tail* in any order?

3   In a coin game you spin three coins. You win if you get three *heads* or three *tails*.

a) Do you think the probability of doing this is:
more than 0.25   less than 0.25   or exactly 0.25?
Have a guess.

b) List all the possible outcomes for one turn at the game like this: How many are there?

    H h *H*
    H h *T*
    H t *H*
    H t *T*
    etc.

c) Your list should tell you that the probability of getting three *heads* is $\frac{1}{8}$.

(i)   What is the probability of getting three *tails*?
(ii) What is the probability of getting three *heads or* three *tails*?
     Notice that you should have chosen 'exactly 0.25' in a).
(iii) In another coin game you spin four coins together. You win if you get three or more *heads*. List all possible outcomes. What is the probability that you will win on any one go?

4   Winston cannot remember how to spell the
    word for this kind of pattern. Here are two
    of his attempts. He can't remember how
    many 's's, how many 'l's, and whether the
    letter in between is an 'a' or an 'e'. Finally,
    in frustration, he guesses a spelling. What
    is the probability that he is correct?

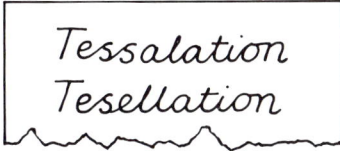

    Tessalation
    Tesellation

▨▨▨▨▨▨▨▨▨▨▨▨ TAKE NOTE ▨▨▨▨▨▨▨▨▨▨

To calculate a probability ● list the equally likely outcomes
                            ● divide the number of equally likely outcomes which are acceptable by
                              the total number of equally likely outcomes.

5   a) How many different routes are there from
       A to B to C?
       List them like this: a then c
                              a then d etc.

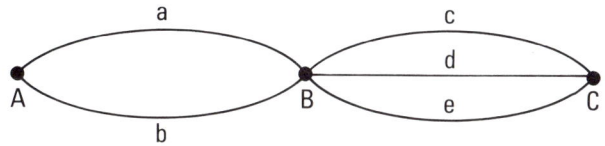

    b) We each choose a route. Explain why the probability that we choose the same route is $\frac{1}{6}$.

    c) Explain why the probability that we choose different routes is $\frac{5}{6}$.

6   You choose two beans from a bag containing two white and two black beans.

    a) Number the beans 1, 2, 3, 4.
       List the possible outcomes, like this: ① ①   ② ②   ① ②

    b) What is the probability that the pair you choose will have different colours?

7   Two dice are rolled together. The circles on the diagram
    represent all the possible outcomes.

    a) How many outcomes are there altogether?

    b) How many of the outcomes give a total score of 4?

    c) What is the probability that you throw a total of 4 with 2 dice?

    d) Alan rolls two dice 360 times. Use your result in c) to
       predict about how many times he will score a total of 4.

    e) About how many times would you expect a total of (i) 2, (ii) 9,
       to be scored in the 360 throws?

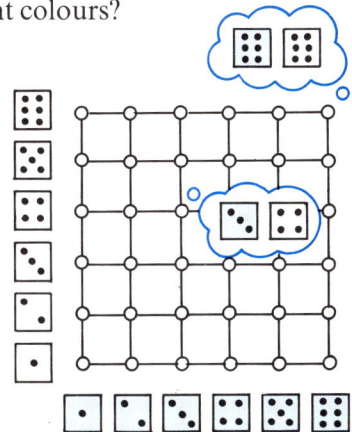

    f) The game is to predict the total score on two dice. Which number would you choose? Why?

B22

8   I spin a coin. You spin a coin.

  a) Explain why the probability that we spin the same result is $\frac{1}{2}$.

  b) What is the probability that we both spin (i) *heads*, (ii) a different result from each other?

Tough luck

1   I take the four aces from a pack of cards, place them face down and shuffle them.
     You pick up one of them, look at it, and replace it.
     I shuffle the cards again. You pick up another card, look at it,
     and replace it. What is the probability that:

  a) your first card was the ace of spades?

  b) your second card was the ace of diamonds?

  c) both cards were the ace of hearts? (This diagram might
      help you to list the equally likely outcomes.)

  d) the cards were the ace of spades and the ace of diamonds in
      that order?

  e) just one of the cards was the ace of clubs?

1st choice   **2nd choice**

heart	heart
spade	spade
diamond	diamond
club	club

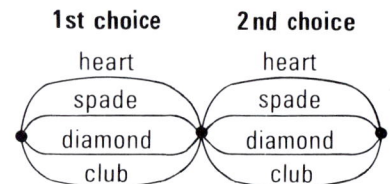

2   In a multiple choice geography test there are ten questions.
     For each question there are five choices of answer.

  a) Tariq guesses the answer to Q1. What is the probability
      that he is (i) correct, (ii) incorrect?

  b) Tariq guesses the answer to Q2. What is the probability
      that he is (i) correct, (ii) incorrect?

  c) What is the probability that Tariq was correct in both Q1 and Q2?

  d) What is the probability that he was incorrect in both Q1 and Q2?

> **Q1**
> Which country has the largest
> population in the world?
> a)India b)Holland c)China
> d)USSR e)USA
>
> **Q2**

B22

3   In a game, two dice are thrown.
     The difference between the two scores is recorded.
     If this is odd, player 1 wins.
     If it is zero or even, player 2 wins.

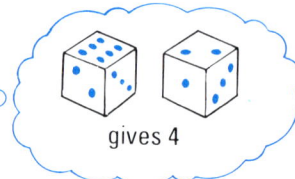

     gives 4

  a) One possible outcome is [ ] [ ] . How many possible outcomes are there altogether?

  b) How many of the possible outcomes give the result 'odd'?

  c) What is the probability that player 1 wins on any throw?

  d) Two players, Red and Blue, play the game 72 times. Red wins if the difference is odd. Blue
      wins if the difference is even or zero. About how many times would you expect:

    (i) Red to win   (ii) Blue to win?

Paying out

1   This is a very simple game of chance. You win if the ball drops
    into the WIN cup.

   a) *Guess* the probability that it will do this.

   b) Here is one route the ball can take to reach the WIN cup.
      How many routes are there altogether into the WIN cup?

   c) How many different routes are there altogether that the
      ball can take?

   d) ... so what is the probability that you win?
      (Was your guess correct in a)?)

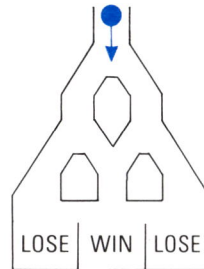

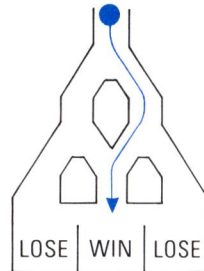

2   This is the game of chance in question 1. It costs 10p a go. You
    win 10p plus your stake money if the ball lands in the middle
    cup. The probability that the ball goes into the middle cup is $\frac{2}{4}$
    or 0.5. (Did you get this right in question 1?)

   a) In 50 goes, about how many times would you expect to win?

   b) About how many times would you expect to lose?

   c) Would you expect to win, lose, or break even over the
      whole 50 goes?

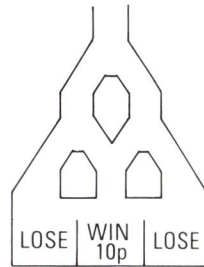

3   Here is the game of chance again. This time the payout has
    been changed. It still costs 10p a go. In 50 tries would you
    expect to win, lose, or break even?
    If you say 'win' or 'lose', write down about how much you
    would expect to win or lose.

THINK IT THROUGH

4   a) Calculate the probability that the ball will fall into:

      (i) cup A   (ii) cup B   (iii) cup C   (iv) cup D.

   b) It costs you 10p a go to play. If you win, you get 10p plus
      your stake money back. You have 40 goes. Would you
      expect to win or lose, and about how much?

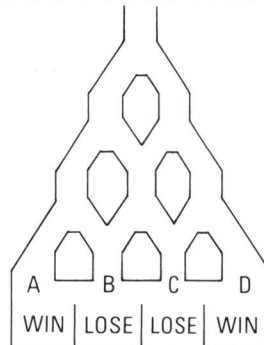

B22

ENRICHMENT

CHALLENGE

1   In this game of chance you have to: • choose one of the two shapes from a bag
                                          • roll the shape.

These are the faces of the tetrahedron: ▲ △ △ △

These are the faces of the cube: ■ ■ ■ ■ ☐ ☐

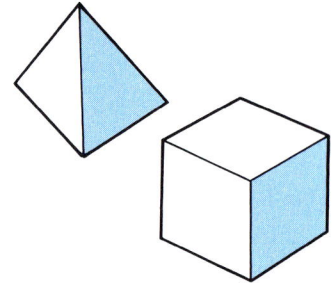

a) Which colour would you predict, white or blue? Why?

b) About how many times in 120 goes would you expect to get:

   (i) a white triangle   (ii) a blue square   (iii) a white face?

Think: about how many times will I choose the tetrahedron? And in about how many of these choices will I roll △ ?

ASSIGNMENT

2   a) Make up a game of chance of your own. Decide how much you should charge for each go, which outcomes are 'win', which outcomes are 'lose', and how much should be paid for each separate 'win' possibility. Make up a tariff board showing the winning combinations and the payments (as on a fruit machine).

> MY GAME
>
> WIN 10p
> WIN 5p
> WIN . . . .

b) Test your game. (Not with real money!) Modify it, if you need to, until you are sure that it:

   • gives a reasonable return to those who play, so that they keep on playing
   • wins for you in the long run.

c) Write a report in which you describe your game and explain how you tested it. Explain how you made sure it met the conditions in b).

B22

CORE

Do you remember . . . ?
28% of £60 is 0.28 × £60,
which is £16.80

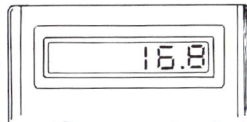

`16.8`

LEISURE and SPORTS

We regret to inform all members that membership fees will increase by 12% next year.

ADULT LEISURE ACTIVITIES

Membership:

Individual	£19.00
Couple	£34.00
Persons over retirement age	£10.00
Child (under 16)	£5.00
16-19 yrs	£6.50

1 a) Calculate how much *you* would pay in membership fees next year.
   b) How much more will a Senior Citizen pay next year than this year?

2 These are some of the items in Hannam's Mail Order catalogue. Calculate the special offer price of each one. Round your results down to the next 1p.

HANNAMS
MAIL ORDER

NEW SEASON SPECIAL OFFERS

5% OFF GOODS
Costing £10 or less

8% OFF GOODS
Costing £10.01 to £25

15% OFF GOODS
Costing £25.01 to £100

25% OFF GOODS
Costing over £100

a)

£7.00

b)

£85

c)

£276

B23

░░░░░░░░░░ CHALLENGE: SELLING HOUSES ░░░░░░░░░░

3   Estate agents charge a fee, called *commission*, for selling houses. For example, Capital and Growth charge 1% of the selling price of the house.

a) What is the commission charge on the £81 500 house?

b) On another sale the commission is £765. What is the price of the house?

c) Swift Sales charge fees of $1\frac{1}{4}$% (1.25%). What is the fee on the £56 000 house?

d) Seekers charge £199 + VAT, no matter what the price of the house. VAT is 15% (of £199). For what range of house prices is Seekers cheaper than:

    (i) Capital and Growth
    (ii) Swift Sales?

Calculating percentages

1 Moira is a sales assistant in a DIY store. One of her jobs is to calculate the price the store should charge for the things it sells.

For example, the store buys electric switches for £1.55 each. This is called the 'cost price'. The store uses a 'mark-up' of 35%. Moira calculates:

$$\boxed{C}\ \boxed{1}\ \boxed{.}\ \boxed{5}\ \boxed{5}\ \boxed{\times}\ \boxed{.}\ \boxed{3}\ \boxed{5}\ \boxed{=}$$

This gives the 'mark-up' to be added to the cost price.

a) Calculate for yourself the 'mark-up' on the electric switches. Round your result down to the next 1p.

b) Next, Moira writes out a price card to advertise the items. Copy and complete it.

☆ **SPECIAL OFFER** ☆
SWITCHES ONLY
£   + VAT

c) VAT is added to the price of a switch when it is sold. VAT is 15%. How much is the VAT on the switch (round the result down to the next 1p)? How much does a customer pay for a switch? (This is called the 'selling price'.)

THINK IT THROUGH

2 These are the prices that the DIY store pays (the cost price) for some of the things it sells. Calculate the price the customer pays (the selling price). Don't forget: the store adds a 35% mark-up, then 15% VAT.

Cost price £3.50 each

Cost price £19

B23

3 Alan works in the same DIY store as Moira.
   This is how he calculates the price to print on the
   price tags:

   `C` `1` `.` `5` `5` `×` `1` `.` `3` `5` `=`

   Switch
   Cost price
   £1.55

a) Use Alan's method to help you to complete this price tag.

b) Check that Alan's method gives the same
   result as Moira's. Write one or two sentences to
   explain why Alan's method works.

   SPECIAL OFFER
   ☆ **SWITCHES ONLY** ☆
   £   + VAT

c) Whose method do you prefer, Alan's or Moira's? Why?

d) Use Alan's method to calculate what to write on the price tag for each
   of these:

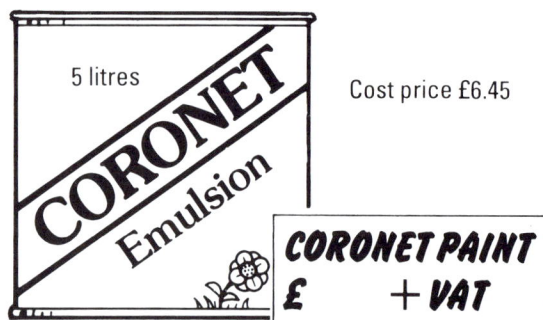

   5 litres
   CORONET
   Emulsion

   Cost price £6.45

   CORONET PAINT
   £   + VAT

   STEP LADDERS
   £   + VAT

   Cost price £11.75

THINK IT THROUGH

e) Alan is working out selling prices for the switches.

   He presses: `C` `1` `.` `5` `5` `×` `1` `.` `3` `5` `×` `1` `.` `1` `5` `=`

   Write one or two sentences to explain why this gives the correct selling price.

f) Use Alan's method in part e) to find the selling price of:

   (i) the tin of paint   (ii) the stepladders.

WITH A FRIEND

4 Moira tries this method for the switch: `C` `1` `.` `5` `5` `×` `1` `.` `5` `0` `=`

   (Moira thinks: 35% mark-up + 15% VAT = 50%, so multiply by 1.5.)
   Discuss with your friend why this does NOT give the correct selling price. Each of you write one
   or two sentences to explain what you decide.

B23

5 Yasmeen is looking for a pair of new shoes in the sale. She finds a pair which were originally £23.

SAVE 15%
ON ALL SHOES
15% OFF all marked prices

She calculates the sale price like this:
$0.85 \times £23 = £19.55$

a) Is she correct?

b) If you say 'yes' in part a) write one or two sentences to explain why her method works.

6 Use Yasmeen's method to find the sale prices in these sales. Round your result to the nearest 1p.

a) **28% OFF** all marked prices
KITCHEN WARE SALE
£34

b) **SAVE 12%** ON ALL SPORTSWEAR
Hobson's Sports Summer Sale
Track Suits Were £28.50

c) **TAPES! TAPES! TAPES!**
CLOSING DOWN SALE
SAVE 70%
Were £5.95 per pack of 4

Present wage   £7000
New wage $£7000 \times 1.24 = £8680$
(100% of £7000 + 24% of £7000 = 124% of £7000)

Original price £250
Sale price $£250 \times 0.82 = £205$
(100% of £250 − 18% of £250 = 82% of £250)

WAGE INCREASE 24% FOR NURSES

SALE   18% OFF all items

B23

Percentage increase and decrease

1   The 'Private cars' bar chart shows that about 280 people in every 1000 own cars in the UK.

$$\frac{280}{1000} = 0.28 = 28\%$$

So about 28% of people in the UK own cars.

a) What percentage of people in these countries own cars:

(i) USA   (ii) Japan?

b) In one country, about 42% of people own cars. Which country?

c) What percentage of people in the UK:

(i) own a telephone
(ii) own a TV set?

STANDARDS OF LIVING

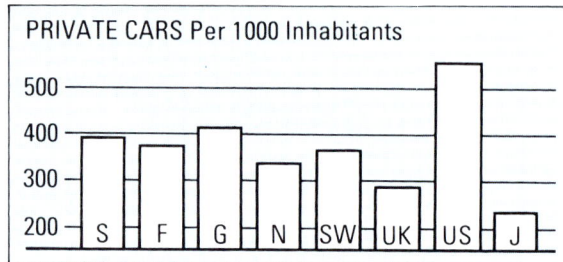

PRIVATE CARS Per 1000 Inhabitants

S	SWITZERLAND	SW	SWEDEN
F	FRANCE	UK	UNITED KINGDOM
G	GERMANY	US	UNITED STATES
N	NETHERLANDS	J	JAPAN

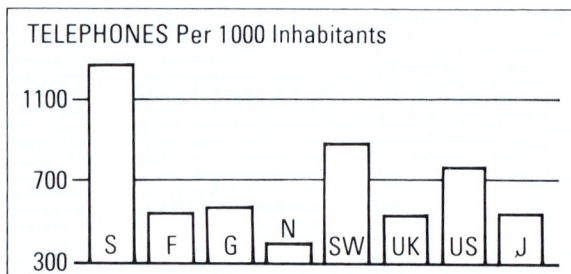

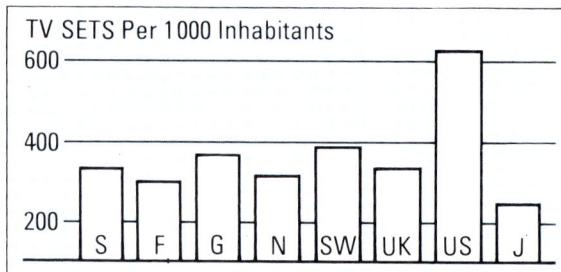

TELEPHONES Per 1000 Inhabitants

TV SETS Per 1000 Inhabitants

d) Which country do you think has the better standard of living – the UK or Sweden? Write one or two sentences to explain why you think this.

2   These charts show the amount of each type of energy consumed in the UK in four different years.

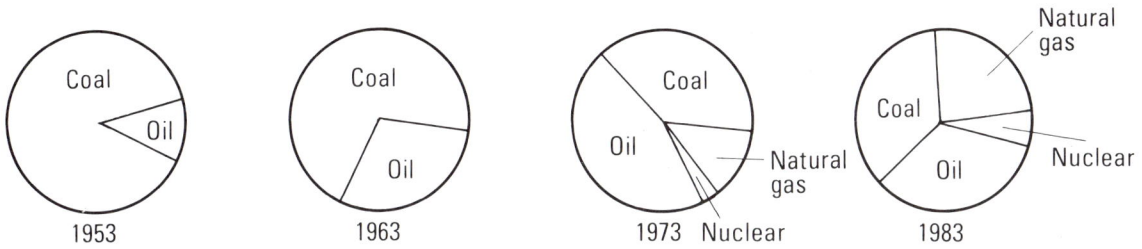

a) Roughly what fraction of the total energy came from oil in 1953?

b) Roughly what percentage is this?

c) The graph below shows how the proportions
    of each type of energy changed between
    1953 and 1973. Copy and complete it.

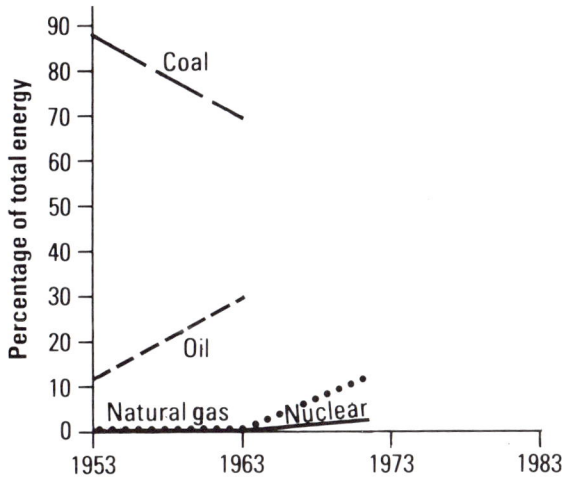

d) Write a short report to say how the types of energy used in the UK changed between 1953 and
    1983. Say which types increased and which types decreased, and by how much.

THINK IT THROUGH

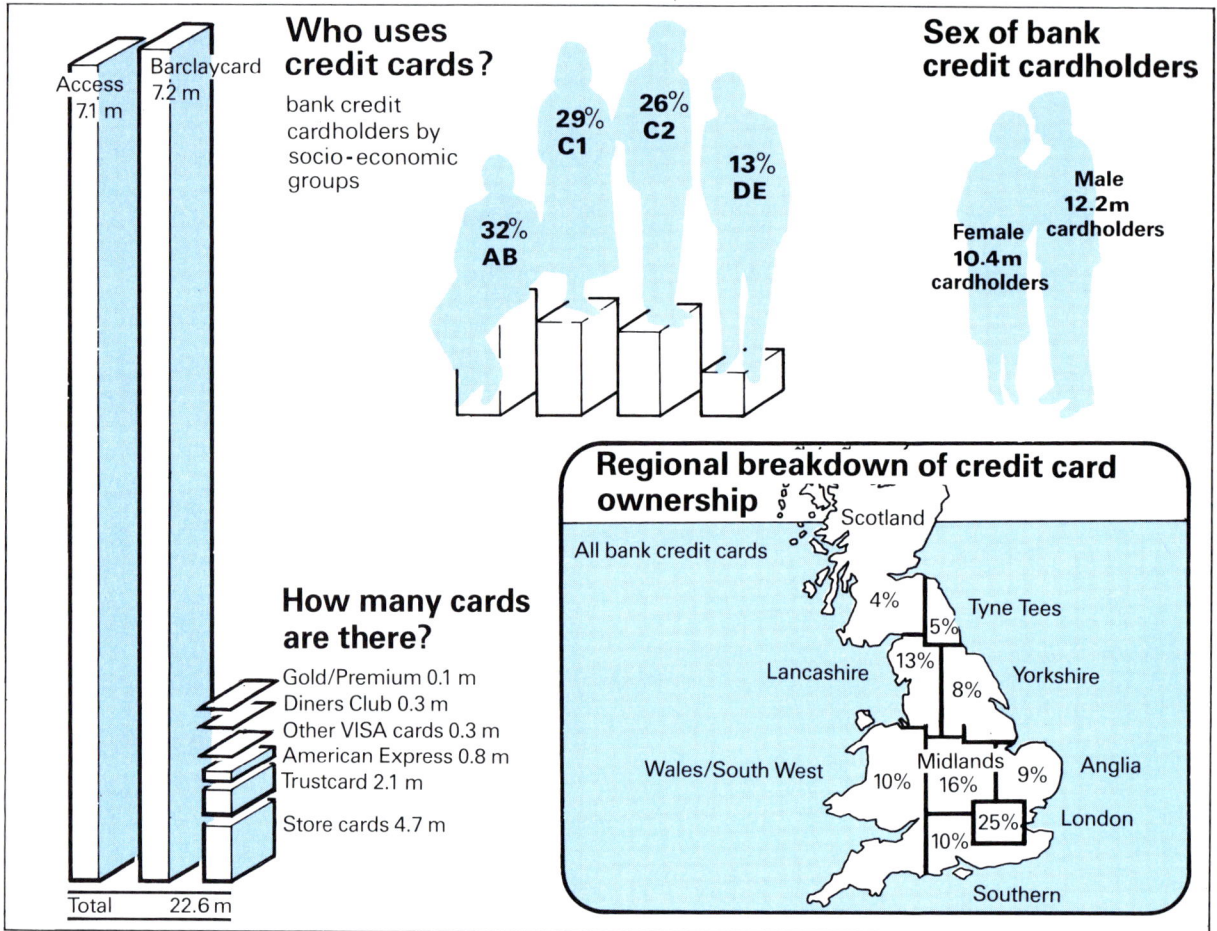

Who uses credit cards?

bank credit cardholders by socio-economic groups

Access 7.1 m

Barclaycard 7.2 m

32% AB

29% C1

26% C2

13% DE

Sex of bank credit cardholders

Male 12.2m cardholders

Female 10.4m cardholders

How many cards are there?

Gold/Premium 0.1 m
Diners Club 0.3 m
Other VISA cards 0.3 m
American Express 0.8 m
Trustcard 2.1 m

Store cards 4.7 m

Total   22.6 m

Regional breakdown of credit card ownership

All bank credit cards

Scotland 4%

Tyne Tees 5%

Lancashire 13%

Yorkshire 8%

Wales/South West 10%

Midlands 16%

Anglia 9%

London 25%

Southern 10%

3   The chart gives information about credit cards in the UK in 1984.

a) How many credit cards were there in use in 1984?

b) Roughly, what percentage of the cards were:
    (i) Access, (ii) Barclaycard, (iii) American Express?

c) Roughly, what percentages of the cards were owned by (i) women, (ii) men?

d) 16% of all credit cards belonged to Midlanders. Roughly how many cards is this?

e) Roughly, how many cards were owned by people in Scotland?

f) Which do you think is more likely?

   (i) More than 10% of people in Wales/the South West owned credit cards.
   (ii) Less than 10% of people in Wales/the South West owned credit cards.

   Write one or two sentences to explain your answer.

- The tree has grown 1.5 m since last year.

  Its fractional increase in height is $\dfrac{1.5}{2} = 0.75$.

- So its percentage increase in height is 75% (of last year's height).

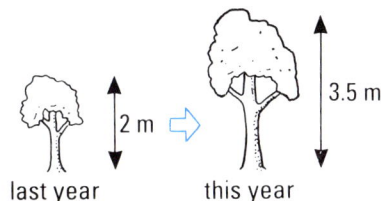

Write each of your results in questions 4 to 10 to the nearest 1 per cent.

4   The train increased its speed from 60 km/h to 90 km/h. What is its percentage increase in speed?

5   The price of a holiday increases from £270 to £300. What is the percentage increase in price?

6   Ranjit's salary increases from £7000 to £7500. What percentage increase is this?

7   Last year 'Homework' built 1200 new homes. This year they hope to build 1500. What percentage increase will this be?

8   A metal bar is 1.76 m long. After heating by a blacksmith it is 1.88 m long. What is the percentage increase in length?

The reservoir normally holds 3000 million litres of water. After a long, dry spell it holds 2500 million litres. The fractional decrease in the amount of water

is $\dfrac{500}{3000} \approx 0.17$ | 0.166666 |

So the percentage decrease is about 17%.

9   At the start of a week a grain store has 500 m³ of grain. At the end of the week it has 475 m³. What is the percentage decrease in stock during the week?

10 The table gives the value (in pounds) of a Nissan Bluebird four-door saloon after 1, 2, and 3 years. By what percentage does its value decrease:

a) during the first year   b) during the second year?

Age			
New	1 year	2 years	3 years
12 000	10 800	8900	7500

B23

11 a) Here are two examples of special offers:
Do you agree with each claim?
Any time you say 'No', explain why.

b) What percentage is offered
free in this carton
of orange?

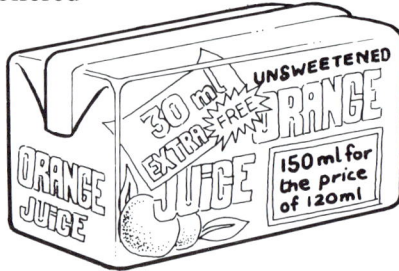

12 Compare each of the following. For each comparison, write a sentence like this: ... is ☐ % as long/heavy as ...

a)   Your longest finger and the length of your hand.

b)   The length of your hand and the length of your arm.

c)   Your weight, and the weight of a shire horse (a shire horse weighs about 1 tonne).

d)   The speed at which you can walk and the speed at which you can run.

CHALLENGE

13 The volume of gas in a gasometer decreased by 20% during one bank holiday weekend. The amount of gas left in the gasometer was 6000 m³. How many cubic metres of gas were there to begin with?

ENRICHMENT

ASSIGNMENT: PRICE CHANGES

1   You have to do the second part of this Assignment about 1 month after the first part.

Part 1

a) Make a list of 20 things which most people buy each week. They should be *essential* items, not *luxury* items.

For example,
Bread
Potatoes
Meat
Petrol
Tea
Detergent
…

Estimate the approximate quantity of each that a family of four buys each week.
Write this on your list: For example, Bread 5 loaves
                            Potatoes 10 kg
                            …

Visit your local shop(s) and find the price of each item on your list. Work out the total cost.

Part 2 (After 4 weeks)

b) Visit your local shop(s) again and find the prices of the items on your list. Work out the total cost. Write a report about what you find.
In your report calculate any individual percentage increases or decreases in price. Calculate also the percentage change in price of your 'shopping basket'.

c) Find out what is meant by 'inflation' and the 'retail price index'. Write about them in your report.

B23

WITH A FRIEND: BIG DIPPER

You need 1 cm squared paper.

1   This is part of a big dipper.
    The car freewheels down the dips
    and up the hills. (It is not driven
    by an engine.)

    a) Decide between you at which point the
       car will be travelling fastest, A, B, C, D,
       E, or F. Write down what you decide.

    b) The speed of the car at point A is
       20 km/h. Its speed at F is 50 km/h.
       Estimate its speed at each of the other
       points.

    c) One of these graphs begins correctly.
       Decide between you which one it is, then
       make a rough sketch of the whole of it.

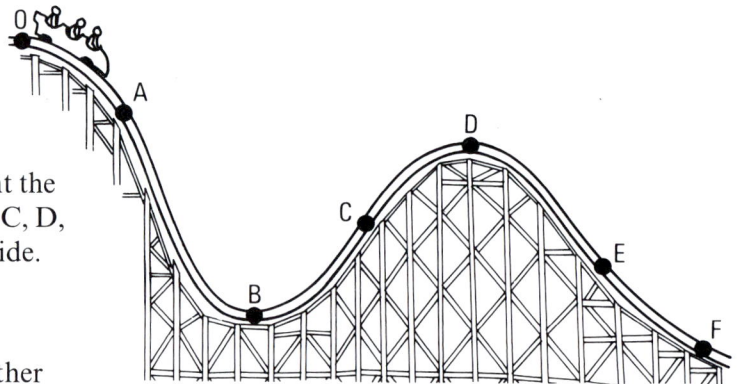

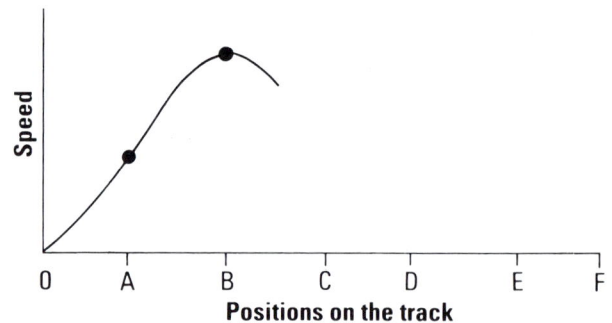

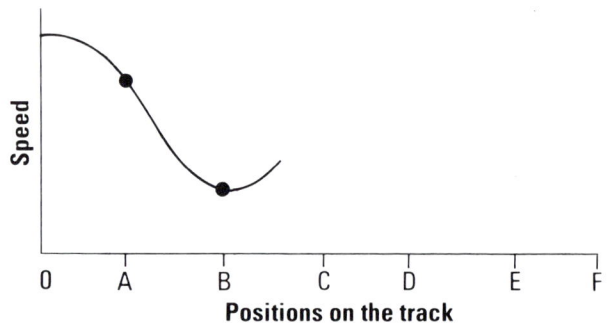

2   Jan Holton lives about 8 km away from the superstore where she does her shopping.

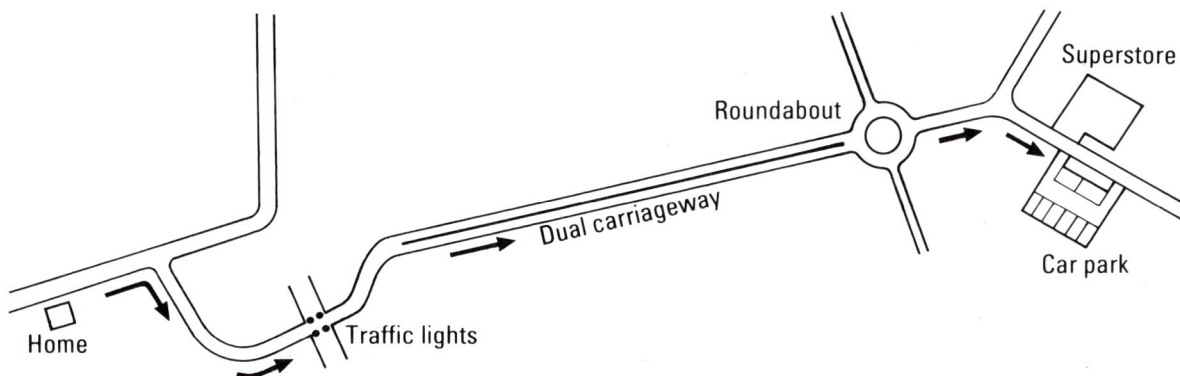

The distance–time graph represents one of her journeys. Labels have been added to show what is happening at different times.

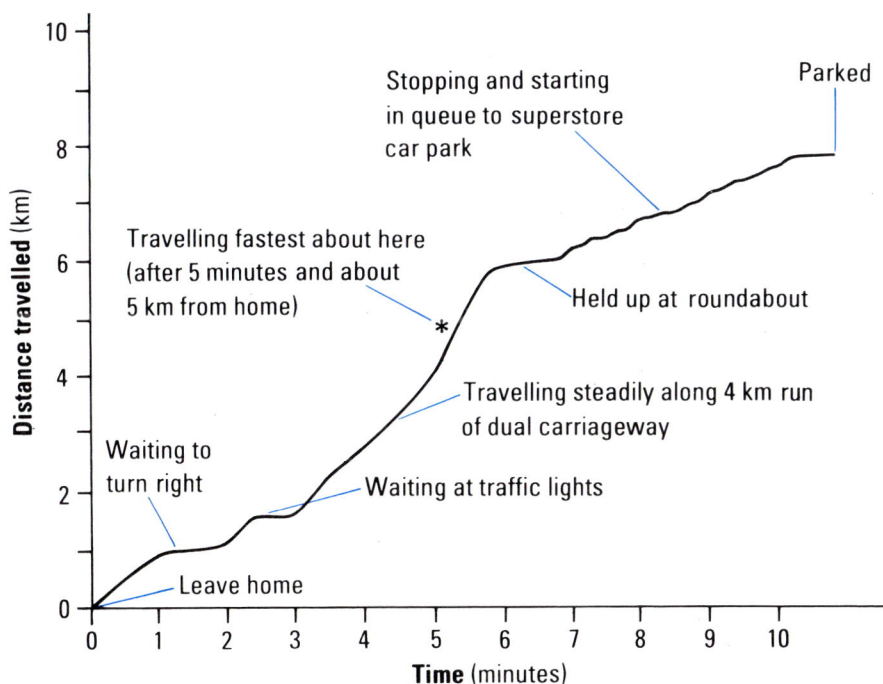

Think of a journey of about 8 kilometres which you might make in a car or a bus. Choose an interesting journey, not one on a long straight road.

Sketch a graph like the one above for your journey. Add labels to your graph, with arrows, to show what is happening at various times.

Put a * to represent the place where your speed is greatest. Label it to show after how many minutes this is, and how far from home you are.

Try to make sure that the slope of your graph is correct for the whole journey.

Make a sketch to show the route you take.

B24

Gradients and graphs

1   The escalator and the lift travel at different speeds. This table shows the distances you travel going up on the escalator.

Time (s)	0	1	2	3	4	5	6	—
Distance (m)	0	$1\frac{1}{2}$	3	$4\frac{1}{2}$	6	$7\frac{1}{2}$	9	—

a) How many metres do you travel each second?

b) The escalator is 10 metres long. Copy and complete the graph for your journey on the escalator.

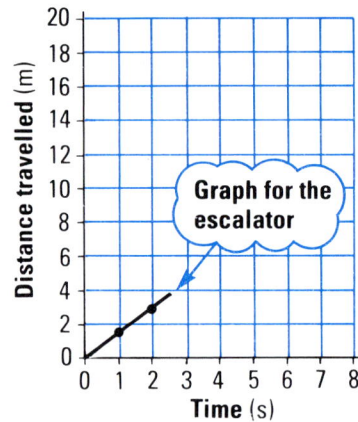

c) Once the lift is moving it travels at a steady 2 m/s. It starts at ground level and travels, without stopping, to Level 4. (It is 4 metres from one level to the next.)
Draw a graph of distance travelled against time for going up in the lift. Start your graph as the lift passes Level 1. Use the same axes as for the escalator.

d) Which graph has the greater slope?

e) On the same axes draw a graph for a moving pavement which carries you 20 metres in 20 seconds.

f) How can you tell from the graph which system is travelling fastest?

2   This is a sketch of the lift and escalator in Brown's Department Store. Use your graphs from question 1.

a) About how long would it take to travel from the ground level to Level 1 using (i) the lift, (ii) the escalator?

b) If there were no stops, about how long would it take to travel from the ground level to Level 3 using (i) the lift, (ii) the escalator?

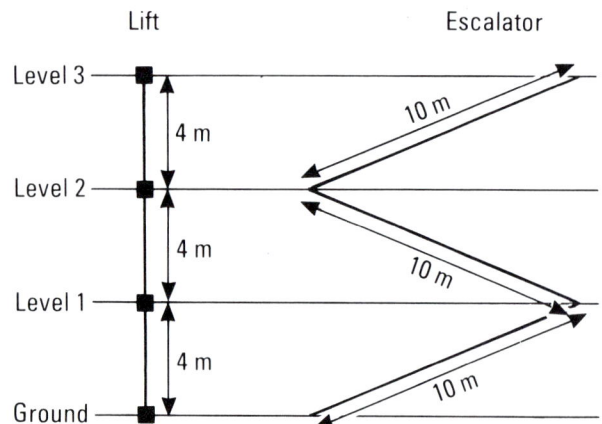

c) On an average journey, the lift stops for
   5 seconds at Level 1 and 6 seconds at
   Level 2. The escalator change-over takes
   3 seconds at Level 1 and 4 seconds at
   Level 2. Copy and complete these
   graphs for the journeys.

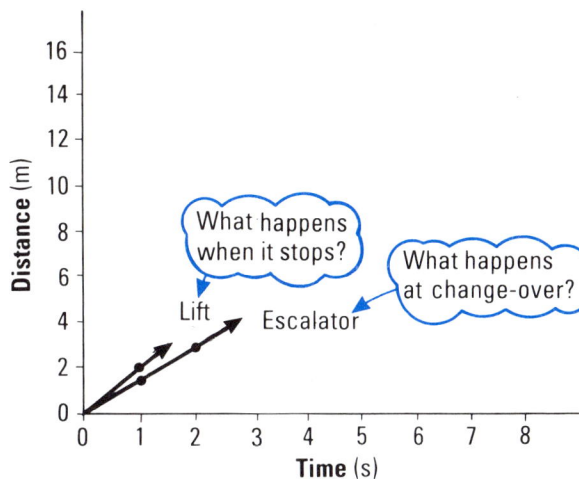

3   The four lines in the diagram are the graphs
    for:

    ● a cruise liner sailing in the Mediterranean
    ● a camel crossing the Sahara desert
    ● Concorde at cruising speed across the
      Atlantic
    ● the Orient Express on the way to Vienna.

    The ship, the animal, the aircraft, and the
    train are all travelling at steady speeds. Write
    down which graph is which.

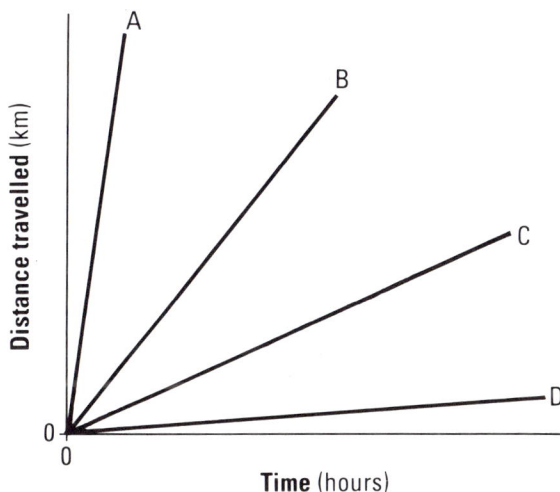

=== TAKE NOTE ===

On distance–time graphs, the greater the slope,
the greater the speed.

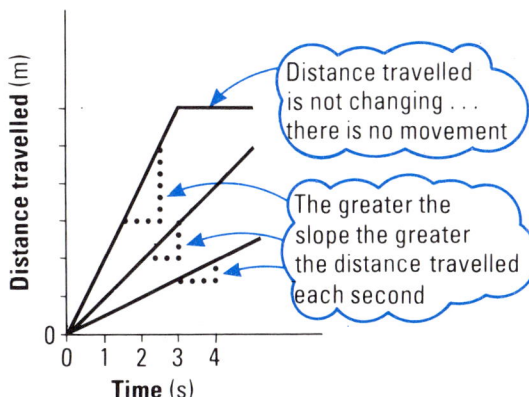

B24

THINK IT THROUGH

4   You need 2 mm squared paper.
A stone is dropped down a mine shaft. The
graph shows the distance it has fallen after 1,
2, 3, and 4 seconds.

a) From the graph we can see that the stone
fell about 35 m during the 4th second.
Roughly, how far did it fall during (i) the
1st second, (ii) the 2nd second, (iii) the
3rd second?

b) Make a rough guess at how far the stone
will fall during the 5th second. (Check
your guess by copying the graph and
continuing it with a smooth curve.)

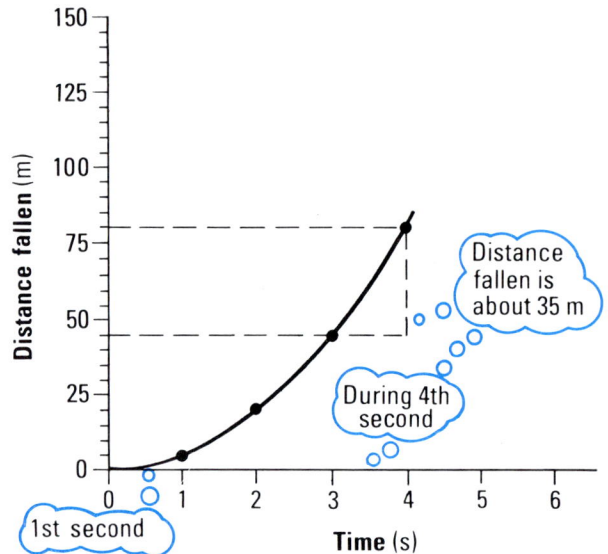

c) As the graph continues its slope gets steeper and steeper. What do you think this tells us about
the speed of the stone? (*Think*: is the stone travelling further in each successive second, not so
far, ... ?) Write one or two sentences to explain why you think this.

5   Write your own *Take note* like the one on the
previous page, but for a curved distance–time graph.

THINK IT THROUGH

6   This graph tells the story of Moira's bath
time.

a) For how many minutes did Moira leave
the taps running?

b) How many more minutes passed before
Moira stepped into the bath?

c) By how much did Moira's presence in the
bath increase the water level?

d) After how many more minutes did she
step out of the bath?

e) Which was faster, the rate at which water
flowed into the bath, or the rate at which
water flowed out of it?

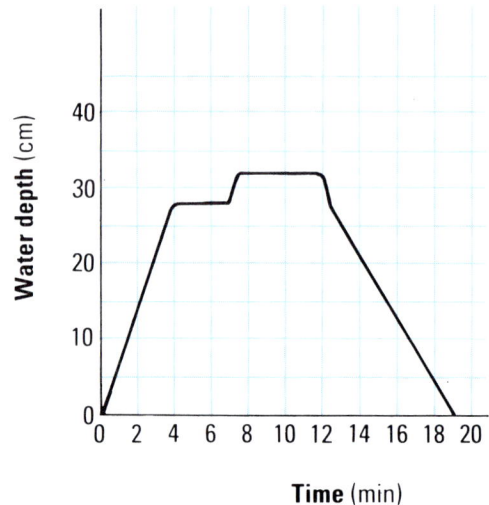

B24

7   The graph represents the progress of two athletes in a 100 m race.

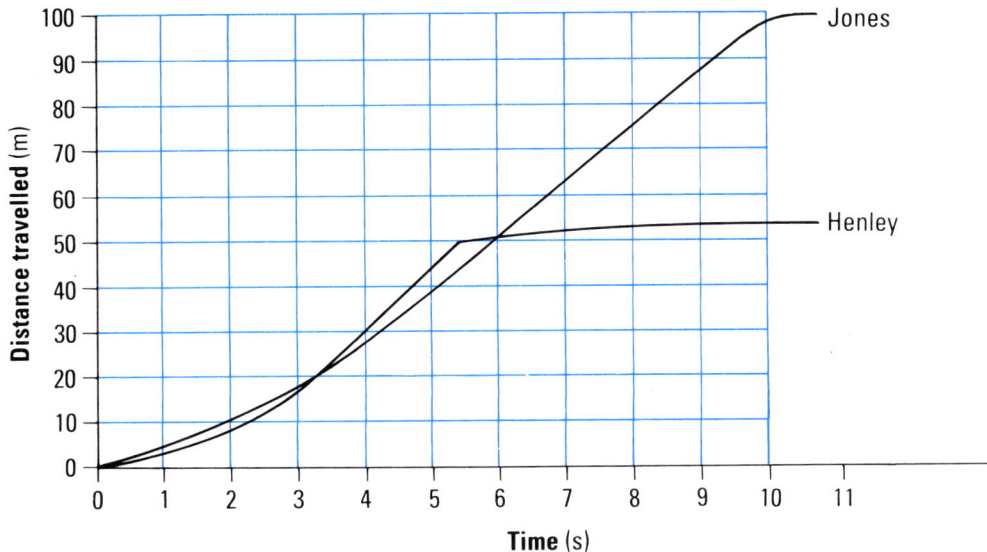

a) This is a newspaper report about the race. Write down all the missing information (A to L).

(A)_____ wins 100 m Challenge.

(B)_____ was fast out of the starting blocks, but (C)_____ was faster still. At the 15 m mark (D)_____ was up by about (E)__ m with (F)_____ struggling to stay with him. Then, a sudden surge by (G)_____ rapidly closed the gap, until by the time they had reached the (H)__ m mark the two athletes were neck and neck. With about (I)__ m to go, (J)_____ looked as though he would win easily, as he pulled away. But then disaster struck. Clutching his right calf muscle he limped helplessly onto the central grass area, leaving (K)_____ to win the race in (L)__ seconds.

b) During the race, who reached the faster speed? Explain how you know.

REVIEW

Prisms are the same shape and size all the way along their length.

These are prisms.

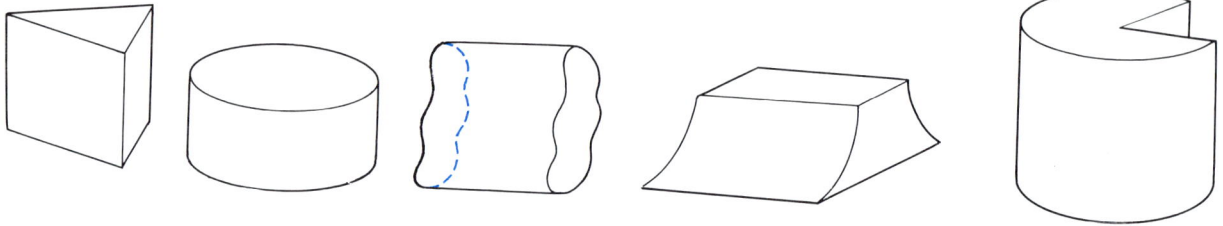

These are not prisms.

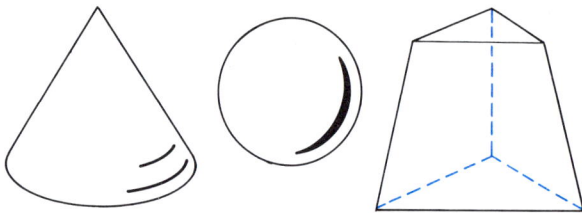

These are cylinders.

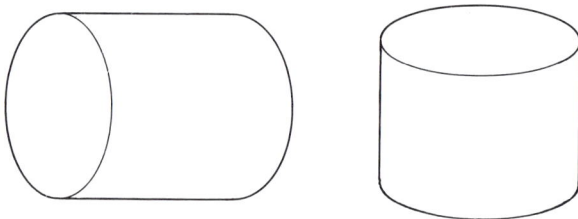

■ Draw two more examples of prisms.

> **Do you remember?**
>
> Volume of prism =
> volume of 1 layer of 1 cm cubes × number of layers

Cylinders are special kinds of prisms – they are prisms with circular cross-sections.

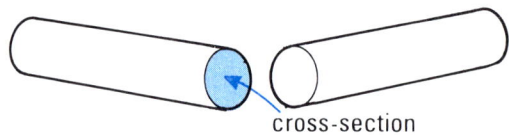

cross-section

This prism has five cubes in the blue layer.
Its volume is $5 \times 6\,\text{cm}^3 = 30\,\text{cm}^3$.

    ↑   ↑
  5 cubes in 6 layers
  each layer of cubes

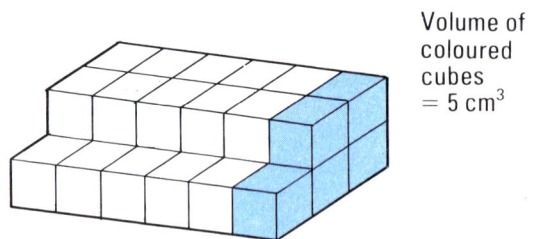

Volume of
coloured
cubes
$= 5\,\text{cm}^3$

■ Sketch a prism whose volume is $20\,\text{cm}^3$.

$6\,\text{cm}^2$ $3\,\text{cm}^2$ $6\,\text{cm}^2$

$3\,\text{cm}^2$ $3\,\text{cm}^2$

The surface area of this prism is
$6+3+6+3+3+3+3+3\,\text{cm}^2 = 30\,\text{cm}^2$.

$3\,\text{cm}^2$ $3\,\text{cm}^2$

■ Sketch a prism whose surface area is $20\,\text{cm}^2$.

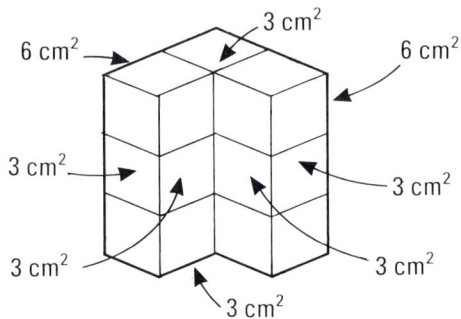

$3\,\text{cm}^2$

CORE

1 A prism is built on this horizontal base. It is made from 1 cm cubes.

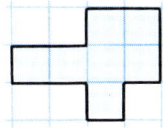

   a) How many horizontal layers are needed if:

   (i) 21   (ii) 28   (iii) $31\frac{1}{2}$

   cubes are used altogether?

   b) The 21-cube prism in part a) looks like this. Draw the other two prisms. (Use 1 cm dotted isometric paper.)

2 A prism is built on a horizontal base. It is made from exactly twenty 1 cm cubes.

   a) Which of these bases can be used, if:

   (i) the cubes are to be kept whole (ii) the cubes can be cut in half?

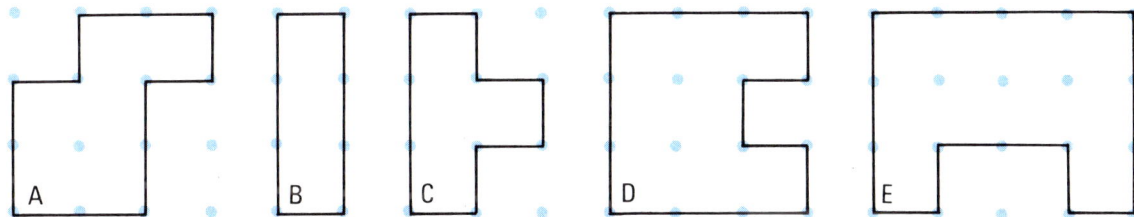

   A   B   C   D   E

   b) How many horizontal layers are needed if base A is used?

3 a) Prisms are built on these horizontal bases. How many 1 cm cubes are needed for each prism, if they are all 3 cm high? Give an approximate number for E.

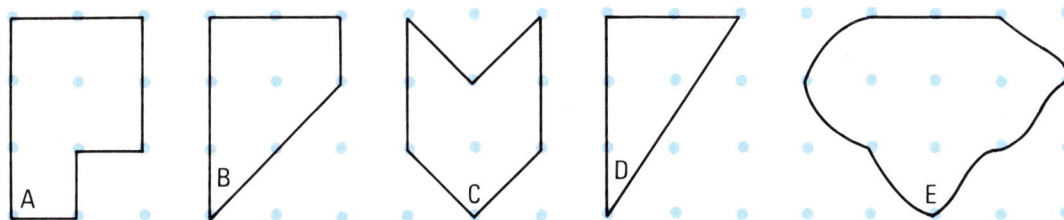

   A   B   C   D   E

   b) Here are sketches of the first two prisms. Copy the sketches and make sketches of the other prisms from roughly the same viewpoint. Use plain paper.

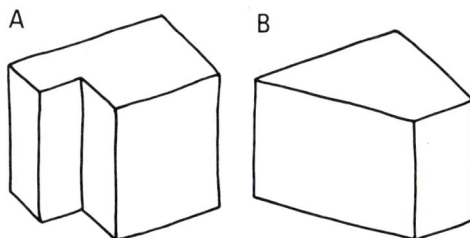

   A   B

B25

4   a) Which of these shapes are prisms?

b) For each one you name as a prism, calculate: (i) its volume   (ii) its surface area.

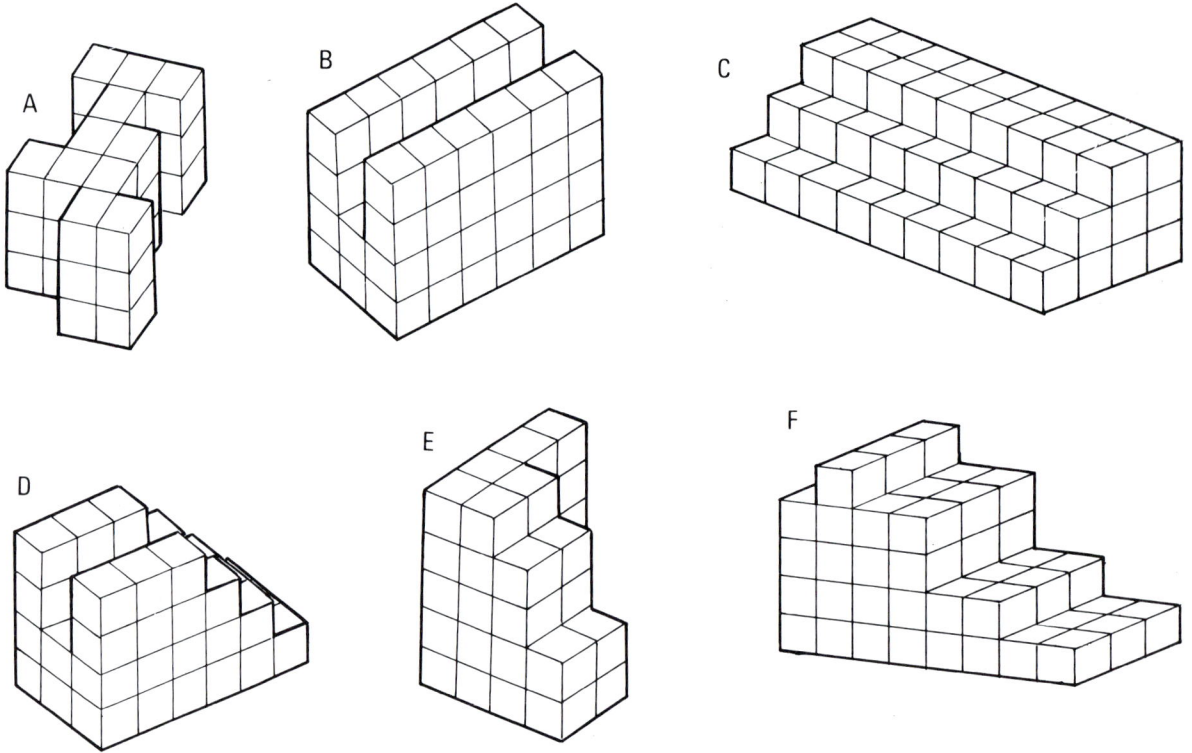

A

B

C

D

E

F

5   Calculate the volume and surface area of each of these prisms:

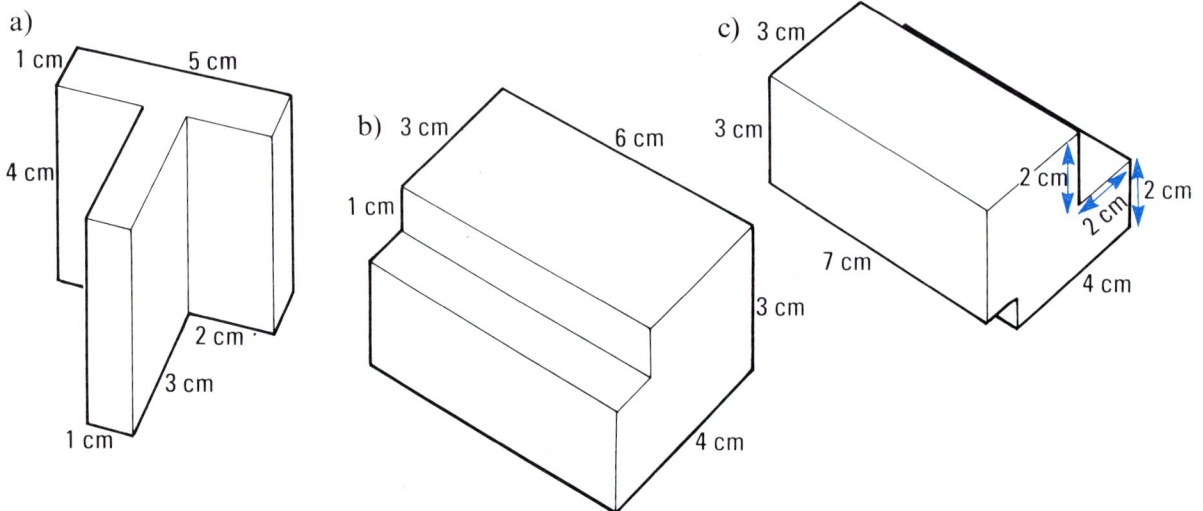

a)

1 cm   5 cm

4 cm

2 cm

3 cm

1 cm

b) 3 cm   6 cm

1 cm

3 cm

4 cm

c) 3 cm

3 cm

7 cm

2 cm   2 cm

2 cm

4 cm

THINK IT THROUGH

6   Prisms are built on these horizontal bases. They are made from *whole* 1 cm cubes and have the same volume. What is the smallest number of cubes needed for each?

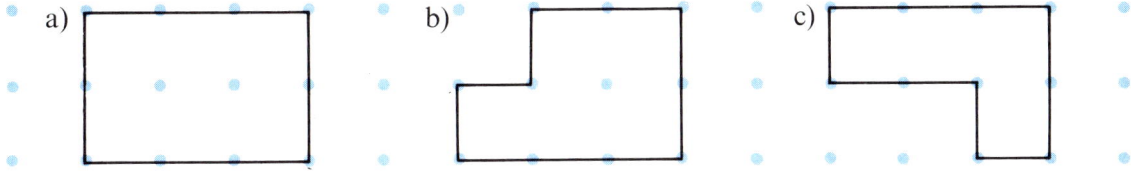

   a)   b)   c)

7   a) A 4 cm high prism is built on this
        horizontal base. What is its volume?

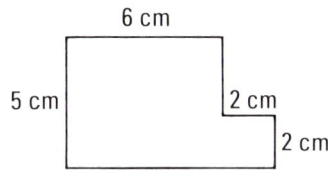

   6 cm

   5 cm   2 cm

                           2 cm

   b) The prism is rearranged to form a cuboid.
       Write down one possible set of values for
       its width, depth and height.

   A 'box shape'

   width

   depth   height

   c) The prism is reshaped to form a cuboid
       with an 8 cm square base. Roughly, what
       is its height?

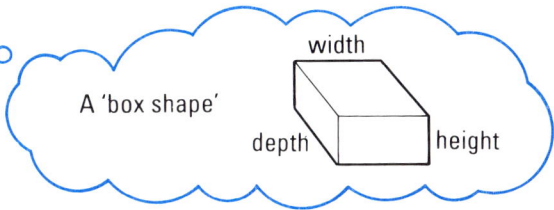

   d) The prism is reshaped again to form a
       triangular prism. Its base is an 8 cm
       equilateral triangle. Roughly, what is its
       height? (Hint: draw the triangle on
       isometric paper.)

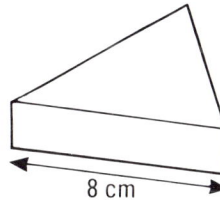

   8 cm

CHALLENGE

8   This Plasticine prism is rolled
    out into a cylinder of the
    same length. Roughly how
    wide is the cylinder?

   4 cm

   4 cm

B25

THINK IT THROUGH

9 a) The area of the shaded face is 9 cm². What is the volume of the prism?

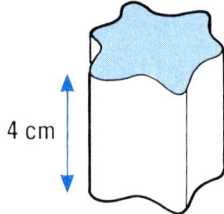

4 cm

b) Here is a pile of 10p pieces. The area of the circular face is about 6 cm². What is the volume of the pile?

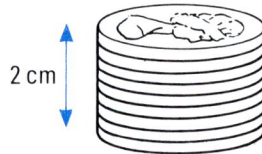

2 cm

c) The volume of this prism is 24 cm³. What is the area of the shaded face?

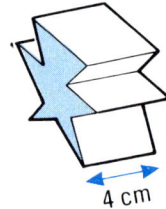

4 cm

10 Copy and complete the *Take note*.

TAKE NOTE

Volume of prism = area of constant cross-section × …

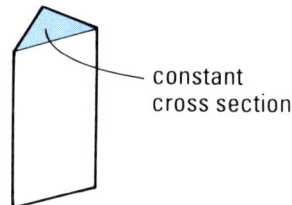

constant
cross section

11 a) Explain why a cylinder is a prism.

b) This cylinder is 8 cm wide and 5 cm high.

(i) Work out the approximate area (A cm²) of a circular face. Use the rule $A = \pi \times$ radius × radius. Choose 3 or 3.14 for π.
(ii) Work out the approximate volume of the cylinder.
(iii) Work out the approximate volume of a cylinder that is the same height but twice as wide.

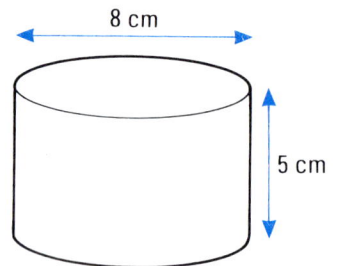

8 cm

5 cm

12 Copy and complete the *Take note*.

TAKE NOTE

$$V = \pi \times r \times \ldots \times \ldots$$

↑   ↖ ↗   ↑

volume of   radius of   height of
cylinder   circular face   cylinder

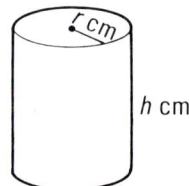

r cm

h cm

B25

ENRICHMENT

1   Melvin brushes his teeth twice a day. This is
a full scale drawing of his toothbrush and
toothpaste tube. He usually covers the full
length of the bristles with toothpaste.
Roughly how long will a 50 ml tube last him?
If the tube costs 30p what does each
brushing cost him?

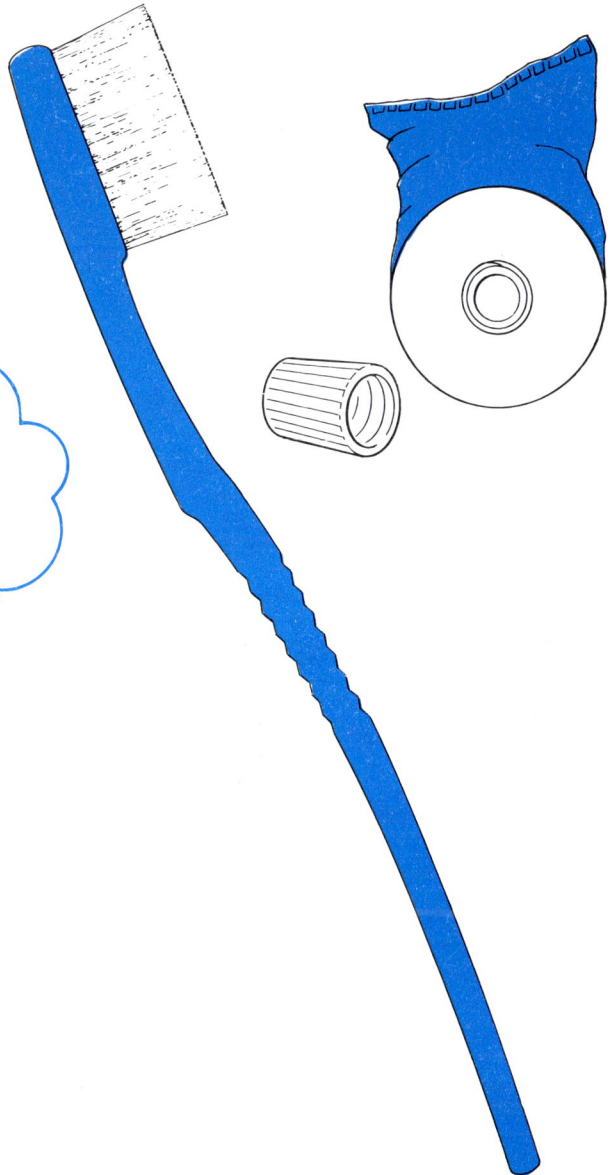

2   Estimate the volume of:

a) your arm
b) your leg
c) your whole body.

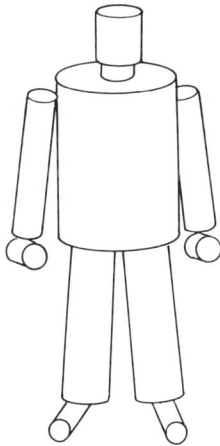

ASSIGNMENT

3   Make two desk tidies like this from card. Your first one should
have a capacity of between 460 and 480 cm^3. Your second one
should have a surface area, including the base, of between 460
and 480 cm^2.
Draw the nets of each of your desk tidies. On your nets write
down the important dimensions. Write a short report to say
how you decided on the dimensions for each of your desk
tidies.

CONSOLIDATION

WITH A FRIEND: POPULATION PYRAMIDS

Mexico

female   male

Australia

female   male

over 74
70–74
65–69
60–64
55–59
50–54
45–49
40–44
35–39
30–34
25–29
20–24
15–19
10–14
5–9
0–4

5   0   5

5   0   5

Percentage of total population

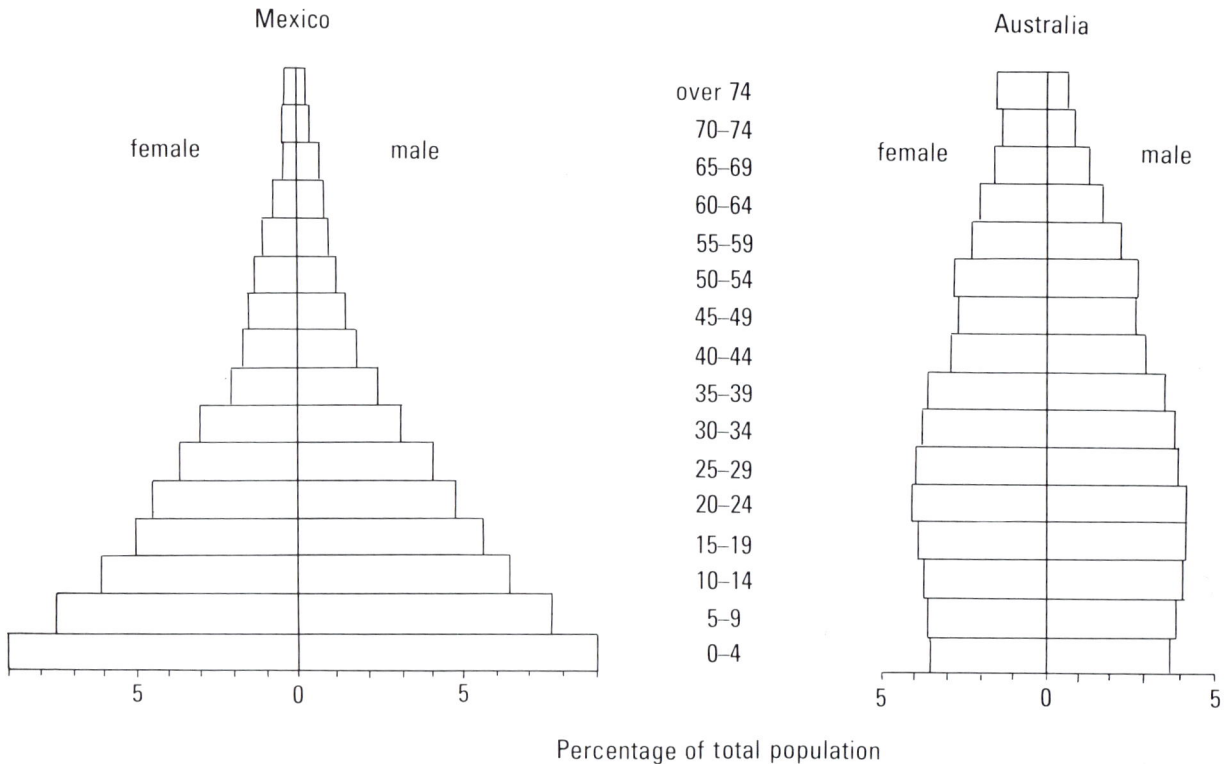

1   Discuss with your friend what the two graphs tell you about **Mexico** and **Australia**. Each of you write one or two paragraphs to explain how the population of **Mexico** differs from the population of **Australia**.

2   This graph appeared in *Which?* magazine. It gives three pieces of information to do with the number of days a car is off the road for repairs and servicing. These are:

- the age of the car
- the number of days it is off the road
- the percentage of all cars affected.

a) Decide between you what information is given by the column marked X. Write down what you decide.

b) Write 'We agree' or 'We disagree' for each of these. If you disagree explain why.

   (i) The older the car, the more days it spends off the road.
   (ii) 7-year-old cars have a very bad record.
   (iii) There are more problems with new cars than with 3-year-old cars.
   (iv) The majority of new cars spend 5 days or more off the road during their first year.

c) You ask ten owners of one-year-old cars how many days their cars were off the road for repairs and servicing during their first year. What would you expect the *total* number of days to be?

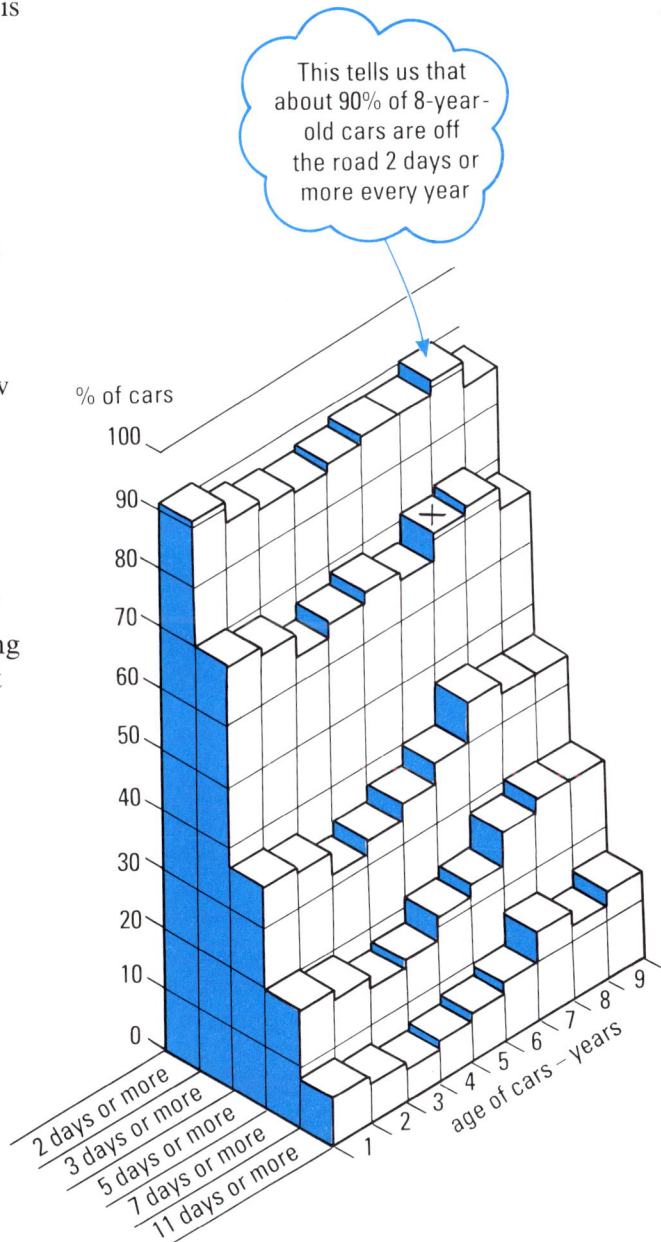

This tells us that about 90% of 8-year-old cars are off the road 2 days or more every year

% of cars

3 This chart shows the times of day when road casualties occurred during 1982.

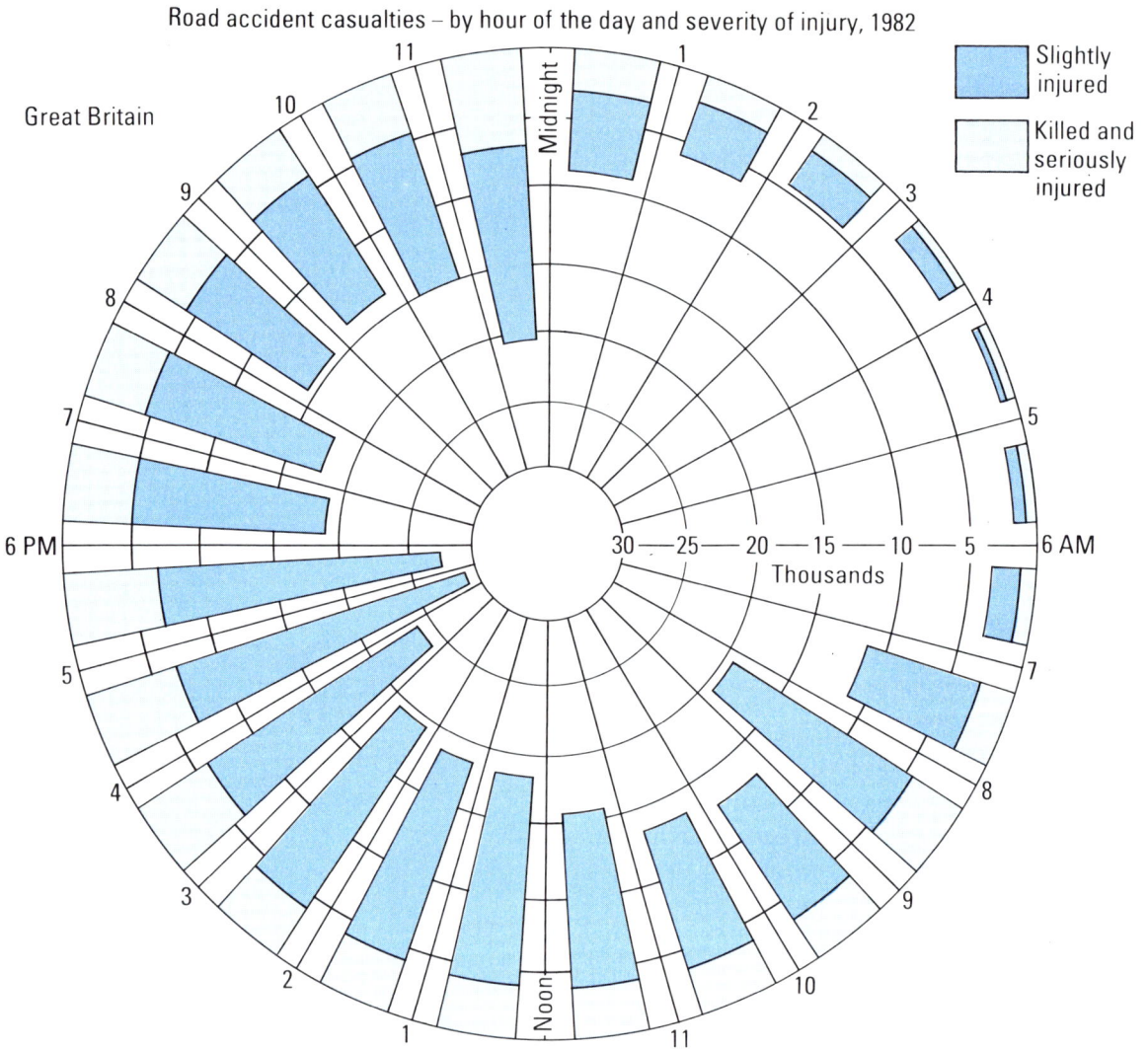

Road accident casualties – by hour of the day and severity of injury, 1982

Great Britain

Slightly injured

Killed and seriously injured

Thousands

a) During which hour did most road accident casualties occur?
b) During which hours were casualty figures below 10 000 for the year?
c) During which hours were casualty figures above 20 000 for the year?
d) During which hours did more than 5000 *serious* accidents occur?

CHALLENGE

e) On 31 January 1983, the wearing of seat belts became compulsory for drivers and front seat passengers of cars and light vans. In the first 5 months after this, serious accidents were down 20%. There were 6125 road accident deaths in 1982. Estimate the total number of deaths during the whole of 1983.

CORE

Pie charts

1 This *pie chart* shows what percentage of their pocket money
   15-year-olds spend, on average, on clothes.

   a) What percentage of their pocket money do 15-year-olds
      spend on things other than clothes?

   b) Roughly, what percentage of your own pocket money do
      you spend on clothes? *Sketch* a pie chart to represent your
      spending on clothes. (Estimate the angle for clothes on
      your chart. You do not have to calculate it accurately.)

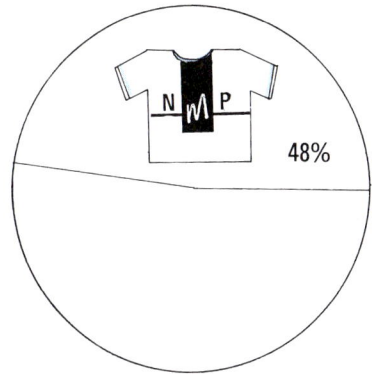

   c) This pie chart shows what percentage of her money a 15-
      year-old girl spends on clothes and entertainment. What
      percentage of her money does the girl *not* spend on clothes
      or entertainment?

   d) Roughly, what is the angle of each sector of the pie chart?

   e) Roughly, what percentage of your own pocket money do
      you spend on entertainment? *Sketch* a pie chart which
      shows your spending on clothes and entertainment. (You
      do not have to calculate the angles accurately.)

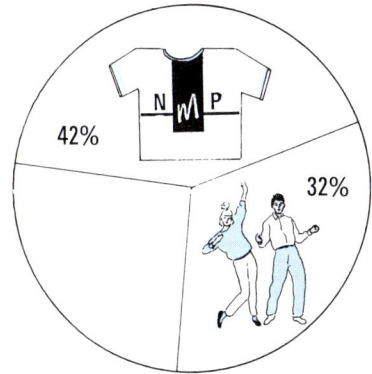

2 This is how a 14-year-old divides up his spending money.

   a) What percentage is spent on:

      (i) clothes   (ii) entertainment?

   b) What percentage is saved?

   c) On the chart, 180° represents 50% spending.

      (i) What does 90° represent?
      (ii) What does 45° represent?

   d) This is how another 14-year-old uses his money:

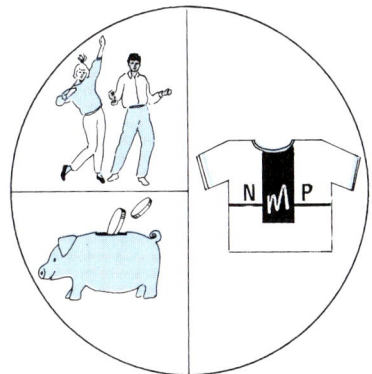

   Clothes 12.5%   Savings 62.5%   Entertainment 25%

   Sketch a pie chart to show the information. On your chart, mark the size of the angle of each
   sector.

B26

3 Ask 10 people in your class 'What do you most like to spread on your toast?' Draw a pie chart like this to represent your results. Make sure that the angles of your chart are *accurate*. (Use your protractor.)

(Hint: Each sector in this chart represents 1 person. How many degrees is each sector? What % does each sector represent?)

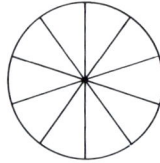

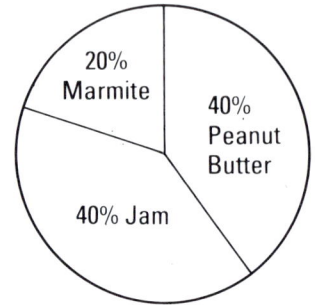

4 Here are the results of three pupils' surveys for question 3:

A
Marmite	5
Jam	2
Peanut Butter	1
Marmalade	2

B
Marmite	4
Jam	1
Peanut Butter	4
Marmalade	1

C
Marmite	5
Jam	3
Peanut Butter	1
Marmalade	1

Here are the three pie charts for the surveys.

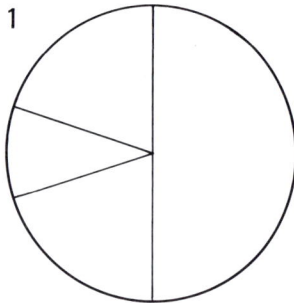

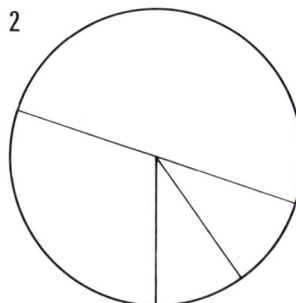

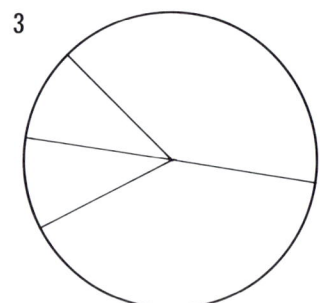

a) Match each pie chart with a survey.

b) How many degrees is the angle for 'Jam' in each pie chart? (Do not use a protractor.)

c) What is the percentage for 'Jam' in each survey?

5 These are the results of a toast survey for 20 people. A pie chart is drawn of the results.

a) What is the angle for:
   (i) Marmite (ii) jam (iii) peanut butter?

b) What is the percentage for each type of spread?

Marmite	8
Jam	5
Peanut Butter	7

Calculations from pie charts

1   To make a good solution for blowing bubbles you need 99% water and 1% washing-up liquid. The pie chart shows this information.

   a) Calculate the angle for *washing-up liquid*. (Remember that 1% of 360° is 0.01 of 360°.)

   b) How many degrees is the angle for water?

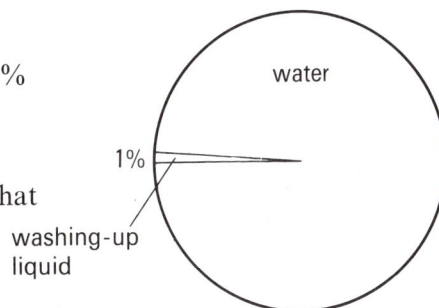

Mixture of water and washing-up liquid needed for bubbles solution

B26

2   a) The pie chart on the left shows that 40% of smokers will die before they reach retirement age. Calculate the angle which represents the 40% deaths. (Do not measure. Remember that 40% of 360° is 0.4 of 360°.)

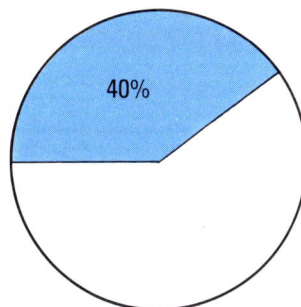

   b) By comparison, only 15% of non-smokers will not reach retirement age. Calculate the angle which represents the 15% deaths.

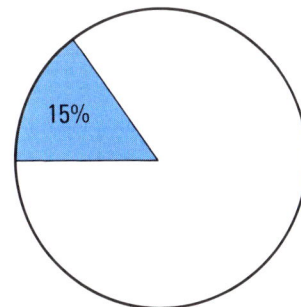

☐ smokers who do not reach retirement age

☐ non-smokers who do not reach retirement age

3   The ratio of female teachers to male teachers in the schools of a city is 2:3.

   a) What percentage of the teachers are (i) female, (ii) male?

   b) A pie chart is drawn to show the information. What is the angle of the chart for female teachers?

TAKE NOTE

40% of teachers in a school live more than 5 miles from the school.

40% of 360° is $0.4 \times 360° = 144°$

The angle of the pie chart for this group of teachers is 144°.

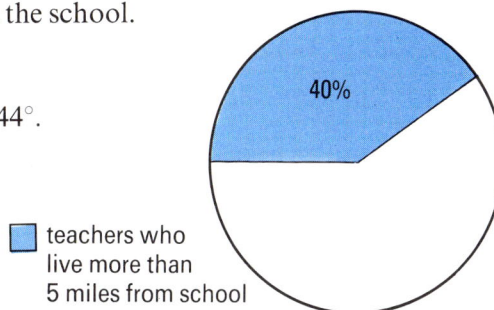

☐ teachers who live more than 5 miles from school

■■■■■■■ ■ THINK IT THROUGH ■ ■■■■■■■

4   Income for the National Trust comes from:

- members' and supporters' fees
- investments
- grants
- property income.

The pie chart shows what percentage came from each source in 1987. Calculate the angle of the chart for each source of income.

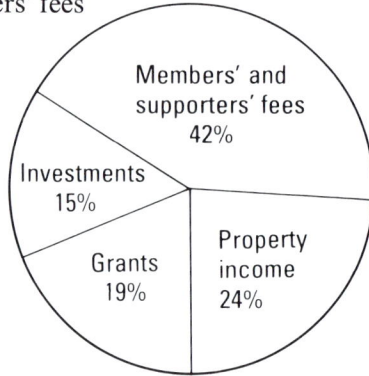

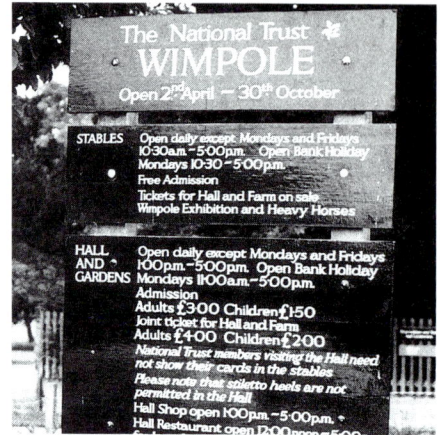

Pie-chart survey

1   Do each of these questions in your head.

a) 400 people were asked which they preferred, butter or margarine.
    200 said 'Butter'.
    100 said 'Margarine'.
    100 said 'No preference'.
    Which of the pie charts correctly represents the information?

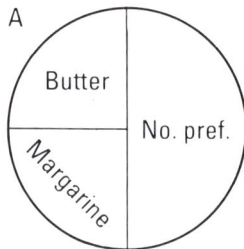

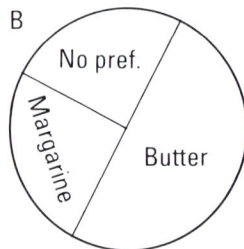

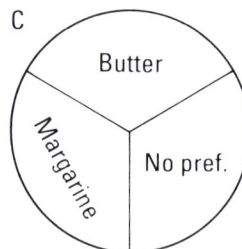

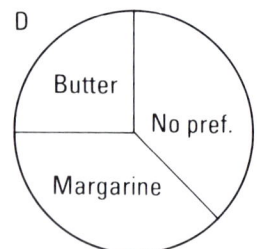

b) 10% of passengers on an aircraft are UK citizens. A pie chart can be drawn to show the information. Which of these is the correct angle for UK citizens?

    (i) 10°   (ii) 36°   (iii) 100°

c) 120 dog owners were asked which of three pet foods their dogs liked best. The pie chart shows the results. How many said 'Lassie'?

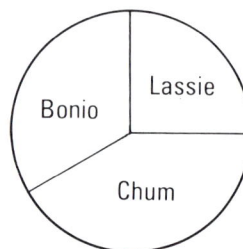

d) A survey of 1000 homes asked for the country of origin of the family's main TV set. The pie chart shows the results. How many TV sets were made in Japan?

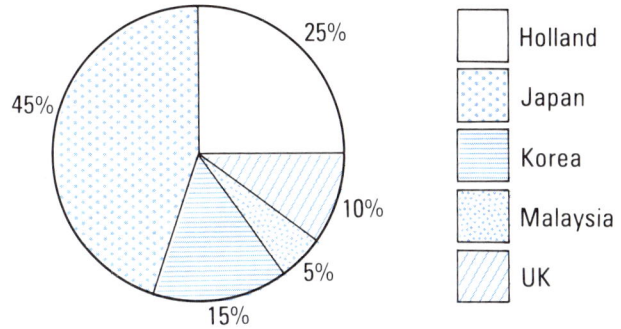

25% Holland
45%
Japan
10%
Korea
5%
Malaysia
15%
UK

B26

2   In 1987 10.8% of the world's population lived in Africa. On a pie chart showing world population, what angle is the sector for Africa?

3   In a survey of 450 cars, 310 are found to be made by Ford. On a pie chart showing the results, what angle is the sector for Ford?

SURVEY

4   Think of a survey you would like to do. Here are some ideas.

- Types of shoes worn by teenagers.
- Favourite group, band, or singer.
- Types of cars owned by teachers.

Carry out the survey. Record your results in a table and in a pie chart. (Make it large and attractive. Give it a title.) Write a report about your survey.

REVIEW

Triangles ABC, DEF, GHI are similar (their angles are equal).

The ratios of corresponding sides are equal:

$$\frac{AB}{BC} = \frac{DE}{FE} = \frac{HI}{GI}$$

$$\frac{CB}{CA} = \frac{FE}{FD} = \frac{GI}{GH}, \text{ and so on.}$$

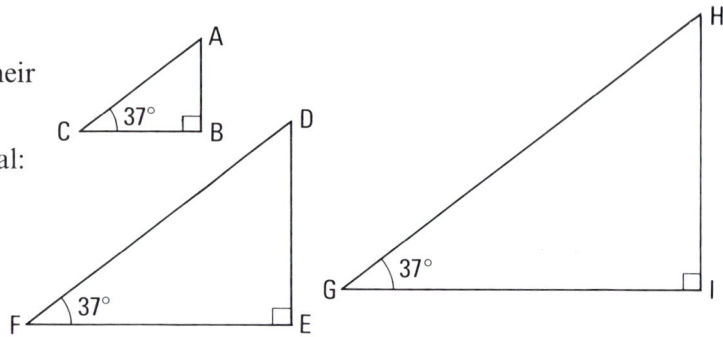

■ Copy and complete these sets of equal ratios:

a) $\dfrac{AC}{AB} = \dfrac{FD}{?} = \dfrac{?}{?}$

b) $\dfrac{AB}{?} = \dfrac{?}{FD} = \dfrac{?}{?}$

Angle	TM/MN Ratio (to 3 DP)	TM/TN Ratio (to 3 DP)	NM/TN Ratio (to 3 DP)	Angle	TM/MN Ratio (to 3 DP)	TM/TN Ratio (to 3 DP)	NM/TN Ratio (to 3 DP)	Angle	TM/MN Ratio (to 3 DP)	TM/TN Ratio (to 3 DP)	NM/TN Ratio (to 3 DP)
0	0.000	0.000	1.000	31	0.601	0.515	0.857	61	1.804	0.875	0.485
1	0.017	0.017	1.000	32	0.625	0.530	0.848	62	1.881	0.883	0.469
2	0.035	0.035	0.999	33	0.649	0.545	0.839	63	1.963	0.891	0.454
3	0.052	0.052	0.999	34	0.675	0.559	0.829	64	2.050	0.899	0.438
4	0.070	0.070	0.998	35	0.700	0.574	0.819	65	2.145	0.906	0.423
5	0.087	0.087	0.996								
				36	0.727	0.588	0.809	66	2.246	0.914	0.407
6	0.105	0.105	0.995	37	0.754	0.602	0.799	67	2.356	0.921	0.391
7	0.123	0.122	0.993	38	0.781	0.616	0.788	68	2.475	0.927	0.375
8	0.141	0.139	0.990	39	0.810	0.629	0.777	69	2.605	0.934	0.358
9	0.158	0.156	0.988	40	0.839	0.643	0.766	70	2.747	0.940	0.342
10	0.176	0.174	0.985								
				41	0.869	0.656	0.755	71	2.904	0.946	0.326
11	0.194	0.191	0.982	42	0.900	0.669	0.743	72	3.078	0.951	0.309
12	0.213	0.208	0.978	43	0.933	0.682	0.731	73	3.271	0.956	0.292
13	0.231	0.225	0.974	44	0.966	0.695	0.719	74	3.487	0.961	0.276
14	0.249	0.242	0.970	45	1.000	0.707	0.707	75	3.732	0.966	0.259
15	0.268	0.259	0.966								
				46	1.036	0.719	0.695	76	4.011	0.970	0.242
16	0.287	0.276	0.961	47	1.072	0.731	0.682	77	4.331	0.974	0.225
17	0.306	0.292	0.956	48	1.111	0.743	0.669	78	4.705	0.978	0.208
18	0.325	0.309	0.951	49	1.150	0.755	0.656	79	5.145	0.982	0.191
19	0.344	0.326	0.946	50	1.192	0.766	0.643	80	5.671	0.985	0.174
20	0.364	0.342	0.940								
				51	1.235	0.777	0.629	81	6.314	0.988	0.156
21	0.384	0.358	0.934	52	1.280	0.788	0.616	82	7.115	0.990	0.139
22	0.404	0.375	0.927	53	1.327	0.799	0.602	83	8.144	0.993	0.122
23	0.424	0.391	0.921	54	1.376	0.809	0.588	84	9.514	0.995	0.105
24	0.445	0.407	0.914	55	1.428	0.819	0.574	85	11.430	0.996	0.087
25	0.466	0.423	0.906								
				56	1.483	0.829	0.559	86	14.301	0.998	0.070
26	0.488	0.438	0.899	57	1.540	0.839	0.545	87	19.081	0.999	0.052
27	0.510	0.454	0.891	58	1.600	0.848	0.530	88	28.636	0.999	0.035
28	0.532	0.469	0.883	59	1.664	0.857	0.515	89	57.290	1.000	0.017
29	0.554	0.485	0.875	60	1.732	0.866	0.500				
30	0.577	0.500	0.866								

1 These tables give the ratios $\dfrac{TM}{MN}, \dfrac{TM}{TN},$ and $\dfrac{NM}{TN}$ for different values of x from 0 to 89.

a) When x is 40 what is the ratio: (i) $\dfrac{TM}{MN}$ (ii) $\dfrac{TM}{TN}$ (iii) $\dfrac{NM}{TN}$?

b) In a triangle the ratio $\dfrac{TM}{MN}$ is 0.84. What is x?

c) In another triangle the ratio $\dfrac{TM}{MN}$ is 0.43. What can you say about x?

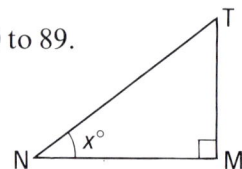

▰▰▰▰▰▰▰▰ ACTIVITY ▰▰▰▰▰▰▰▰

2   You need a scientific calculator which has these keys: [sin] [cos] [tan]
    Make sure it is in DEGREE MODE.
    (Ask your teacher if you are not sure about this.)

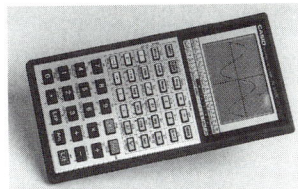

    a) Press [C] [6] [0] [tan] Write down the result.

    b) Try with the numbers 10, 20, and 30 after [C]

    c) Now compare your results with the tables in question 1.
       Write one or two sentences to explain the meaning of the display when [tan] is pressed.

    d) Investigate the [sin] and [cos] keys. Write one or two sentences to explain what these
       cause the calculator to display.

3   Use the [sin] [cos] and [tan] keys.
    For the triangles, find these ratios correct to 3 DP:

    a) $\dfrac{PQ}{QR}$ b) $\dfrac{PQ}{PR}$ c) $\dfrac{QR}{PR}$ d) $\dfrac{LN}{LM}$

    e) $\dfrac{MN}{LN}$ f) $\dfrac{MN}{LM}$ g) $\dfrac{ST}{UT}$ h) $\dfrac{UT}{ST}$ i) $\dfrac{UT}{US}$

B27

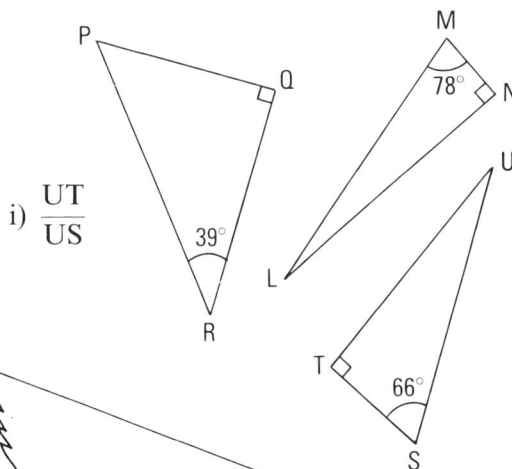

4   The ratio $\dfrac{h}{d}$ in the diagram of the tree is 0.46.
    Search with your calculator to find the value
    of x correct to 1 DP.
    (For example, press [C] [3] [7] [tan]
    How close is 37° to giving the correct ratio?)

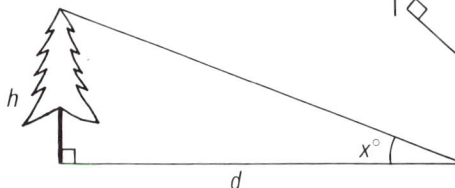

5   The ratio $\dfrac{k}{l}$ for the ladder leaning against the
    wall is 0.32. Search with your calculator to
    find the value of y correct to 1 DP.

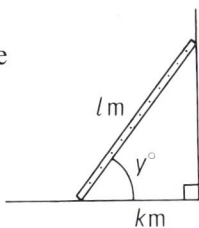

6   The diagram shows a section of a roof.

    a) Find these ratios correct to 2 DP:

       (i) $\dfrac{DC}{AC}$ (ii) $\dfrac{AC}{AB}$

    b) Using your results in a), search with your
       calculator to find the values of x and y
       correct to the nearest whole number.

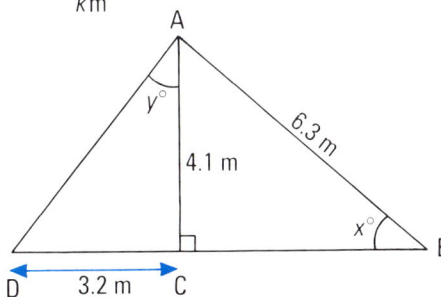

TAKE NOTE

Mathematics which deals with angles and distances is called *trigonometry*.

[sin] [cos] and [tan] are called *trigonometrical ratios*.

We use them to help us to calculate distances and angles
in right-angled triangles.

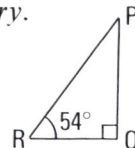

When we press

[C] [5] [4] [tan] the calculator displays the ratio $\dfrac{PQ}{QR}$. We write $\tan 54° = \dfrac{PQ}{QR} = 1.376$ (3 DP).

[C] [5] [4] [sin] the calculator displays the ratio $\dfrac{PQ}{PR}$. We write $\sin 54° = \dfrac{PQ}{PR} = 0.809$ (3 DP).

[C] [5] [4] [cos] the calculator displays the ratio $\dfrac{QR}{PR}$. We write $\cos 54° = \dfrac{QR}{PR} = 0.588$ (3 DP).

B27

7 a) As the value of x increases does the ratio $\dfrac{AB}{BC}$ increase or decrease?

   b) Use your answer in a) to decide which
   is larger, $\tan 30°$ or $\tan 80°$.

   c) As the value of x increases does the ratio $\dfrac{CB}{CA}$ increase or decrease?

   d) Use your answer in c) to decide which is larger, $\cos 30°$ or $\cos 80°$.

   e) As the value of x decreases does the ratio $\dfrac{AB}{AC}$ increase or decrease?

   f) Use your answer in e) to decide which is larger, $\sin 30°$ or $\sin 80°$.

   g) Check your decisions in b), d), and f) with your calculator.

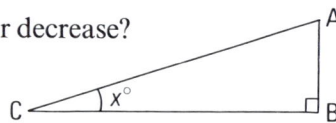

WITH A FRIEND: TRIG. RATIO GUESSING GAMES

8 There are three different games: one for sin, one for tan, and one for cos. Play each one.

sin game   Choose an angle between 0° and 89° (only whole numbers). Ask your friend to guess
           the sin of the angle. An answer within 0.1 of your calculator display scores 1 point.
           Now your friend chooses an angle. And so on. The first one to 5 points wins the
           game.

cos game   Play the same game, but this time ask for the cos of the angle. Again the first one to
           5 points wins the game.

tan game   Play the same game for tan. This time the scoring is a little more complicated. For
           angles between 0° and 45°, the answer must be within 0.1 of the calculator display.
           For angles between 45° and 89° the answer must be within 0.5 of the calculator
           display. The first one to 5 points wins the game.

Calculating the lengths of sides

■■■■■■■■■■■■■ TAKE NOTE ■■■■■■■■■■■■■

You can memorize the meaning of sin, cos, and tan like this:

$$\sin x° = \frac{\text{opposite}}{\text{hypotenuse}} \qquad \cos x° = \frac{\text{adjacent}}{\text{hypotenuse}} \qquad \tan x° = \frac{\text{opposite}}{\text{adjacent}}$$

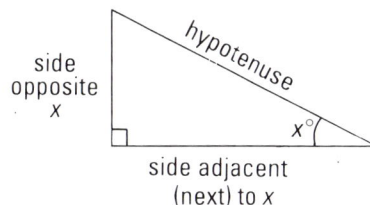

side opposite x

side adjacent (next) to x

(Sin is short for sine, cos for cosine and tan for tangent.)

1 a) Check that to calculate the height of the telegraph pole we can write this equation:

$$\tan 40° = \frac{h}{12}$$

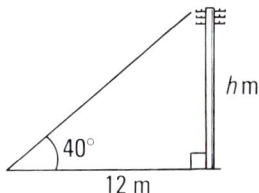

h m

40°

12 m

b) Check that tan 40° is 0.84 (2 DP). Copy and complete this solution.

$$0.84 = \frac{h}{12}$$

$$\times 12 : \quad ? \times 12 = h$$

$$h = ?$$

2 When we calculate lengths using sin, cos, or tan we work with equations like those in question 1. Here are some more examples. Copy and complete the solution of each:

a)
$$0.86 = \frac{k}{7}$$
$$\times 7 : 0.86 \times ? = k$$
$$k = ? (2SF)$$

b)
$$\frac{k}{2.1} = .36$$
$$\times ? : k = .36 \times ?$$
$$k = ? \ (2SF)$$

c)
$$0.5 = \frac{k}{0.98}$$
$$\times 0.98 : k = ? \times ?$$
$$k = ? \ (2SF)$$

d)
$$\frac{5}{k} = 12$$
$$\times k : 5 = ? k$$
$$\div 12 : k = ? (2SF)$$

e)
$$\frac{3.7}{k} = 0.46$$
$$\times ? : 3.7 = 0.46 ?$$
$$\div ? : \quad k = \frac{?}{?}$$
$$k = ? (2SF)$$

f)
$$0.52 = \frac{20}{k}$$
$$\times k : 0.52 k = 20$$
$$? : k = \frac{?}{?}$$
$$k = ? (2SF)$$

3 a) Copy and complete this equation about the chimney: $\tan 24° = \dfrac{h}{?}$

b) Copy and complete this solution:

$$\frac{h}{150} = 0.45 \ (2SF)$$
$$\times 150: \quad h = 0.45 \times ?$$
$$h = ? (1SF)$$

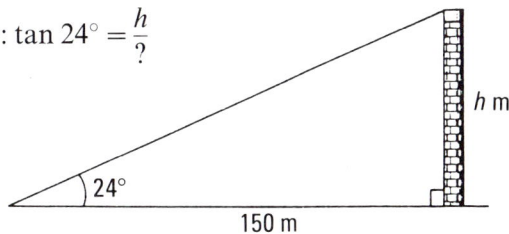

h m

24°

150 m

4   Copy and complete this solution, to find
    the distance d m from the crane cab to the hook:

$$\cos 32° = \frac{d}{10}$$

$\times\,?:\quad 0.848 \times\,? = d$

$$d =\ ?\quad (1\,SF)$$

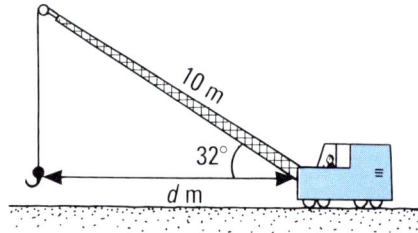

5   Use the same method as that in questions 3 and 4 to find d (correct to 2 SF) in each of these
    situations:

a)

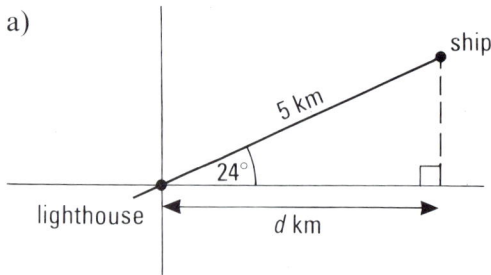

b)

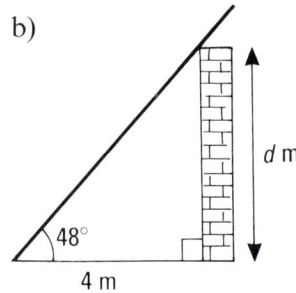

6   In each example a distance in a triangle is calculated by finding the value of k. Copy and complete
    each solution:

a) $\dfrac{PQ}{3.8} = \square\ 27°$

$= \square\ (3\,SF)$

$\times 3.8:\quad PQ = 0.510 \times 3.8$

$= \square\ (2\,SF)$

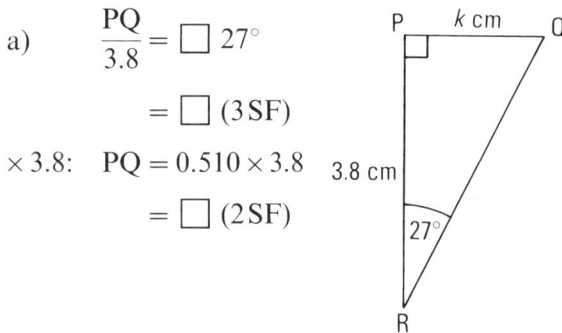

b) $\dfrac{MN}{9.4} = \cos \square$

$= \square\ (3\,SF)$

$\times 9.4:\quad MN = 0.839 \times \square$

$= \square\ (2\,SF)$

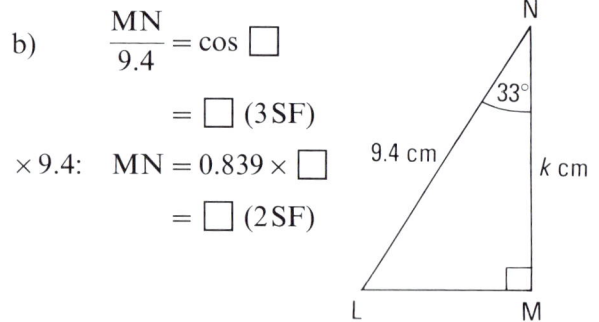

c) $\dfrac{AC}{10.7} = \square\ 72°$

$= \square\ (3\,SF)$

$\times 10.7:\quad AC = 0.951 \times 10.7$

$= \square\ (2\,SF)$

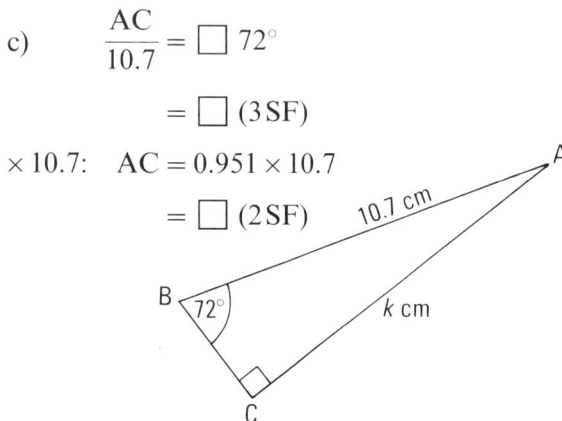

d) $\dfrac{4.8}{QR} = \square$

$= \square\ (3\,SF)$

$\times QR:\quad 4.8 = QR \times \square$

$\div\,?:\quad \dfrac{4.8}{1.11} = QR$

$QR = \square\ (2\,SF)$

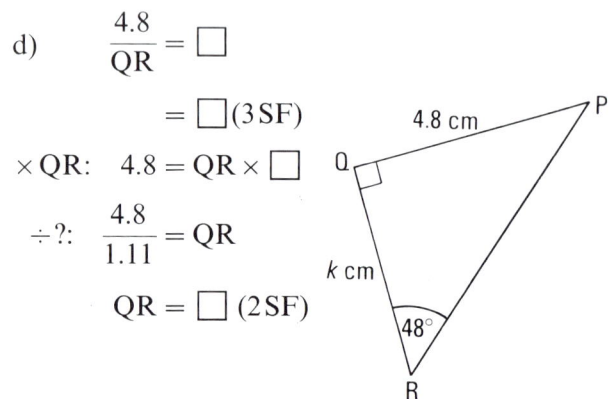

e) $\dfrac{7.6}{LN} = \Box$ 82°

$= \Box$ (3SF)

$\times$ LN: $7.6 = \Box \times LN$

$\div$?: $\dfrac{7.6}{0.990} = LN$

$LN = \Box$ (2SF)

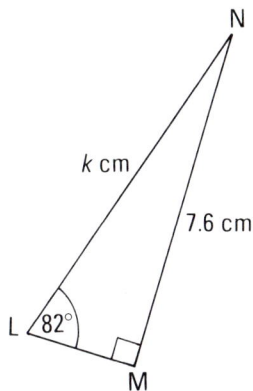

f) $\dfrac{9.3}{AB} = \Box$

$= \Box$ (3SF)

$\times$?: $9.3 = \Box \times AB$

$\div$?: $\dfrac{9.3}{\Box} = AB$

$AB = \Box$ (2SF)

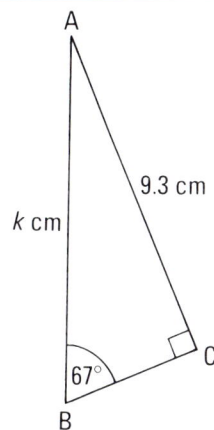

THINK IT THROUGH

7 Find the distance marked ? in each situation.

a)
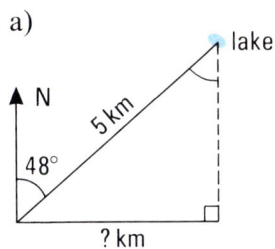

b)
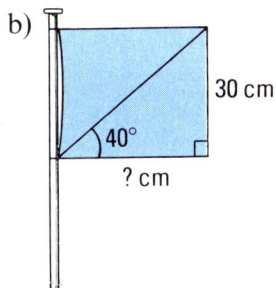

c)
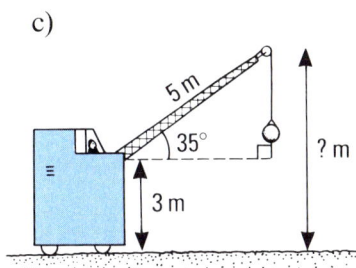

8 a) In the diagram of the roof section, how many degrees is angle CAD?

   b) Calculate (i) AD, (ii) CD.

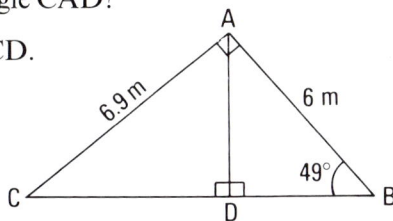

9 Calculate the values of a and b in each part correct to 2 SF.

a)
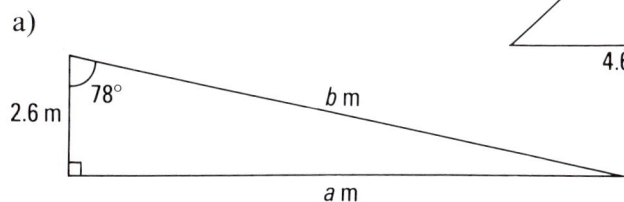

b)
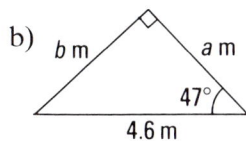

c)
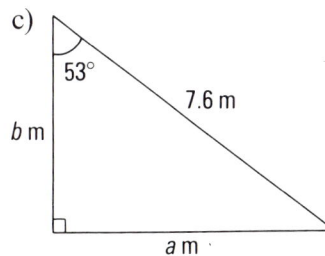

ENRICHMENT

1   In the diagram, find:
    a) AC   b) AD (Remember Pythagoras.)

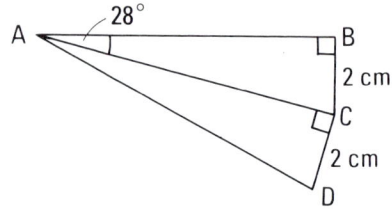

2   The diagonals of a rectangle are 8 cm long. They meet at an angle of 32°. How long and how wide is the rectangle?

3   a) Find tan 45°.
    b) Make a sketch of a triangle to help you to explain why tan 45° *must* be exactly 1.

4   a) Find cos 60°.
    b) Sketch the equilateral triangle. Use your sketch to help you to explain why cos 60° *must* be $\frac{1}{2}$.

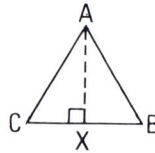

5   ABCDEF is a regular hexagon.

    a) Explain why AOB is an equilateral triangle.
    b) AB = 2 cm. Calculate the shortest distance between AB and ED. (Pythagoras again.)

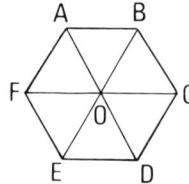

6   ABCDE is a regular pentagon.

    a) How many degrees is angle ABC?
    b) Calculate the length of AC.

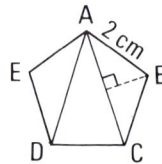

7   In the diagram of the cuboid ∠AHE = 30° and ∠CHG = 40°. HC = 4 cm. What are the dimensions of the cuboid?

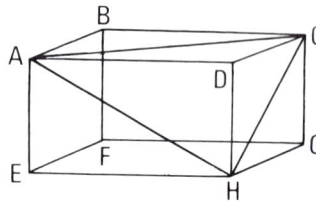

8   The area of triangle ABC is 20 cm². How long is a) XB, b) XY?

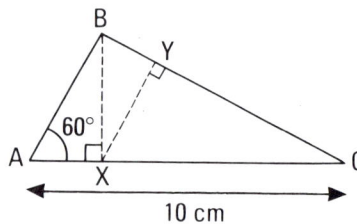

CORE

1 a) Write down an equation for this problem:

Three identical silver cups together weigh 80 g more than one of the cups itself. What does each cup weigh?

Use P g for the weight of a silver cup.

b) Solve your equation. Check that the value of P you find satisfies your equation *and* the problem.

2 a) Try to find the solution to this problem:

Two identical bronze statues and a sculpture of a pair of horses' heads together weigh 24 tonnes. What is the weight of each statue, and what is the weight of the horses' heads sculpture?

Write down what you notice.

b) You will have realised that there is not enough information to enable us to find the weights.
But we can list some of the possibilities.
Check that these are possible:

Statue 4 tonnes   Horses' heads 16 tonnes
Statue 10 tonnes   Horses' heads 4 tonnes

Write down three more possibilities.

c) Use N tonnes for the weight of each statue and H tonnes for the weight of the horses' heads sculpture. Copy and complete this equation:

$\boxed{} N + \boxed{} = 24$

d) If N is 9, what is H?

e) If H is 9, what is N?

f) Write one or two sentences to explain the connection between the graph on the right and the problem.

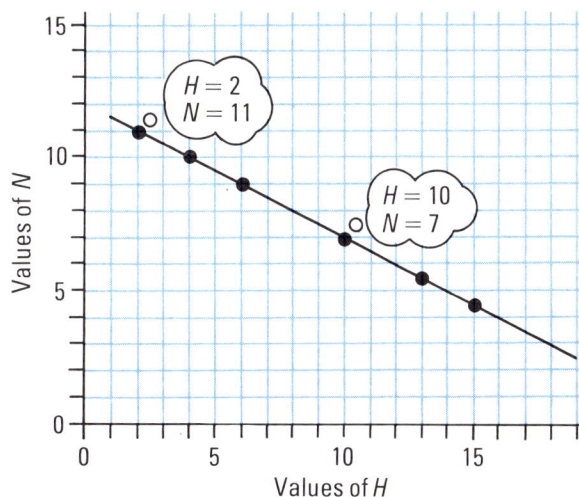

Graph labels: Values of N (vertical axis), Values of H (horizontal axis), $H = 2$ $N = 11$, $H = 10$ $N = 7$

The equation $2N + H = 24$ has two 'unknowns', N and H.

It has many possible solutions: $(2, 11)$, $(10, 7)$, ...

$$\uparrow \ \uparrow \quad \uparrow \ \uparrow$$

$H \ N \quad H \ N$ The solutions can be plotted as a graph.

B28

3   Check that one solution to the equation $x - 2y = 10$ is $x = 12, y = 1$.
    Write down three more solutions.

4   a) Here is another problem for which we
        cannot decide the solution:

        A gold necklace is 4 cm longer than a
        silver necklace. How long is each necklace?
        One possible solution is:

        gold necklace   silver necklace
        24 cm   20 cm

        Write down two more possible solutions.

    b) Use p cm for the length of the gold
        necklace and q cm for the length of the
        silver necklace. Write down an equation
        connecting p and q.

    c) Two possible solutions to the problem are shown on the chart. Draw a line to represent
        some other possible solutions.

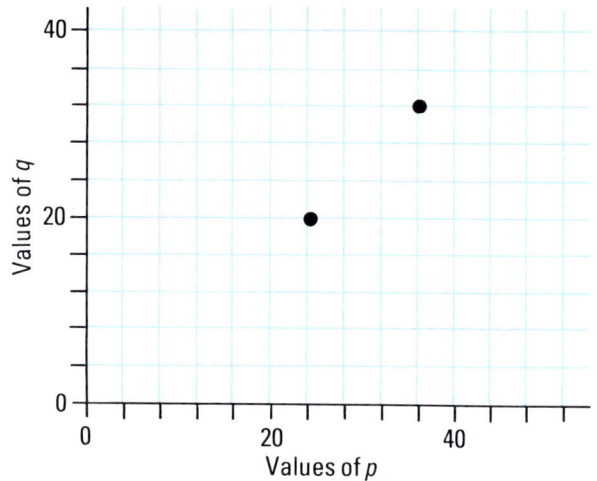

5   Write a problem of your own, like those in questions 2, 3, and 4, which can have many
    solutions. Draw a line on a chart to represent all the possible solutions.

6   a) Copy and complete the equation
        represented by the balance:

        $\boxed{} + 2 = N$

        | $3T$ kg | 2 kg | | N kg |

    b) Check that these are possible solutions to
        your equation:

        $T = 1, N = 5 \qquad T = 2\tfrac{1}{2}, N = 9\tfrac{1}{2}$

    c) Write down three more solutions.

    d) Copy the chart. Draw a line to represent
        all the solutions to your equation that fit
        on the graph.

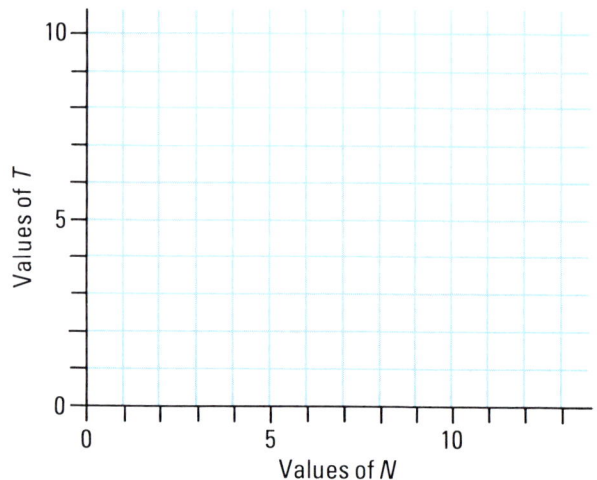

7 a) Write down an equation to represent each of these balances.

   b) Draw a chart to represent the solutions of each equation.

(i)

| x kg | | y kg |

(ii)

| x kg | x kg | | y kg | 1 kg |

Simultaneous equations

1 Rupinder's boss says 'Bring me a mixture of ten large and small buttons'.

   Rupinder could bring 9 large and 1 small
   or 8 large and 2 small
   or ...

   a) Copy the axes of this chart. Mark with dots, ●, all the possible combinations of large and small buttons which Rupinder could bring.

   b) Copy and complete this equation for Rupinder's problem:

   L ? S =   ?
   ↑   ↑
   number of   number of
   large buttons small buttons

   c) Just as Rupinder decides she will take 6 large and 4 small buttons, the boss says, 'I want twice as many small ones as large ones'. Can Rupinder do what the boss wants? Explain your answer.

   d) (i) Copy and complete this equation for the boss's second request:

       $2 \square = S$

      (ii) Check that $S = 6$, $L = 3$ is a solution to your equation in (i).

      (iii) On the same axes you drew in a), mark a line of crosses to represent the solutions to the equation in (i). Is there a point that is both 'dotted' and 'crossed'?

   e) Your graphs show that it is impossible for Rupinder to carry out her boss's orders. Explain how they do this.

2 a) Write down four possible solutions for each of these
      equations:

   (i) $a + b = 10$ (ii) $b = 2a$

   b) Copy the axes. Draw lines to represent all the solutions to
      the two equations that fit on the graph.

   c) Is there a pair of values for a and b which is a solution to
      both equations? If you say 'Yes' write down the values.

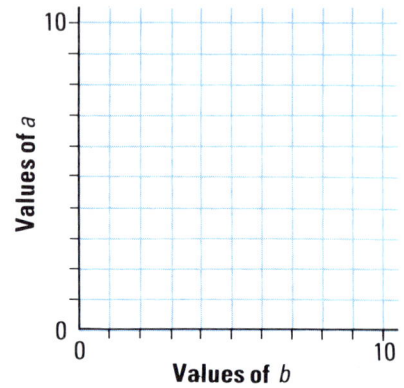

THINK IT THROUGH

3 Question 2 has something in common with question 1. Write a short paragraph to explain what it
   is. In your explanation say why the problem in question 2 c) has a solution, while the problem in
   question 1 does not.

B28

4 a) Here are two equations and four solutions of each. Complete the solutions in each list:

                    value of x value of y
                         ↘   ↙
   $x + y = 10$ Solutions $(0, 10)\ (3, ?)\ (5, ?)\ (?, 8)$
   $2x + 1 = y$ Solutions $(0,\ \ 1)\ (4, ?)\ (?, 7)\ (?, 11)$
                         ↗   ↖
                    value of x value of y

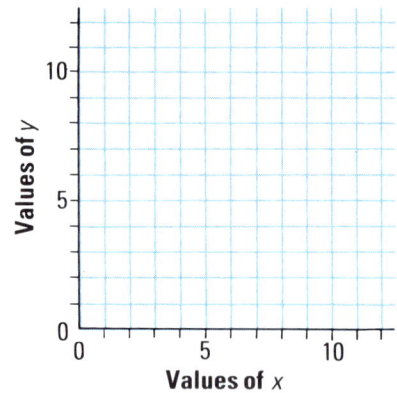

   b) Copy the axes.   (i) Draw a line to represent
                            solutions of $x + y = 10$.
                       (ii) Draw a line to represent
                            solutions of $2x + 1 = y$.

   c) Which values of x and y satisfy both equations?

TAKE NOTE

When we search for values which satisfy two
equations together, we say we are solving
the equations *simultaneously*.
We call the equations *simultaneous equations*.

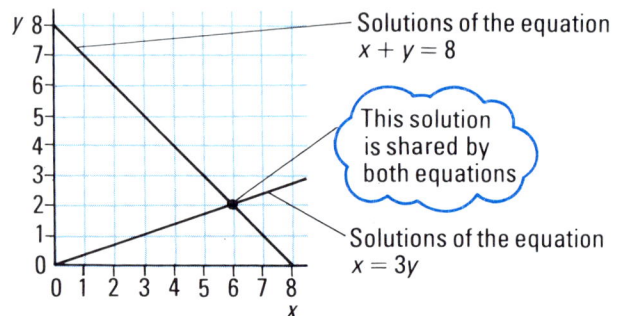

Solutions of the equation
$x + y = 8$

This solution
is shared by
both equations

Solutions of the equation
$x = 3y$

5 a) Check that $x = 2$, $y = 1$ is a solution of each of these
   equations:

   (i) $2x = y + 3$ (ii) $x - y = 1$

   b) Copy and complete these solutions of equation (i):

   ($^-4$,) ($^-3$,) ($^-2$,) ($^-1$,) (1,) (2,) (3,) (4,)
      ↑   ↖
   value value
    of x of y

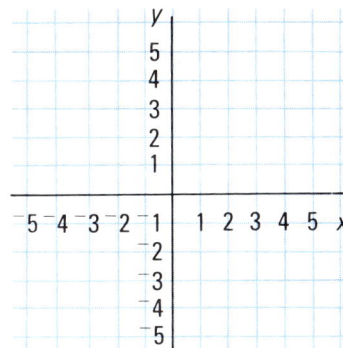

   c) Copy and complete these solutions of equation (ii):

   (, $^-5$) (, $^-4$) (, $^-3$) (, $^-2$) (, $^-1$) (, 0) (, 1) (, 2)

   d) Draw lines to represent solutions of each equation. Check from your graph that $x = 2$, $y = 1$ is
      a solution of both equations.

   e) How can you tell from your graph that there is not another solution which the equations
      share? Write one or two sentences to explain.

6 a) Copy and complete this equation
      connecting a and b in rectangle A.

      $a + b = ?$

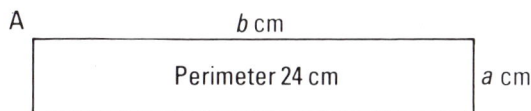

   b) Copy and complete this equation for
      rectangle B.

      $2a + b = ?$

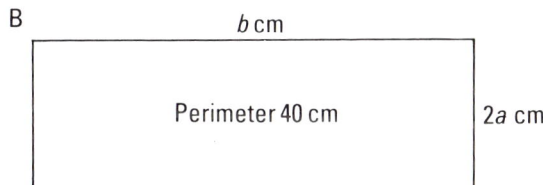

   c) Write down four possible pairs of values
      for a and b in rectangle A.

   d) Write down four possible pairs of values
      for a and b in rectangle B.

   e) Draw lines on the graph to represent the
      solutions to the equations in a) and b).
      Write down the solution which satisfies
      both equations. Write one or two
      sentences to explain what this solution
      tells us about the rectangles.

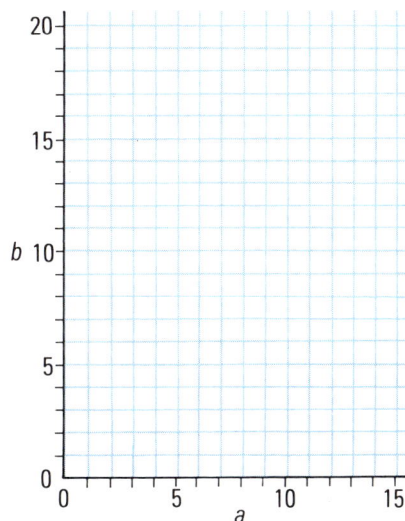

A

b cm

Perimeter 24 cm a cm

B

b cm

Perimeter 40 cm $2a$ cm

B28

7 a) Use the graphs to help you to find the common solutions to these pairs of equations:

   (i) $x = y$ and $x + y = 6$
   (ii) $x = y$ and $x + y = {}^-6$
   (iii) $y = 5x$ and $x = y$
   (iv) $y = 5x$ and $x + y = 6$

   b) Use the graphs to help you to explain why the equations:

   $$x + y = 6$$
   $$\text{and } x + y = {}^-6$$

   do not have a common solution.

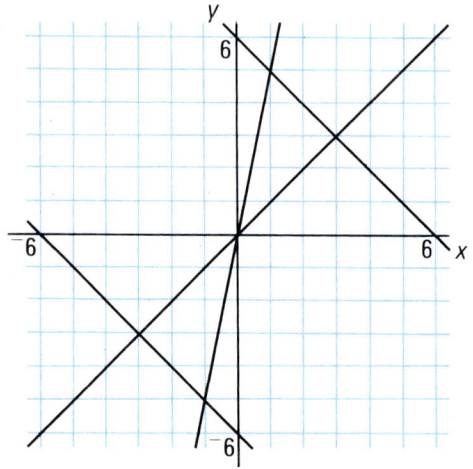

━━━━━━━━ CHALLENGE ━━━━━━━━

8 Solve the simultaneous equations:

   a) $x - y = 5$ and $x + 2y = 5$ b) $2x - y = 4$ and $2x + 4y = 9$

B28